Mineral Resource Evaluation II:
Methods and Case Histories

Geological Society Special Publications
Series Editor A. J. Fleet

GEOLOGICAL SOCIETY SPECIAL PUBLICATION NO. 79

Mineral Resource Evaluation II:
Methods and Case Histories

EDITED BY
M. K. G. Whateley & P. K. Harvey
Department of Geology
University of Leicester, UK

1994
Published by
The Geological Society
London

THE GEOLOGICAL SOCIETY

The Society was founded in 1807 as The Geological Society of London and is the oldest geological society in the world. It received its Royal Charter in 1825 for the purpose of 'investigating the mineral structure of the Earth'. The Society is Britain's national society for geology with a membership of 7500 (1993). It has countrywide coverage and approximately 1000 members reside overseas. The Society is responsible for all aspects of the geological sciences including professional matters. The Society has its own publishing house which produces the Society's international journals, books and maps, and which acts as the European distributor for publications of the American Association of Petroleum Geologists and the Geological Society of America.

Fellowship is open to those holding a recognized honours degree in geology or cognate subject and who have at least two years relevant postgraduate experience, or who have not less than six years relevant experience in geology or a cognate subject. A Fellow who has not less than five years relevant postgraduate experience in the practice of geology may apply for validation and, subject to approval, may be able to use the designatory letters C. Geol (Chartered Geologist).

Further information about the Society is available from the Membership Manager, The Geological Society, Burlington House, Piccadilly, London W1V 0JU, UK.

Published by The Geological Society from:
The Geological Society Publishing House
Unit 7
Brassmill Enterprise Centre
Brassmill Lane
Bath BA1 3JN
UK
(*Orders*: Tel. 0225 445046
 Fax 0225 442836)

First published 1994

The Geological Society 1994. All rights reserved. No reproduction, copy or transmission of this publication may be made without prior written permission. No paragraph of this publication may be reproduced, copied or transmitted save with the provisions of the Copyright Licensing Agency, 90 Tottenham Court Road, London W1P 9HE, UK. Users registered with Copyright Clearance Center, 27 Congress St., Salem, MA 01970, USA: the item-fee code for this publication is 0305-8719/94 $7.00.

British Library Cataloguing in Publication Data
A catalogue record for this book is available from the British Library
ISBN 1-897799-06-3

Typeset by Bath Typesetting Ltd
Bath, England

Printed in Great Britain by
Alden Press, Oxford

Distributors

USA
 AAPG Bookstore
 PO Box 979
 Tulsa
 Oklahoma 74101-0979
 USA
 (*Orders*: Tel. (918)584-2555
  Fax (918)584-0469)

Australia
 Australian Mineral Foundation
 63 Conyngham St
 Glenside
 South Australia 5065
 Australia
 (*Orders*: Tel. (08)379-0444
  Fax (08)379-4634)

India
 Affiliated EastWest Press PVT Ltd
 G-1/16 Ansari Road
 New Delhi 110 002
 India
 (*Orders*: Tel. (11)327-9113
  Fax (11)326-0538)

Japan
 Kanda Book Trading Co.
 Tanikawa Building
 3-2 Kanda Surugadai
 Chiyoda-Ku
 Tokyo 101
 Japan
 (*Orders*: Tel. (03)3255-3497
  Fax (03)3255-3495)

Contents

Preface vii

Definitions
RIDDLER, G. What is a mineral resource? 1
ARMITAGE, M. G. & POTTS, M. F. A. Some comments on the reporting of resources and reserves 11
JAKUBIAK, Z. & SMAKOWSKI, T. Classification of mineral reserves in the former Comecon countries 17

Data
GRIBBLE, P. D. Fault interpretation from coal exploration borehole data using SURPAC2 software 29
HATTON, W. INTMOV; a program for the interactive analysis of spatial data 37
NATHANAIL, P. Reserve assessment of a stratified deposit with special reference to opencast coal mining in Great Britain 45

Deposit variability
NOTHOLT, A. J. G. Phosphate rock: factors in economic and technical evaluation 53
BELL, T. M. & WHATELEY, M. K. G. Evaluation of grade estimation techniques 67
DOWD, P. A. Optimal open pit design: sensitivity to estimated block values 87
SCOBLE, M. & MOSS, A. Dilution in underground metal mining: implications for grade control and production management 95
SIDES, E. J. Quantifying differences between computer models of orebody shapes 109

Finance
GORMAN, P. A review and evaluation of the costs of exploration, acquisition and development of copper and gold projects in Chile 123
O'LEARY, J. Mining project finance and the assessment of ore reserves 129

Case histories
DOWD, P. A. The optimal design of quarries 141
AL-HASSAN, S. & ANNELS, A. E. Geostatistical estimation of manganese oxide resources at the Nsuta Mine, Ghana 157
ANNELS, A. E., INGRAM, S. & MALMSTROM, L. Structural reconstruction and mineral resource evaluation at Zinkgruvan Mine, Sweden 171
ARTHUR, J. & ANNELS, A. E. The application of geostatistical techniques to *in situ* resource estimation in the sand and gravel industry 191
CAMERON, R. I. & MIDDLEMIS, H. Computer modelling of dewatering a major open pit mine: case study from Nevada, USA 207
CRUMP, L. A. & DONNELLY, R. Opencast coal mining; a unique opportunity for Clee Hill Quarry 219
BARRY, J., GUARD, J. & WALTON, G. Database management at the Lisheen deposit, Co. Tipperary, Ireland 233
MITCHELL, C. J. Laboratory evaluation of kaolin: a case study from Zambia 241
O'LEARY, J. Cia Minera Los Pelambres: a project history 249

Preface

The theme of this volume is Mineral Resource Evaluation. The chapters cover a wide range of activities in this field and they describe some of the methods that are currently in use to help in the evaluation of mineral resources, including exploration drilling, sampling, resource estimation, mine design, financial evaluation and mine sampling and grade control. Case histories of mineral resource evaluation are also described, with examples from all over the world, including Canada, USA, Chile, Ghana, Sweden, Zambia, Ireland, and of course, UK.

One notable aspect of this volume is the number of chapters which deal with mineral resource definitions. A mineral resource is a means to an end not an end in itself. The aim is to develop a mine which will maximize the Net Present Value (NPV) for the present shareholders. It is therefore important to define resources sufficiently accurately to convince the banks to lend the necessary 75–80% of the money required for the capital expenditure to develop the mine. It would appear that the banks are the final arbiters and maybe it is they who should be telling the mining companies how they should be describing their resources. Perhaps the mining industry, through their various institutions, in conjunction with the financial organizations should set up an international commission to recommend a standard nomenclature with recognized definitions. As recommendations they would not be enforceable, but companies, Governmental organizations and individuals could then compare their preferred way of describing resources to the recommended nomenclature and at least sensible and meaningful comparisons could then be made.

The editors are particularly grateful to those companies and organizations who provided support for this volume either directly or indirectly through their demonstrations of software, publications or services. In particular, they would like to acknowledge the following:

Golder Associates Ltd
Datamine International
Hall & Watts Systems Ltd
Pergamon Press
J. H. Reedman & Associates Ltd
Northern Exploration Services
Crowe, Schaffalitzky & Associates Ltd
Cambridge University Press
Water Management Consultants Ltd

Michael K. G. Whateley
Peter K. Harvey

What is a mineral resource?

G. P. RIDDLER

British Geological Survey, Keyworth, Nottingham NG12 5GG, UK

Abstract: Mineral resource evaluation is one of a whole spectrum of quantitative methods which have been used since biblical times for the purpose of improving the process required to aid problem-solving and to increase the quality of strategic management decisions at all levels. Such evaluation leads to the classification of mineral resources. It is important to have a clear knowledge of how mineral resources are classified and what the classification nomenclature actually means to ensure decisions are based on a sound understanding of criteria applied.

As interest for investment in mineral resource development spreads into Eastern Europe and the former Soviet Union, the need for some form of harmonization in terminology, nomenclature, criteria, as well as the approach to the classification process for the definition of mineral resources has again been highlighted.

The approach to quantitative studies is discussed, and guidelines are given that may improve mineral reserve and resource estimates so that implementation failure of mining projects can be avoided.

Approaches to the classification of mineral resources are reviewed along with the uses made of such mineral resource information.

The argument is put for a standard approach to mineral resource classification and consistency of international nomenclature with a discussion of some of the related problems. Principles are established which may suggest a way forward illustrated by a recent major resource assessment project.

There has been much debate about the nomenclature used when classifying mineral resources. This has continued in the mining press recently following the publication of the latest attempt at defining resources and reserves by the Institution of Mining and Metallurgy (IMM 1991).

Mining has become a global business meaning that individual multinational mining companies operate in many different countries. Mineral resource development has in recent years opened up to such companies in the former Soviet Union and Eastern Europe. The approach to mineral resource classification, criteria and nomenclature still varies considerably from country to country (Armitage & Potts, this volume; Jakubiak & Smakowski, this volume). In this era of global communication there is a need for some form of standardization.

It has always been important to distinguish between reserves and resources. This paper deals particularly with resource classifications because resource assessment is a primary function of the Minerals Group at the BGS. Some reference to mineral reserves classification is made in this paper to demonstrate their relationship to mineral resources but mineral reserve classification is essentially the responsibility of mineral deposit developers and should be carried out at the time an investment is made.

For the purposes of this paper the distinction between the classification of resources and reserves is made as follows. The term *resources* refers to mineral deposits which can or at some time in the future may be mined economically. The term *reserves* is generally considered to represent that portion of resources the presence of which is *geologically assured* and can be *mined economically now*. Determination of reserves would normally require a multidisciplinary feasibility study. Since the level of data required for determining reserves and resources differs and the two classifications are used for different purposes, the separate terms are fully justified.

Quantitative studies

Mineral resource evaluation and classification is one of a whole spectrum of quantitative studies which have been an adjunct to decision making for a long time. When considering how the approach might be harmonized between countries it is worth considering the findings of operational management research projects completed outside the minerals industry on the approach to quantitative studies (Huxham 1987).

This research suggests a staged approach

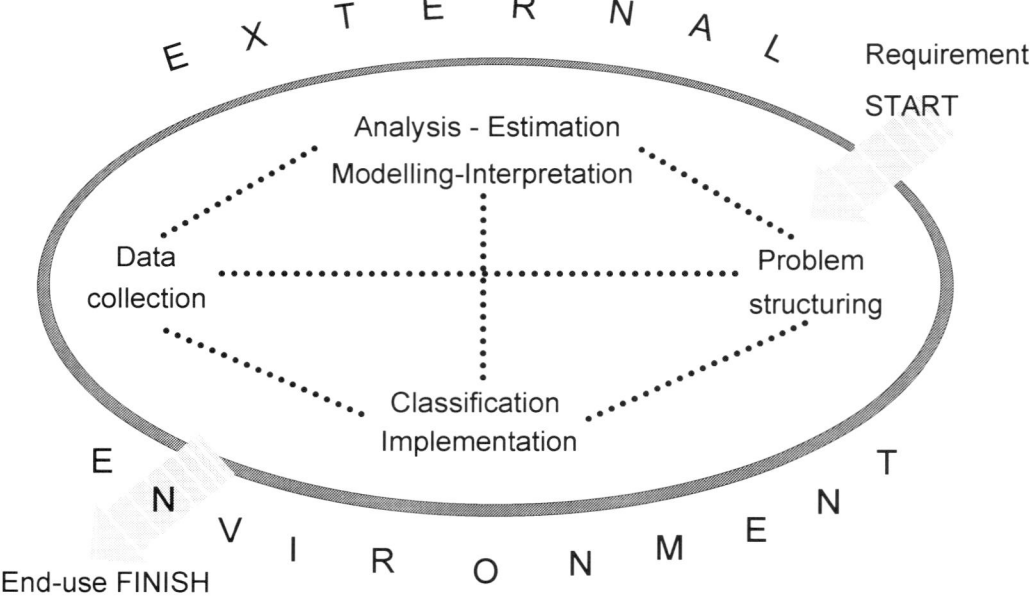

Fig. 1. The main stages of interactive quantitative studies.

which provides a logical framework. Some relatively straightforward, obvious, but often neglected guidelines emerge indicating how such studies may improve the potential for success in project implementation. These guidelines indicate how the quality of mineral resource evaluation and classification may be improved so as to avoid implementation failures at the project development stage.

All quantitative studies start with the definition of the requirements of the study which are governed to a great extent by the proposed end-use. There are four main stages to quantitative studies, data collection, problem structuring, data analysis/estimation/modelling/interpretation, and implementation/classification, all of which are interactive (Fig. 1).

Data collection has been recorded since biblical times, the Book of Numbers, appropriately enough, describing population census, used at the time for taxation or military service. In the minerals industry we have become familiar with the use of sampling for assay and mineralogy by channelling, trenching or drilling as the means of data collection. Of course while this may be adequate for resource classification, additional qualitative and quantitative data on the external environment such as capital, labour, commodity markets, local economics, planning and development policy, infrastructure and other criteria would have to be added for reserve classifications (Riddler 1988).

Problem structuring. This is the stage where the end-use must be considered and appropriate data collection put in hand and can involve a wide range of specialized multidisciplinary input. The optimal approach will be chosen through consideration of various alternatives. It is probable that in many cases a particular development project will not proceed as a result of the output from this stage.

Data analysis/estimation/modelling/interpretation. This can involve a broad range of techniques being applied to the data including statistics, simulation and linear programming.

It is important that such techniques are properly defined and explained because non-specialists can be led to assume that the sophistication of a statistical technique ensures the reliability of the inferences it provides. The right questions should be asked and the answers questioned as more information comes to hand through the interactive process.

An early example of this stage comes from the work of John Graunt, a seventeenth century draper, who was probably the world's first statistician. He was the first to be recorded as making reasoned estimates on the basis of a specified method mainly because of a lack of accuracy in previous classification in the data. The requirement for his study had the added impetus in that the end-user was the King who wished that a weekly watch be kept on the rise in Plague deaths in the city of London so that he could receive early warning to decamp to the relative safety of the countryside (Kennedy 1983).

In the minerals industry, this stage is applied to the estimation of reserves and resources. Use of the results of this by non-specialists such as accountants, lawyers and planners is increasing. Since the estimations have a direct bearing on the commercial viability of mineral development projects, and there is now legal precedent for such viability to be taken into account in planning decisions in UK, it is important that the limits of these estimations are clearly understood.

Implementation/classification. Once data have been collected and estimates made, reserves and resources can be identified and classified. Implementation of mineral developments based on data models and this classification can then take place.

It is well known that implementation of mineral development projects can fail because inadequate attention is paid to some factor or other during the process of estimation and classification of reserves and resources through the inadequacy of the approach or the level of data used to distinguish between them.

Avoiding project implementation failure

Based on such a staged approach to quantitative studies, operational management research findings indicate that by following certain guidelines at the *data/estimation/modelling/interpretation* stage, implementation failures may be avoided.

The guidelines highlighted by the research can be related to minerals industry activities as follows.

Ensure that all relevant lines of communication relating to input data are kept open. For example, geologists may have the major input to a resource assessment but other disciplines such as metallurgy, mining engineering, mineralogy, environmental science or economics come into play when reserves are being estimated.

Involve those who commissioned the study or will use any product of the analysis phase. The end-user should be involved because the requirements of the study may change as it evolves. For example, local planning authorities may impose local constraints and conditions on a development and it would be prudent to find out what these may be at the earliest opportunity so that resource and reserve estimates may take these into account.

Keep the implementation phase in mind from the start. That is, what is the end-use proposed for the study, who is to be using it, where and when? For example, some of the common uses of mineral resource and reserve evaluation data are given in Table 1.

Table 1. *Some uses of mineral resources evaluation data*

Company annual reports

Strategic planning

- exploration
- development
- mineral supply
- location of resources of critical metals

Investment planning

- bankable documents
- asset valuation

Awareness of mineral endowment

Attracting inward investment

Decide on the emphasis and level of detail. For example, is a resource assessment or a reserve estimation required? The reserve estimation requires much more detailed information.

Decide on the suitability of effort on data analyses for the purpose at hand. For example, detailed mineralogy may be required for reserve estimation on complex enriched copper deposits but not necessarily for resource assessments.

Keep models simple and be aware of their inadequacies and shelf-life. For example, about AD 50, the Roman author Pliny records a site specific model for classification of reserves at one prospect, using the orientation of the deposit as a criterion. He stated that all easterly trending deposits were profitable (Agricola 1556). This was probably perfectly adequate 2000 years ago for a specific local small-scale operation but is perhaps not a model which would have widespread application now.

As technology and knowledge advance, models necessarily become more complex, but they should still be clear, understandable to the user and continuously reviewed.

The qualitative *external environment* (Fig. 1) can have a significant impact on the outcome of a quantitative study. If for example it changes or is not accounted for during the data analysis/estimation/modelling/interpretation stage of the

Table 2. A comparison of major resource classification systems

	US Dept of the interior 1973	IGS 1982	Canada, Dept of energy, Mines and Resources	USSR Geological Directorate	Germany	UN Committee on Natural Resources	IMM 1991	Australian AIMM AMIC 1988
↑	Measured			Explored (ABC1+abc1)	Proved		Measured mineral resource	Measured
	Indicated	1	Demonstrated		Probable	R1	Indicated mineral resource	Indicated
Increasing geological assurance	Inferred	2	Surmised	Prospective (C2+c2)	Possible	R2	Mineral potential	Inferred
	Hypothetical	3	Speculative	Predicted	Prognostic	R3		
	Speculative							
Diminishing economic potential	Economic/recoverable	E	Economic	Balance (ABC1C2)	Mineable	E	Proved/probable mineral reserve	Proved/probable reserve
	Paramarginal	S				Marginal S		
↓	Submarginal		Subeconomic	Out of balance (abc1c2)	Potential	Subeconomic		

study then inadequacy and failure at implementation will result, rendering the whole quantitative study useless.

A good example of an external environment factor on minerals industry activity is the effect that designated planning constraint areas may have on the classification of mineral resources or reserves.

In the UK constraints on mineral related developments include national parks, urban areas, green belts, sites of special scientific interest, national nature reserves, areas of outstanding natural beauty, agricultural land and so on. These all affect planning consent decisions. The impact of such constraints is illustrated by UK fuller's earth reserves. Reserves (1991) with planning consent amount to 0.78 million tonnes which represents only about 28% of what could be classified as a reserve. 72% of the 'available' fuller's earth is without planning consent and must be classified as a resource as this amount although it may be mined *economically*, is not available to be mined *now*.

The point here is that no matter how sophisticated the model for classifying reserves may be it is only those reserves with planning permission that can sustain production now and it is only these that can be truly classified as reserves.

By adopting some form of standard approach, methodology or framework that can be recognized and understood by various users along the lines indicated by the operational management research into the approach to quantitative methods, some of the more obvious problems can be anticipated. Resource and reserve estimates and classification may be improved therefore and this will assist the successful implementation of mining projects by increasing the quality of strategic management decisions at all levels.

Approach to classification of resources and reserves

Resources

For the mineral supply process to continue it is important to have a knowledge of available resources in particular regions. At a company level such information is critical for decision making on worldwide exploration programmes and for strategic planning decisions which address resource replenishment requirements (Riddler 1988).

In view of the importance of these decisions with regard to the normally high level of investment required, a clear knowledge of how the mineral quantities are estimated, how classifications are defined and what these mean is essential. The problem is that there are many different national approaches to the classification of mineral resources which are inconsistent and of differing standards.

To illustrate this point a comparison of major resource classification systems is given on Table 2. The opening up of Eastern Europe and the former Soviet Union adds another dimension to nomenclature and classification criteria (Jakubiak & Smakowski, this volume). Such a variety of nomenclature for what may be equivalent classifications of reserves and resources begs the question 'What is a mineral resource?'.

While the process involved in the *data analysis/estimation/modelling/interpretation* stage has to vary since the data are site/technology/time specific, there is certainly scope for a harmonization of the nomenclature used to describe the resultant classified categories of mineral resources after that stage is complete.

Reserves

Because of environmental planning constraints and the uncertainty over the issue of extraction permits, the classification of sustainable reserves is getting more difficult.

For example, in UK the proposed Mineral Planning Guideline (MPG6) for the aggregates industry which is currently at the consultation stage provides two options, the status quo and Option 2, which proposes a 20% reduction in aggregate production, a halving of the Company land banks from 10 to 5 years (that is, halving reserves with planning consent).

Option 2 could be expected to force up prices and encourage the use of aggregates from superquarries in Scotland or elsewhere as well as recycling and use of waste materials. On the other hand it raises questions as to how much of these alternative resources could be converted into reserves because of external factors such as planning and environmental constraints in the alternative supply areas as well as transportation costs.

Special interest groups can stop mineral development projects after permitting, so should reserves only be classified after production commences?

Classification of reserves therefore has to be a dynamic process, with volumes switching in and out of the demonstrated resource according to prevailing conditions at the time. Regardless of such classification, it should be possible to envisage standardization of nomenclature.

Classification and nomenclature: present and future

So what is happening to address the problem of such a varied nomenclature (Table 2)? Codes, guidelines and definitions for mineral resources and reserves abound and there is still much scope for moving towards a common nomenclature.

The USBM/USGS between 1973 and 1980 (McKelvey 1973; USGS 1980) produced a resource/reserve classification system and subsequently in 1986 produced guidelines which were broadly adopted by the minerals industry. Variations of this have followed.

In 1988 (with an update in 1992), the Australian IMM and the Australian Mining Industry Council published their code followed by Guidelines in 1990 (Aus IMM 1988, 1990, 1992). In 1991, the SME published broadly similar guidelines (SME 1991).

At the Prospectors and Developers International Convention in Toronto in April 1993 the CIM announced that it was to produce its report on classification by mid-1993.

The CIM has consulted many stakeholders. Its system of classification will be designed to:

- cover economic/non-economic criteria;
- make allowance for commodity price fluctuation;
- have flexibility to allow reserves to move in and out of resource categories;
- encompass all external environmental criteria;
- base resource/reserve classification on a feasibility study;
- utilize terminology from the Aus IMM, IMM, SME and AIME.

Together with the Aus IMM and the SME which are the professional bodies in currently significant mining countries, the CIM are moving towards a standardized system of nomenclature which will have *measured, indicated,* and *inferred* resources and *proven* and *probable* reserves. Along with the attendant guidelines these appear to meet most requirements including US Stock Exchange Commission regulations. Another common feature is that all systems require a multidisciplinary feasibility study for the conversion of resources into reserves with the consideration of economic, mining, metallurgical, marketing, environmental, social, governmental factors, as well as statements on grade and dilution.

This of course raises the question as to who judges the quality of the study and determines when there is enough information.

The latest initiative is being implemented under the auspices of the CMMI (Council of Mining and Metallurgical Institutions) which has set up a working party to seek to establish a set of definitions of reserves and resources for international use.

Resource assessment case history

The major input to resource assessments can reasonably be expected to come from geologists. This case history relates to phosphorite deposits in the northern region of Saudi Arabia. The strata comprised a layered carbonate sequence with three major phosphorite horizons. The area assessed was 107 000 km^2.

Data were collected from outcrop, trenching and drilling.

Approach

The general approach adopted for this resource assessment included:

- compilation of historical and new drilling and trenching data;
- preparation of contour plans of thickness, grade, accumulation and overburden thickness;
- definition of mining parameters
    overburden thickness
    grade
    deposit thickness
    stripping ratio;
- a literature search on resource assessment nomenclature; the definitions used were based on the most common relevant usage, and derive from the mineral endowment concept of Harris and Agterberg (1981) and the resource classification system used by USBM/USGS (1980)—the former avoided economic issues and the latter was widely used in the world mineral industry; the USBM/USGS system had been in use for some time and its advantages and limitations are realized and reported in the literature—one such limitation being the lack of indication of mineral availability; this system, however, was primarily designed for resource assessment;
- the definition of geological and chemical terms;
- establishment of criteria for the resource classification such as
    minimum grade (based on beneficiation technology)

Table 3. *Definition of resource assessment terms*

Mineral endowment is defined as the total accumulation of elemental metal in a specified region which
- is above the crustal average
- has a minimum grade (q)
- has a minimum specified quantity (t) of mineralized material at grade (q)
- has a depth less than (h) metres

where q, t and h are based on current mining practice and are the criteria used as a basis for the resource assessment.
[For the phosphorite case history, $q = 15\%$ P_2O_5; $t = 1\,\text{m}$; $h = 50\,\text{m}$.]
The total accumulation is known or believed to occur, is in a form that can be extracted physically and has no economic criteria for exploitation.

Mineral endowment = total resources + mineral occurrences.

Total resources
- are concentrations of metals and minerals that are known or believed to occur
- meet the thickness, grade and depth criteria for the resource assessment
- could be commercially exploited in future say within 20 to 30 years given the necessary infrastructure

Mineral occurrences
- are concentrations of metals and minerals that are known or believed to occur
- can be extracted physically
- fall below the thickness, grade and depth criteria for the resource assessment
- are therefore unsuitable for exploitation in the foreseeable future

Identified resources
- are concentrations of metal/mineral that are known to occur to an acceptable level of confidence
- meet the thickness, grade and depth criteria for the resource assessment
- could be commercially exploited in future
- sampled at distances not greater than a specified maximum

Demonstrated resources
- are concentrations of metal/mineral that are known to a high level of confidence
- meet the thickness, grade and depth criteria for the resource assessment
- are divided into a measured, reserve base or indicated categories depending on stringency of mining criteria
- are sampled at distances not exceeding a specified minimum based on geostatistical drilling

Reserve base
- is that portion of the demonstrated resource which meets the thickness, grade and depth criteria for the resource assessment
- has a stripping ratio of less than 5:1
- includes economic, marginally economic and sub-economic categories, subdivisions that can be made when economic criteria are established
- represents that part of the identified resource with the greatest potential for current exploitation

Indicated resources
- are concentrations of metal/minerals that are known to a high level of confidence
- meet the thickness, grade and depth criteria for the resource assessment
- could be commercially exploited in future
- are sampled at distances between the specified minimum and maximum

Undiscovered resources
- are those concentrations of metal/mineral that are believed to occur based on sparsely located or no sampling points
- are thought to meet the resource assessment criteria
- could be commercially exploited in future
- are divided into hypothetical and speculative resources

Hypothetical resources
- are concentrations of metal/mineral believed to occur based on sparsely located sampling points in known areas
- are thought to meet the resource assessment criteria of thickness, grade and depth
- could be commercially exploited in future
- are sampled above the specified maximum but not more than three times that maximum

Speculative resources
- are concentrations of metal/mineral believed to occur based on geological extrapolation
- are thought to meet the thickness, grade and depth criteria for the resource assessment
- could be commercially exploited in future
- are sampled at distances greater than five times the specified maximum and are believed to exist

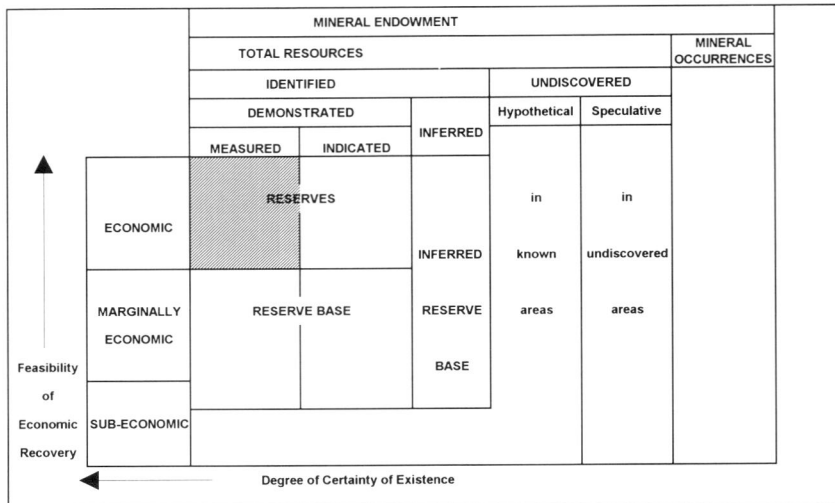

Fig. 2. Phosphorite project: definition of resource assessment terms.

maximum depth
minimum thickness;
(this was achieved by comparing operating mines in similar geographical and political environments)
- cross-traverse drilling for geostatistical analysis to establish range (continuity) to classify demonstrated resources;
- definition of resource assessment terms.

Definition of resource assessment terms

The nomenclature used in the resource assessment is shown in Fig. 2. More detailed definitions of the resource assessment terms are given in Table 3.

The resource assessment produced tonnages and average grades for each category of the classification which follows the logical hierarchy of the classification for each of the three phosphorite horizons in the region (Fig. 3). These estimates were accompanied by maps showing the spatial distribution of each class (Riddler et al. 1986).

The main principles of any estimates are contained in the USBM/USGS (1980) definitions which warrant that 'resources must be continuously reassessed in the light of new geological knowledge, of progress in science and technology and of shifts in economic and political conditions'. The current trends towards usage of systems based on the USBM/USGS system is clearly emerging as one way forward, and its application is demonstrated by this case history.

No reserves were estimated, but a full feasibility study has now been carried out on the reserve base.

The role of the British Geological Survey

In the early 1980s BGS (IGS as it was then) prepared a classification for mandatory corporate use influenced by the nomenclature proposed by the UN Committee on Natural Resources (Table 2) (HMSO 1982). This system has not been widely adopted as it failed, according to the IMM, to address the requirements of certain end-users particularly those seeking bankable documents.

The BGS now use as a general basis, the resource classification proposed by USBM/USGS (1980) in all discussions of mineral resources while at the same time monitoring, contributing towards and adopting in due course a consistent international system of classification and nomenclature. This assists the BGS in the provision of its mineral resource information products now in preparation on a consistent basis. These include an Industrial Minerals Map of the UK which will incorporate information such as ball clay resources in SW England. The BGS is also producing a Metallogenic Map of UK and compiling a Mineral Occurrence Database which will include resource information such as the gold occurrences in UK.

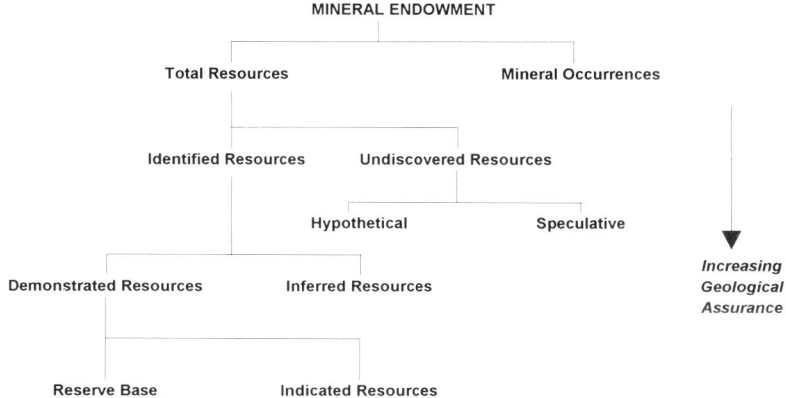

Fig. 3. Hierarchy of the resource assessment produced for each phosphorite member.

The way forward

More information is required to classify mineral reserves than mineral resources which involves greater expense. Companies therefore will only define mineral reserves required to maintain production over a fixed forward period, which is relatively short. Mineral reserves are therefore perceived by the public as being short term, limited and running out, which is not the case, because they usually form only a small part of the overall mineral resource inventory. This is a view expressed by the International Council on Metals and the Environment (Crowson 1992).

ICME point out that the confusion in the distinction between mineral reserves and resources has led to the misguided public perception that the world is running out of its mineral resources. This is leading to international policies which put more stress on protection and conservation of mineral resources rather than supporting more economical, environmentally efficient methods of exploration, production, use, recycling and disposal.

In conclusion, therefore, the importance of standardizing mineral resource and reserve classification nomenclature must be stressed so that it may be more easily understood by users and the public perhaps by adopting an approach similar to that suggested in this paper.

It is time for the minerals industry worldwide through its professional bodies, to pursue and co-ordinate efforts and arrive at a common approach to the classification and a consistent nomenclature for mineral resources so that the public no longer has to ask the question 'What is a mineral resource?'.

This paper is published with the permission of the Director of the British Geological Survey.

References

AGRICOLA, G. 1556. *De Re Metallica*. [Translated by H. C. Hoover & L. H. Hoover, Dover Publications Inc, New York, 1950].
ARMITAGE, M. G. & POTTS, M. F. A. Some comments on the reporting of resources and reserves. *This volume.*
AIMM/AMIC 1988/1992. *Australian Code for Reporting of Identified Mineral Resources and Ore Reserves.* Report of the Joint Committee of the Australasian Institute of Mining and Metallurgy and the Australian Mining Industry Council.
—— 1990. *Guidelines to the Australian Code for Reporting of Identified Mineral Resources and Reserves.* The Joint Committee of the Australasian Institute of Mining and Metallurgy and the Australian Mining Industry Council.
CROWSON, P. 1992. *Mineral Resources: The Infinitely Finite.* The International Council on Metals and the Environment.
HARRIS, D. P. & AGTERBERG, F. P. 1981. The appraisal of mineral resources. *Economic Geology,* 75th Anniversary Volume, 897–938.
HMSO 1982. *Select Committee on the European Communities: Strategic Minerals:* Session 1981–82 20th Report.
HUXHAM, C. 1987. *Quantitative Methods Unit 1.* University of Strathclyde Graduate Business School.
IMM 1991. *Definitions of reserves and resources.* The Institution of Mining and Metallurgy.
KENNEDY, G. 1983. *Invitation to Statistics.* Basil Blackwell, Oxford.
MCKELVEY, V. E. 1973. *Mineral Resource Estimates and Public Policy.* US Geological Survey Professional Paper, **820**.

RIDDLER, G. P. 1988. *Corporate Strategy for Mineral Exploration*. MBA Dissertation, University of Strathclyde Graduate Business School.

——, VAN ECK, M., ASPINALL, N. C., McHUGH, J. J., PARKER, T. W. F., FARASANI, A. M. & DINI, S. M. 1986. *Sirhan–Turayf Phosphate Project—An Assessment of the Phosphate Resource Potential of the Sirhan–Turayf Region*. Saudi Arabian Deputy Ministry for Mineral Resources Technical Record **RF-TR-06-2**.

SME 1991. *A Guide for Reporting Exploration Information, Resources and Reserves*. SME Planning Committee: Mining Engineering.

USGS 1980. *Principles of a resource/reserve classification for minerals*. US Geological Survey Circular, **831**.

Some comments on the classification of resources and reserves

M. G. ARMITAGE & M. F. A. POTTS

Steffen Robertson and Kirsten UK Ltd, Summit House, 9–10 Windsor Place, Cardiff CF1 3BX, UK

Abstract: Resource and reserve classification systems are used by stock markets, investors, mining companies etc. in order to make their decision making apparently more soundly based. In view of the subjectivity of geology and diversity of mining methods, in producing and adhering to such systems those charged with the responsibility of estimating reserves run the risk of inadvertently misleading these people and of risking their own integrity. This is not to say that we should not have classification systems at all, but more that they need to reflect the reality as much as the requirement. There are additional problems at present because of the number of classification systems currently in use and the widely different uses to which they are commonly put. The worldwide acceptance of a single terminology of resource and reserve classification would go some way to reducing these. Different classification systems may be required for different applications. In addition, a move to a more systematic quantitative assessment of confidence is proposed which may reduce some of the problems inherent in such an imprecise science.

The Institution of Mining and Metallurgy (IMM) has recently (1991) produced definitions and guidelines for the reporting of resources and reserves. The system of resource and reserve classification proposed joins a growing list of similar such systems already in use. The two most commonly used systems at present are that proposed in 1980 jointly by the Bureau of Mines (USBM) and the US Geological Survey (USGS), and that proposed in 1989 by the Australasian institution of Mining and Metallurgy (AIMM) and the Australian Mining Industry Council (AMIC). Other systems in regular use include those proposed by the Association of Professional Engineers of Ontario (APEO), the Society for Mining, Metallurgy and Exploration (SMME), and the United States Securities and Exchange Commission (SEC), respectively.

All these systems use slightly different terminologies, but, more confusingly, give slightly different definitions for the main terms common to each, i.e. 'resource' and 'reserve', and also 'measured', 'indicated' and 'inferred'. The result is that figures assigned to these terms in international literature, feasibility studies, and consultants and company annual reports are unclear unless the system being used is stated. This is, unfortunately, rarely the case.

More fundamental problems rest with the diversity of mineral deposits, the progressive stages of exploration and exploitation, the different mining methods, and the varying degrees of confidence in continuity and grade of mineralization, that the systems attempt to describe. In attempting to cover so much with so few categories the systems depend heavily on the judgement of the user. The constraints imposed by the systems often force reporters of reserves to use hybrid terminology or to ignore them altogether. More importantly, even with the best and most honest will in the world, and even if the system being used is stated, the results are open to misinterpretation and potential misuse.

This paper describes the workings of the USBM/USGS, AIMM/AMIC and IMM systems and discusses the differences between them and the potential problems with their use and interpretation. The USBM/USGS and AIMM/AMIC systems are discussed as they are probably the most widely known of the systems currently in use; the IMM system because it is the latest offering by a professional body. The paper does not seek to comment as to which of these is the 'better' system; they are all similar in style and preference is largely individualistic. Instead it concentrates discussion on specific problems facing both the user, in trying to adhere to any such system, and the interpreter (who may only have a limited knowledge of geology or mining) in trying to understand the meaning of the resulting classification. Although the paper stops short of proposing an alternative system, it does introduce a new approach to the subject which the authors consider would reduce some of these problems.

Current classification systems

USBM/USGS classification

This system, proposed in USBM & USGS (1980), is largely based on 'The McKelvey Box' of McKelvey (1972). The main terminology is summarized in Table 1. It covers everything from as yet undiscovered deposits to 'blocked-off' stopes. It also covers deposits not economic at present but which may become economic in the future. It can therefore be used to classify anything from a particular mine's economically mineable tonnage, to a country's mineral potential. 'Identified Resources' are blocks of ground for which grades, tonnage and geological continuity have been estimated, 'Subeconomic Resources' are parts of 'Identified Resources' which studies have shown could not be economically extracted at the time of determination, and 'Undiscovered Resources' are as yet undiscovered unspecified deposits projected to be present based on geological interpretation and guesswork. 'Reserves' are specific blocks of ground containing mineralization deemed to be economic to mine at the present day. 'Reserves', and 'Subeconomic Resources', are further classified as 'measured', 'indicated', or 'inferred' dependent on the degree to which the geological continuity has been confirmed. Briefly, for 'measured' the continuity is 'so well defined that the size, shape and mineral content are well established'; while for 'indicated' the sites available for inspection are 'too widely or otherwise inappropriately spaced to outline the ore completely or to establish its grade throughout'. 'Inferred' is estimated from 'assumed continuity or repetition for which there is geological evidence' but few samples or measurements.

AIMM/AMIC classification

Since its introduction this system, AIMM & AMIC (1989), has been the main challenger to the USBM/USGS system for international recognition. The system is similar in that it distinguishes between resources and reserves based on economic analysis (specifically on the results of feasibility studies), and between subcategories of both of these based on perceived geological continuity. It is dissimilar in precise definitions and in that it attempts only to address specific identified mineral occurrences and not 'global' resources. A 'Resource' is an 'identified *in situ* mineral occurrence' which has 'reasonable prospects for eventual economic exploitation'. An 'Ore Reserve' is that part of a 'Resource' that feasibility studies have shown 'could be recovered economically under conditions realistically assumed at the time of reporting'.

'Measured Resources' have 'confirmed continuity', while 'Indicated Resources' have a 'reasonable indication of continuity'. 'Inferred Resources' are estimates where the available data are of insufficient coverage to enable the 'geological framework to be confidently interpreted, and the continuity of mineralization to be predicted'. 'Measured Resources' require a 'firm understanding of the geology and controls of mineralization' and even 'Indicated Resources' 'assume continuity of mineralization'.

A 'Proved Ore Reserve' is that portion of a 'Measured Resource' that studies have shown is economically mineable even after account is taken of dilution, while a 'Probable Ore Reserve' is the equivalent portion of an 'Indicated Resource'. Table 2 shows how the system works in practice.

Table 1. *USBM/USGS system*

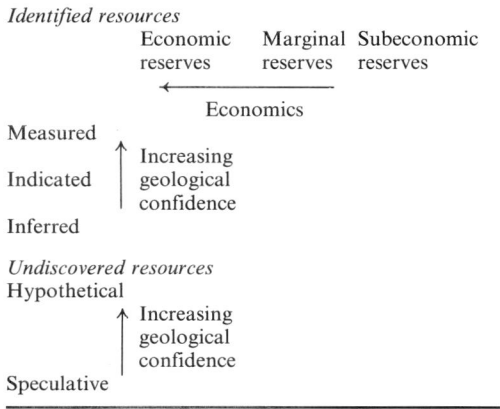

Table 2. *AIMM/AMIC system*

Ore reserve	Resource
1A Proved ore reserve ←	2a Measured resource
	↑ Increasing geological confidence
1B Probable ore reserve ←	2B Indicated resource
	↑ Increasing geological confidence
	2C Inferred resource
Economic input ←	
3 Pre-resource mineralization	

The guidelines to the AIMM/AMIC system also refers to an additional category 'Pre-Resource Mineralization'. This relates to specific mineral occurrences where there is as yet insufficient data to enable it to be classified as a 'Resource'.

IMM classification

These definitions and reporting guidelines, proposed in IMM (1991), are the product of a working party set up in 1989, the aim of which was to impose rules on, and maintain the credibility of, the reporting of resources and reserves in the UK. The London Stock Exchange has this year made it a condition of listing that companies adhere to these guidelines in the same way as the Australian Stock Exchange has for some time with the AIMM/AMIC system.

The IMM definitions are clearly based on the AIMM/AMIC system but there are subtle, yet significant, differences between the two. Both restrict their attention to specific identified mineral occurrences only, both distinguish between resources and reserves as a function of the result of feasibility studies and both subdivide resources based on perceived geological continuity. They differ, however, in that their sub-division of reserves which in the IMM system is a function of the depth of the economic studies carried out as well as the perceived geological continuity, not solely geological continuity as is the case with the AIMM/AMIC system. Also, in the IMM system both resources and reserves may be *in situ* or 'mineable' dependent on whether the quoted tonnages allow for mining dilution and losses. In addition the 'inferred' category of resource has been dropped. Table 3 gives the terminology and the workings of the system.

Table 3. *IMM system*

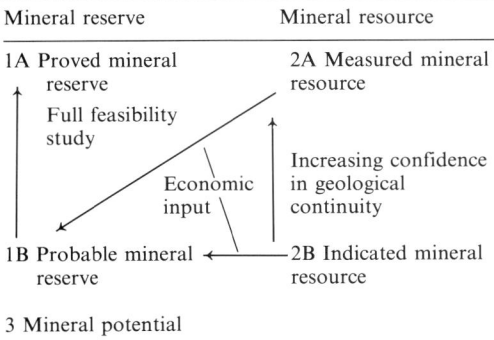

Briefly, a 'Measured Mineral Resource' is defined as that portion of a 'Mineral Resource' where 'the geological character, continuity, grades and nature of the material are so well defined that the physical character, size, shape, quality and mineral content are established with a high degree of certainty'. An 'Indicated Mineral Resource' is simply that portion where the 'sites used for inspection and sampling and measurements are too widely or inappropriately spaced to enable the material or its continuity to be defined or its grade throughout to be established'.

A 'Proven Mineral Reserve' is that portion of a 'Measured Mineral Resource' which a full feasibility study (sufficiently detailed to enable a decision on implementation to be made) has shown to 'justify extraction at the time of determination'. A 'Probable Mineral Reserve' is that portion of a 'Measured' and/or 'Indicated' Resource' which economic studies, not of the detail of a full feasibility study, have similarly shown to justify extraction.

The system also contains a 'Mineral Potential' category which corresponds to the 'Pre-Resource Mineralization' category of the AIMM/AMIC system.

Discussion

Problems with terminology

The same terms are used by all the three systems discussed here but within each system they have slightly different definitions. USBM/USGS 'resources' may be uneconomic and even undiscovered, while AIMM/AMIC and IMM 'resources' must be both identified and potentially economic. AIMM/AMIC 'reserves' take account of mining dilution, but this is not necessarily the case with IMM 'reserves'. Also IMM 'proven reserves' are ready to mine, while AIMM/AMIC 'proven reserves' may still not have the backing of a feasibility study and may simply be 'resources' roughly discounted at an early exploration stage to allow for likely mineability induced losses and dilution. Also the IMM 'resource' subdivision is more subjective than the AIMM/AMIC subdivision which, for example, requires 'Measured Resources' to have 'confirmed continuity' as opposed to simply being known with 'a high degree of certainty'. This situation is further complicated if the terminology of other systems not discussed in detail here are also considered. Those proposed by the Association of Professional Engineers of the Province of Ontario

(APEO) and the Society for Mining, Metallurgy, and Exploration (SMME), for example, use similar terms (e.g. ore reserve, geological reserve, in situ reserve and mineable reserve; and inferred, indicated and measured resources, and probable and proven reserves respectively) but in a slightly different way again.

The result is that reported resources and reserves are at best unclear and at worst misleading unless the system being used is stated. As mentioned in the introduction to this paper, this is rarely the case as can be seen in a quick scan through any of the many mining publications. The problem would clearly be best addressed by the adoption of one system, such as has been done by the Australian Stock Exchange and more recently the London Stock Exchange. There is little doubt that adoption of the AIMM/AMIC system by the Australian Stock Exchange has improved the standard of the reporting of resources and reserves in that country as can be seen in mining publications, in feasibility documents and in company annual reports.

In this regard, the authors randomly selected Annual Reports from 15 UK listed companies all of which reported and discussed resources and reserves. Of these only two indicated which classification system they had used (in both cases the AIMM/AMIC system), two had instead described and used their own systems (for example type 1 reserve, type 2 reserve etc.) and the remainder gave no indication of resource or reserve definitions at all.

Mining is a worldwide activity and many companies run, or fund, operations in different continents as well as different countries. Yet though the problem has been addressed by several national institutions and to some extent locally controlled, there can be little doubt that the terminology used worldwide for the reporting of resources and reserves is in a confused state. It is the opinion of the authors that what is required is one internationally accepted system of resource and reserve classification.

Problems with usage

The USBM/USGS, AIMM/AMIC and IMM systems can be used to classify deposits in the process of being evaluated. They can be used to classify tonnages of prospective mines from the early exploration stage to the production of a full feasibility document. In addition the USBM/USGS system covers unexplored and even unknown deposits and this can be used to classify a country's mining potential.

In addition to the above it is important, from a planning point of view, that a mine is able to quantify the tonnages of ore that are in the different stages of preparation for mining i.e. 'ore blocked out', and 'ore developed'. This was covered by abandoned terminology such as 'ore in sight' which was in use when diamond drilling was not the prevalent exploration technique but is not covered by modern systems such as those discussed in this paper here which all now use their extremes at the exploration phase (using previous systems there would be no 'proven' or 'possible' ore until mining commenced). This is a weakness of such systems as despite this they are still commonly used on operating mines for this specific purpose. This results in the situation that blocks of ground classified as 'proven' in feasibility documents are reclassified as 'probable' on the commencement of mining, and/or in the generation of locally adapted hybrid systems of classification.

The use of such a few categories to cover a wide variety of mineral deposit types and mining methods causes additional problems. The classification of vein gold deposits is a very different business to that of limestone deposits. It is hard to see how a deep sub-outcropping shear zone gold deposit, which is planned to be mined from a shaft, can have any 'proven' or 'measured' tonnage even using the newer definitions of the systems discussed here until it has been extensively exposed in underground development. This is supported by the major difficulties with geological continuity recently experienced by mines such as Big Bell in Australia, and Goldstream, Nickel Plate and Silbak-Prem in Canada (H. G. Taylor, pers. comm.). This despite the owners of these properties having completed sufficient studies to attract the development capital required.

It is, however, also hard to envisage many investors risking money in a venture that at the end of its full feasibility study has no 'Proved Mineral Reserve' or 'Proved Ore Reserve'. It is necessary therefore for classification systems to take account of the ability of different deposits to achieve the various requirements of different categories while an alternative would be to have different systems for different types of deposits and mining methods.

In summary, therefore, it is regarded by the authors as important that any proposed international classification system should cover the requirements of all the uses to which it is likely to be put and not just the requirements of deposits during the build-up to mining. It should also be able to be used to classify meaningfully, and reflect confidence in the profitability of mining of, all deposit types using any mining style. At the same time it must still be simple, and capable of being clearly understood by all those who are likely to use it.

Problems with subjectivity

The three systems discussed here subdivide to a similar degree. Each uses a handful of phrases to describe confidence in the presence and economics of exploiting any type of non-petroleum mineral deposit (though the AIMM/AMIC system has a separate terminology for coal deposits) using any type of mining method. In so doing each has trodden a fine line between over-categorization and under-distinction. The result in all cases is a subjective system that relies heavily on experience and knowledge in its use. All the systems contain 'grey' areas within which different geologists, in trying to adhere to any one system (even using the same information), would classify the same body differently dependent on their view of the geological continuity. This is not a function of the respective ability of different geologists but rather because geology is not a subject that can be easily boxed and because with mineable reserves one is predicting not measuring. It is consequently hard to make realistic comparisons for quoted resources and reserves for different deposits unless they have all been produced by the same person or team or they are supported by details regarding the geology and the extent of sampling. Clearly to simply report a resource or reserve figure and a corresponding classification is insufficient, and when used in this way, because of the subjectivity of geology and the broadness of the current systems, a classification system may become a liability. Despite this, the facility to compare deposits and rank anticipated benefits and returns on investment is a major reason for having the classification in the first place.

Suggestions that geostatistical techniques on their own can quantify error and form the basis of classification are flawed in most cases. Certainly they can quantify grade uncertainty and to a limited extent geological uncertainty, but only down to the spacing for which there are data (i.e. the drillhole spacing). Where an underground mine is being planned this is often insufficient to determine the variations in geometry and grade at the detail required to predict mineability on a stope scale.

There will always be a subjective element in the reporting of resources and reserves, yet in placing too much emphasis on broad classification systems we run the risk of both losing good information and of implying a confidence to the resulting classification that cannot be justified. The effect of this can be reduced by the inclusion of details regarding the geology and sample coverage. More fundamentally, however, as such detail will be left out at some point and anyway is not meaningful to all parties concerned, there may be some room for reducing the problem by either increasing the number of categories or alternatively by making classification systems more end-use specific. Planned underground mines for example have different requirements regarding geological continuity than open pit mines. An alternative would be to introduce a way in which uncertainty regarding geological and grade continuity and its relevance to the proposed mining method could be ranked in a more methodical manner. This is discussed more in the next section of this paper.

Summary

Three main causes of concern with current classifications of resources and reserves have been discussed. These are:

- that there are several classification systems currently in use all of which use the same, or very similar terminology, but all of which give these terms significantly different definitions;
- that the most commonly used classification systems were not designed to cover all the situations in which they are now used;
- that, by attempting to categorize an area as subjective as geology, a certainty and distinction is implied that cannot be justified nor would be likely to be repeated if the same work was undertaken by different people.

These areas of concern can cause problems enough when viewed in isolation but when combined, however, the problems multiply. For example, if a proven, probable and possible reserve is reported without reference to either a given system or the amount of supporting evidence, it can simply mean that in the view of the writer the first reported tonnage is considered better known than the second which is in turn better known than the third. In reading the report, not only do we not know on what basis the given tonnages were allocated to the various categories, we do not even know what these categories mean.

The first two areas of concern could be reduced if a single terminology for resource and reserve classification which covered all the requirements of governments, mining companies, banks and shareholders could be internationally agreed. Different classification systems may be required for different applications but if so these should be compatible not contradictory. The third area of concern is inherent to the subject. A suggested method for reducing this subjectivity is given in the following section.

An alternative approach to classification

Most proposed classification systems, as well as defining terms and categories, also highlight those aspects that should be considered by 'the responsible person' in both calculating and categorizing the deposit grades and tonnages. These include the reliability of the geological data and interpretation; the perceived geological continuity and deposit type; the drilling technique and core recovery; the type of sampling, the sampling coverage and the reliability of the sampling method; the sample preparation and sample analyses; the variability of the assays; the planned mining method and impact on reserves of mineability and selectivity; the impact of any geotechnical problems; the processing route and expected recovery; the total costs and consequent likely cut-off grade; and land ownership aspects and mineralized extensions.

A major problem for 'the responsible person' when categorizing a calculated tonnage and grade is in quantifying the effect, and relative importance, of all these on the confidence in the figures calculated. In shear-zone gold deposits the major uncertainty may be geological continuity; in complex massive sulphide ores and refractory gold ores, the recovery may be more uncertain; in highly folded and faulted ores then mineability may be the key. Where 'the responsible person' has experience of many different deposit types there is more chance that more accurate estimates of these effects will be made. Where 'the responsible person' has more restricted experience then there is more potential for large errors of judgement.

A solution may be to incorporate these aspects more directly into the classification process. The authors envisage a system in which each perceived area of uncertainty is given a quantitative rating reflecting its potential impact on the calculated resource or reserve. The system would need to give examples for each rating. For example, regarding continuity a shear-zone gold deposit in an Archaean greenstone belt intersected by drillholes only would be an example of a higher uncertainty rating than a chromite seam cropping out on the Bushveld Complex or the Great Dyke. The respective ratings could then be cumulated (in a manner reflecting their relative importance) and the sum used as the basis for categorization. A degree of subjectivity would still remain, but in asking all the relevant questions, and giving examples for each rating, the system would force all issues to be addressed and enable 'the responsible person' to estimate better how the various factors relate to his/her deposit compared to how they do at other deposits elsewhere. The resulting classification should then be more robust. There may also be an additional benefit here in that the classification system could be used to target the exploration strategy better to the requirements of the type of deposit and planned mining method.

The production of such a system would require input from experienced workers in all fields of relevance to resource and reserve classification. The result may be a reduction in the subjectivity of the process and an improved ability to compare resource estimates produced by different groups.

References

AUSTRALASIAN INSTITUTE OF MINING AND METALLURGY 1989. *Australasian Code for Reporting Identified Mineral Resources and Ore Reserves.* Report of the Joint Committee.

MCKELVEY, V. E. 1972. Mineral resource estimates and public policy. *American Scientist*, **60**, 32–40.

THE INSTITUTION OF MINING AND METALLURGY 1991. *Ore and Reserves Working Party Report.*

US BUREAU OF MINES & US GEOLOGICAL SURVEY 1980. *Resource/Reserve Classification System.* US Geological Survey Circular C831.

Classification of mineral reserves in the former Comecon countries

Z. JAKUBIAK[1] & T. SMAKOWSKI[2]

[1] *Consulting Geologist, 73 Roxburgh Road, London SE27 0LE, UK*
[2] *Mineral Economy and Energy Research Centre, Polish Academy of Sciences, Ul. Wybickiego 7, 31-261 Krakow, Poland*

Abstract: Classification of reserves in the former Comecon countries is based on two reference axes: the abscissa represents the degree of reserve identification and the ordinate shows the possibility of economic utilization. Accordingly, reserves are classified as documented and prospective and as economic and uneconomic.

Documented reserves are divided into categories A, B, C_1 and C_2. The first two of these categories relate only to developed and blocked reserves, with category A being in practical terms restricted to reserves under advanced exploitation, often at a stage nearing exhaustion. Category C_1 comprises reserves which have been identified and examined to such an extent as to enable a positive definition of their suitability for exploitation; these reserves are often in early production phases. Category C_2 relates to reserves which have been identified and documented at a preliminary stage only. Terminology of prospective reserves differs from country to country. In the former USSR and in the majority of other countries that adopted this classification, prospective reserves are divided into categories P_1, P_2 and P_3. However, in the classification used in Poland, on which this paper is based, prospective reserves are divided into prognostic reserves in categories D_1, D_2 and D_3 and theoretical reserves in category E, categories D_1 and D_2 being equivalent to categories P_1 and P_2 respectively and combined categories D_3 and E roughly equivalent to category P_3. Included in category D_1 are reserves either in areas adjoining documented reserves classified in category C_2 or reserves that occur at depths considered to be uneconomic. Assigned to category D_2 are reserves inferred from surface showings and indirect indications and quantified on the basis of statistical analysis of reserves known in similar mineralized structures in the proximity, whilst category D_3 includes reserves inferred from indirect indications alone. Finally, theoretical reserves E are deducted only from regional structural, lithological, metallogenic and other considerations.

Reserves in categories D_2 and above are divided along the ordinate into economic and uneconomic on the basis of the so-called criteria of balance defined by authorities in charge of the project(s). In centrally-planned economies the selection of these criteria generally reflected prevailing economic policies and priorities rather than profit considerations. Another set of criteria called the criteria of workability is then applied to economic reserves to define mineable reserves and these are called industrial reserves.

Put in the context of the IMM definitions, categories A and B are equivalent to measured mineral resources, categories C_1 and C_2 are broadly comparable to indicated mineral resources, whilst prospective reserves straddle the fields covered by indicated resources and mineral potential. Industrial reserves are comparable to the mineral reserve category as defined by the IMM definitions.

Historical background

Principles of the mineral reserve classification used in the former Comecon countries were developed in the Soviet Union in the early 1940s. The idea was that reserves should be assigned to clearly defined categories combining measures of confidence of reserve identification with criteria of economic feasibility to enable the State to keep balance books of mineral reserves for planning purposes. Legislation was passed in the Soviet Union (in 1941) and in other Comecon countries to enforce the system. Additional laws, instructions and guidelines were laid down in an attempt to standardize procedures and methods of reserve delineation and evaluation and the type of required documentation. All mineral reserve inventories had to be approved by the State and then entered in balance books of reserves. It has to be borne in mind that this practice has always been a matter of inventory rather than profit.

With the adoption of the system in other Comecon countries, it was inevitable that additions and amendments had to be made in various countries to adopt the system to differing local conditions and requirements. Generally, such amendments were minor and concerned only those categories of reserves that were defined with a lesser degree of confidence, i.e. potential reserves. However, effects of these additions and amendments have accumulated over the years eventually creating significant divergences both in the terminology and in the classification criteria. Numerous meetings and conferences were held in an attempt to standardize the approach to potential reserves, the most important of which was probably the conference in Leningrad in 1976, but they have failed to achieve their desired effect. Even the USSR, the creator of the system, finally revised its own original classification and introduced an amended system in 1981 (Diatchkow 1993).

Whilst divergencies developed in the classification of potential reserves, definitions of the main four categories of reserves, A, B, C_1 and C_2, and their use have remained more or less as they were originally expressed throughout the former Comecon countries.

In preparing this paper, the writers relied mainly on Polish sources of information.

Definitions

The Polish mining legislation defines the mineral deposit as a natural or artificial mineral concentration exploitation of which can bring an economic benefit (Zółtowski 1964). The economic benefit is not understood as the ability to generate a profit but includes social benefits.

The term 'reserve' denotes a mineral inventory that comprises identified quantities of mineral (as defined above), some of which may be economic and workable at a given time, and partly identified and/or postulated quantities of mineralization which either cannot be recovered by known mining and processing methods or because their geological occurrence is not adequately known. The term 'resource' has not been used despite growing pressure to restrict the use of the term 'reserve' to identified reserves which are subject to economic feasibility studies.

To determine the amount of data necessary to assign reserves to various categories prior to an investment decision, mineral deposits were divided into groups reflecting the form and complexity of mineralization. In the USSR mineral deposits were divided into four groups on the basis of geological structure and complexity of mineralization and into five groups on the basis of a deposit type. In other countries, including Poland and Czechoslovakia, mineral deposits were divided into three groups depending on the size and on variations in form, composition, type of mineralization and grade. These groups were defined as follows.

Group I—deposits characterized by simple geometry and structure, uniform mineralized widths, generally in excess of an economically marginal width, and a fairly uniform distribution of grade and deleterious constituents, e.g. simple undeformed or weakly deformed stratiform deposits or large intrusive massifs of a uniform composition.

Group II—deposits characterized either by simple structure but showing irregular distribution of grade and deleterious constituents or by complicated structure but characterized by regular distribution of grade and deleterious constituents; mineralized widths in such deposits commonly approach the marginal economic width, e.g. stratiform deposits of variable width or irregular mineralization contained in medium and small intrusive bodies.

Group III—deposits characterized by complex geometry, highly variable widths and/or grades and other deposits that do not fall in Groups I and II, e.g. highly deformed and variable stratiform deposits, polymetallic vein deposits, weathering residues, detrital deposits, salt domes, etc.

Principles of reserve classification

The reserve classification system of the former Comecon countries is based on two reference axes (Table 1). The abscissa, from right to left, indicates increasing degree of reserve identification. The ordinate, going upwards, indicates increasing possibility of economic utilization of reserves.

As seen in Table 1, geological reserves, understood as the total mineral reserve inventory, are divided along the abscissa into documented and prospective reserves, with the latter subdivided into prognostic and theoretical reserves. The key elements and definitions are as follows.

Documented reserves. Reserves of which parameters, including quantity, grade, quality and depth of occurrence have been investigated on a specified grid base and which are appropriately documented. The density of the grid is a vital element guiding the scope of investigations required for various categories. Strict regula-

Table 1. *Classification of mineral reserves used in Poland*

		DEGREE OF RESERVE IDENTIFICATION →							
		GEOLOGICAL RESERVES							
		DOCUMENTED				PROSPECTIVE			
						PROGNOSTIC			THEORETICAL
		A	B	C_1	C_2	D_1	D_2	D_3	E
ECONOMIC	WORKABLE (INDUSTRIAL)								
	UNWORKABLE								
	UNECONOMIC	POTENTIAL RESERVES (RESOURCES)							

↑ POSSIBILITY OF ECONOMIC UTILIZATION ↑

tions have been passed in this respect and must be adhered to under all circumstances. Equally strict requirements have been maintained for reporting standards and documentation required for each category. The documented reserves are subdivided into categories A, B, C_1 and C_2. Categories A and B are defined with a very high degree of confidence and fully assessed from the economic point of view. In fact, requirements for the delineation of reserves in category A are impossible to fulfil even at working mines until reserves are nearly exhausted. It is a common practice, therefore, to quote blocked out reserves as A + B. Reserves in categories C_1 and C_2 require further work to improve the determination of their parameters and quality as well as further work to determine the choice of appropriate methods of access, exploitation and processing. This is particularly relevant to C_2 reserves that are determined from isolated openings and/or from widely spaced boreholes or outcrops.

Prognostic reserves. Prospective reserves that are inferred from indirect indications, showings and isolated sampling (Gałkiewicz 1962). They include three categories of reserves:

D_1—reserves inferred from indications (geochemical, geophysical etc) and/or showings and, sometimes, identified by isolated drillholes but not sufficiently to be included in category C_2, e.g. a salt dome identified by geophysical indications and confirmed with a single borehole;

D_2—reserves inferred from indications and surface showings, such as outcrops, aureoles of disseminated mineralization, alteration haloes etc. and quantified on the basis of statistical analysis of reserves known in close proximity;

D_3—unquantified reserves inferred by drawing analogies with distant producing areas on the basis of indirect indications alone.

Theoretical reserves. Prospective and unquantified reserves that are expected to occur in a given geological environment but that have not yet been discovered in the area studied or in its proximity. Possibilities of the occurrence of such reserves are deduced from theoretical considerations taking into account lithological, structural, metallogenic and other relevant data. These reserves are denoted by the letter E.

In the former USSR and in the majority of other countries that adopted this classification, prospective reserves are divided into categories P_1, P_2 and P_3. Prognostic reserves in categories D_1 and D_2 are equivalent to categories P_1 and P_2 respectively and combined categories D_3 and E are roughly equivalent to category P_3.

The divisions along the ordinate apply to documented reserves and to prognostic reserves in categories D_1 and D_2. Factors governing the main division into economic and uneconomic reserves are called criteria of balance and take into account current mining methods and practices, processing technology and economic feasibility criteria. Another set of criteria is

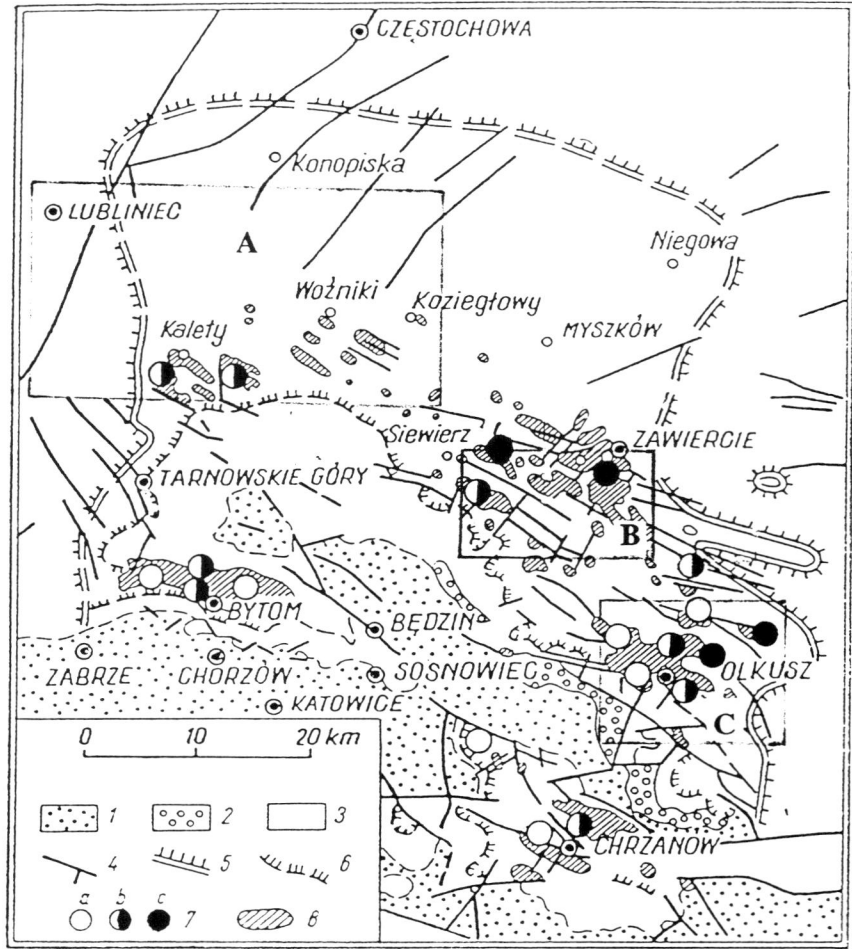

Fig. 1. Areas selected for regional exploration; background map from Przenioslo (1978). 1, Carboniferous; 2, Permian; 3, Triassic; 4, faults; 5, boundaries of ore-bearing dolomite occurrence; (d, dolomite; w, limestone); 6, ore-bearing dolomite outcrops; 7, zinc/lead ratio in the ores (a, more than 5; b, from 5 to 2; c, less than 2), 8, zinc and lead haloes.

applied to economic reserves documented as category B, and occasionally as C_1, to select mineable portions of these reserves under a chosen development option. The key elements and definitions are as follows:

Economic reserves—reserves that meet given criteria of balance;

Uneconomic reserves—reserves which do not meet the above criteria at a given time but have a reasonable potential of becoming economically viable in the foreseeable future;

Workable (industrial) reserves—part(s) of economic reserves selected for extraction in the course of the mine design; their calculation takes into account mining losses and dilution;

Unworkable reserves—part(s) of economic reserves not included in workable reserves.

A variety of other terms, which are not part of this classification, are in use for specific purposes in various countries. The scope of this paper does not allow for any further elaboration on this subject.

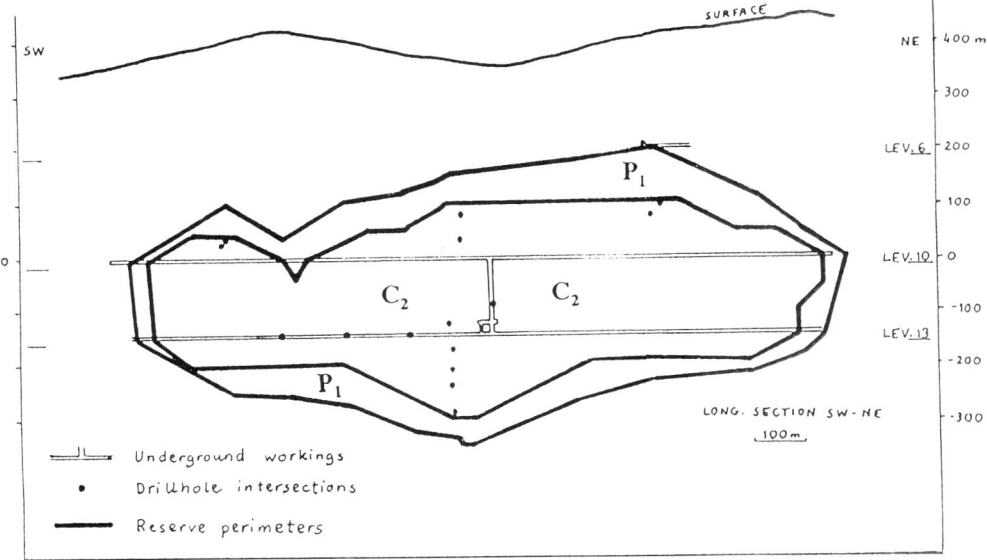

Fig. 2. Delineation of reserves in the Silver Vein, Rožňava, Slovakia (courtesy of Ing J. Popreňak).

Technical appraisal

A process of a mineral deposit delineation begins with investigations of large geological units with the view to identifying the most promising portion of the area. This is the stage at which a geologist can stretch his imagination to deduce what is called theoretical reserves or perhaps can even risk his reputation a little by announcing a discovery of a D_3 reserve. For instance, a geological unit that is known as Palaeozoic Cracovides and runs along the eastern margin of the Upper Silesian Coal Basin was, in the 1970s, identified as a potential host for porphyry-type Cu–Mo mineralization. The areas highlighted as having theoretical and D_3 reserves included Pilica (100 km^2), Bebło (65 km^2) and Zawiercie–Myszków (80 km^2).

In the succeeding phases of the mineral deposit appraisal the following points must be addressed:
(1) form and composition of the deposit;
(2) perimeters of the deposit;
(3) type and quality of the mineral;
(4) geological, hydrogeological and geotechnical conditions that can affect mining;
(5) estimate of geological reserves (global reserve) with subdivisions into economic and uneconomic reserves;
(6) occurrence of accompanying minerals and deleterious constituents.

The following scope of work is required for the various categories of reserves.

Categories D_2 and D_3—preliminary regional exploration involving desk studies to compile and analyse geological information relevant to the mineral potential of the area under consideration and field reconnaissance to confirm reported or suspected showings of mineralization. Postulated reserve areas are shown on small or medium scale maps and parameters are established on the basis of statistical analysis of corresponding parameters of reserves known in the same geological unit or in a corresponding unit of another, not too distant, area. The outlined procedure is well illustrated by exploration for Pb–Zn deposits in Upper Silesia and adjoining regions of southern Poland. Lead–zinc mining in Upper Silesia has a long history going back to the thirteenth century. Geology is well documented and it is well established that Pb–Zn mineralization follows definite horizons within extensive zones of dolomitization in the Lower Shelly Limestone of mid-Triassic age; the extent of dolomitization, therefore, demarcates the extent of the prospective area (Fig. 1). Initial estimates of prognostic reserves covered the whole area of dolomitization where the main ore-bearing horizon was known to be at a depth of less than 1000 m. The marginal criteria selected for that programme included a cut-off grade of 3% Zn + Pb and a minimum width

Table 2. Requirements for density of observation points for mineral reserve evaluation in Poland, after Żółtowski (1964), updated by the Central Geology Authority regulations of 1980

Mineral deposit	Group	Distance between observation points in metres			
		C_2	C_1	B	A
Hard coal, sapropel, bituminous shale	I	3000–4000	1500–3000	1000–1500	Workings < 500 m apart
	II	1500–3000	1000–1500	500–1000 incl. min. one working	Workings < 300 m apart
	III	1000–1500	500–1000	Workings 250–500 m apart	Workings < 200 m apart
Brown coal	I	1000–2000	500–1000	250–500	
	II	500–1000	250–500	125–250	Exploitation
	III	250–500	125–250	75–125	
Iron ore, sulphur	I	2000–5000	500–2000	200–500	Workings on 2 sides 100–200 m apart
Barite, phosphates	II	500–2000	200–500	100–200	Workings on 2 sides 50–100 m apart
	III*	200–500	100–200	Workings on 2 sides 50–100 m apart	Workings on 3 sides 50–100 m apart
Copper ores	I	2000–5000	800–2000	300–800	Workings on 2 sides 200–300 m apart
	II	1500–3000	500–1500	200–500	Workings on 2 sides 100–200 m apart
	III	500–1500	200–500	Workings on 2 sides 100–200 m apart	Workings on 2 sides 50–100 m apart
Lead–zinc ores, nickel ores, tin ores	II	400–600	200–400	100–200	Exploitation
	III	200–400	100–200	75–100	
Halite	I	One workings per 9–12 km^2	2000–3000	Workings on 2 sides 1000–2000 m apart	Workings on 2 sides 200–600 m apart
	II	One working per 4–6 km^2	1000–2000	Workings on 2 sides 600–1000 m apart	Workings on 2 sides 100–300 m apart
	III†	One working per 1–3 km^2	500–1000	Workings on 2 sides 100–300 m apart	Exploitation
Potassium and magnesium salts in salt domes		Drillholes 1000–1500 m apart	Workings on 2 sides 200–400 m apart	Workings on 2 sides 100–200 m apart	Exploitation
Limestone, gypsum, anhydrite, magnesite	I	3–5 workings per 0.5 km^2	300–600	150–300	
	II	300–600	150–300	75–150	Exploitation
	III	200–300	100–200	Workings on 2 sides 50–100 m apart	Workings on 2 sides 25–50 m apart
Clays for fire-resistant products and decorative ceramics	I	One working per 0.25–0.5 km^2	200–500	100–200	Exploitation
	II	200–400	100–200	50–100	
Clays for building ceramics and cement industry	I	300–500	200–300	100–200	Exploitation
	II	200–300	100–200	50–100	
Road and building stone	I	3–5 exposures or workings per 1 km^2	300–600	150–300	
	II	5–8 exposures or workings per 1 km^2	150–300	100–150	Exploitation
	III	150–300	75–150	50–75	
Sand and gravel	I	250–350	150–250	75–150	Exploitation
	II	150–250	75–150	50–75	
Glass sand, foundry sand	I	250–500	150–250	75–150	
	II	150–250	75–150	50–75	Exploitation
	II	50–100	25–50	15–25	

* The lower range distance applies to barite deposits.
† Group III applies only to magnesite deposits.
The maximum distances should never be exceeded.

of 2 m. A factor of 0.1 was used to derive tonnages, based on experience showing that one in ten drillholes in similar areas is likely to intersect mineralization satisfying these criteria. Three areas, shown in Fig. 1 as A, B and C, were selected for more detailed exploration. Prognostic reserves in Area A were assigned to category D_2 and prognostic reserves in Areas B and C, adjoining documented reserve areas and already explored to a degree, were assigned to category D_1. All prognostic reserves inferred between depths of 500 m and 1000 m were assigned to category D_2.

Category C_2—preliminary appraisal based on: data obtained from geological maps; examination of outcrops and existing workings; geophysical interpretation; assays and small-scale sample testing; and, if justified, preliminary technological and engineering studies. Boundaries of the deposit are delineated by geological mapping and geophysical surveys and interpolated and/or extrapolated from isolated outcrops, workings and drillholes. The degree of identification is such that various interpretations of the deposit are possible. If applicable, protective pillars should be outlined and reserves contained in them estimated as a proportion of the total reserves. Hydrogeological, engineering and other conditions should be outlined in general terms. An example from Slovakia (Fig. 2) shows a longitudinal section of a complex Cu–Ag vein with subordinate antimony and mercury evaluated by diamond drilling and underground development on the spacing required for the delineation of reserves in category C_1. However, as the variability of width and grade proved to be more complex than anticipated, reserves could only be classified in category C_2; a rim around it was annotated as a P_1 resource, a step below category C_3.

Category C_1—appraisal assuring approximate elucidation of points (1) to (6) and all other factors that are likely to affect the possibility of mining the deposit. Reserve perimeters are delineated from outcrops, workings and drillholes laid out on a suitable grid, with verification carried out to check geological interpretations and analytical results.

Category B—deposit delineation enabling unambiguous interpretation of points (1) to (6). The appraisal is based on geological data collected on an infill grid, systematic analytical work, pilot plant testing and, in the case of shallow or already exploited deposits, also on trial mining and processing.

Category A—detailed deposit delineation and unquestionable elucidation of all factors that are relevant to exploitation and treatment. The appraisal must be based on trial production or ongoing production and on process testwork on bulk samples. Reserve blocks must be bordered by outcrops, drillholes and workings at strictly specified intervals.

It is essential that the delineation of reserves is conducted on a specified grid. Detailed guidelines in this respect have been legislated (Table 2). Although based on experience and common sense, a forced adherence to these guidelines has been criticized and resisted. Even stricter legal requirements regulate the type and scope of documentation required for each category of reserves, defining even such details as scales of geological maps required for each category of reserves (Żółtowski 1964).

A very positive feature of this classification is the fact that the divisions between various reserve categories have been designed with the specific aim to control a decision tree in the sequential process of mineral exploration, evaluation and mine development. Thus, delineation of an area of prognostic reserves D requires a decision whether to undertake detailed exploration, which, if successful, results in the delineation of reserves in category C_2; this, in turn, requires a decision whether to undertake preliminary evaluation. If successful, this evaluation work results in upgrading all or part of the reserve to category C_1 and prompts a decision whether to undertake detailed evaluation and to select parts of the deposit for such investigation. Conceptual studies of the deposit development and reserve utilization as well as technical and economic preview of a mine design are carried out as part of the work programme. This phase, if successful, results in the delineation of reserves in category B, which justifies the last pre-production phase involving the selection of workable reserves on the basis of a mine design and a long-term production scheduling.

Criteria of balance

Most of the current reserve inventories in the former Comecon countries predate 1992. Feasibility criteria that defined conditions at which these mineral reserves could be regarded as being suitable for commercially justified exploitation are called the criteria of balance (Żółtowski 1964; Kozubski 1965). They were formulated either for groups of deposits sharing the same characteristics or for single deposits by teams of advisers working for ministries and other government departments responsible for the development of projects. Work of such advisers was regulated by a complex set of official

Table 3. *Feasibility criteria for copper deposits in the Lubin–Głogów district of Poland (from Wanielista & Butra 1991)*

Depth interval (m)	Cut-off grade (%Cu)	Minimal average metal content in a drillhole intersection or in underground profile (including Ag) (%Cu)	Minimal metal content in a deposit or in a documented area (including Ag) (%Cu)
<600	0.6	1.0	1.7
601–1200	0.7	1.1	2.0
1201–1600	0.7	1.2	2.2
1601–2000	0.8	1.3	2.4

instructions, directives and guidelines designed to standardize the criteria used. In Poland, it was accepted to use the so-called simplified criteria for the definition of undeveloped economic reserves in categories D_2, D_1 and C_2 and, occasionally, in C_1 and the so-called detailed criteria for the definition of undeveloped economic reserves in categories C_1, B and A and for the definition of all categories of reserves in working mines (Rutowski 1967). At various times preference was given either to criteria based on capital expenditure requirements and payback period, i.e. mine development would go ahead when the documented reserve was adequate to generate revenue for the capital expenditure repayment (Żółtowski 1964), or the criteria based on parameters of investment effectiveness and a marginal reserve value (Wanielista 1976) or to criteria based on the so-called cost limits represented by maximum production costs when getting the mineral from domestic sources (Popreňak 1993a, b) or to criteria determined by the condition that the output value per unit must be at least equal to current unit production costs (Wanielista & Butra 1991). All these criteria are now being superseded by the use of discounted cash flow techniques.

Criteria established in the 1960s were often too lenient resulting in large funds having been authorized for the evaluation of deposits of dubious economic potential, i.e. low-grade iron ores in central Poland or small brown coal deposits in the west of Poland. Increasingly more stringent criteria have been applied since those days with a result that they have often been tougher than criteria used for comparable projects in free market economy countries.

Regardless of the type of criteria used, the results are converted to several easily understood parameters, the most important of which include:

- minimum quantity of economic reserve;
- minimum average grade and/or maximum average content of deleterious constituents;
- sample grade cut-off and maximum allowed contents of deleterious constituents;
- weighted grade cut-off and maximum allowed contents of deleterious constituents per intersection;
- requirements for physico-mechanical and chemico-technological characteristics of a mineral deposit;
- minimum width;
- maximum depth of occurrence for mineral deposits considered for underground mining and maximum stripping ratio for deposits considered for opencast mining;
- maximum thickness of barren and low-grade intercalations.

Examples

(1) Criteria of balance developed for copper deposits in the Lubin–Głogów district of Poland, established by a team working on behalf of the Minister of Smelting in January 1978, were translated into three copper cut-off grades shown in Table 3 and a minimum mineralized intersection width of not less than 2 m (Wanielista & Butra 1991).

(2) Hard coal deposits in the Upper Silesian Coal Basin are included in economic reserves if they occur at a depth of less than 1250 m and satisfy a number of other criteria, including *inter alia* a minimal caloric value of 3000 kcal kg^{-1}, a marginal width of 0.8 m, less than 1.3% S, less than 20% ash, thickness of barren intercalations less than 5 cm.

Criteria of workability

Criteria of workability are determined for economic reserves in specific deposits that are under exploitation or being considered for development. These criteria are designed on the following grounds:

Table 4. Correlation between prognostic and documented reserves of brown coal

Deposit	Prognostic reserves 1981 (10^3 tonnes)	Documented reserves 1991 (10^3 tonnes)	% Change
Bilczew	2,890	5,092	+76.5
Cybinka	100,600	237,487	+136.1
Chelmce	70,000	44,348	−36.6
Gostyn	1,105,000	1,988,830	+80.0
Laczki	236	1,820	+671.2

- method of the deposit development;
- technical constraints of selected mining methods;
- technical constraints of selected processing and beneficiation methods;
- technological and product quality norms;
- environmental constraints of the mining, benefication and the use of the mineral and its products and conditions of disposal and utilization of waste;
- detailed economic analysis, involving forecast mining, processing and marketing costs and the forecast product price range over the relevant period;
- utilization of associated minerals and protection of unworkable and uneconomic reserves.

The criteria are defined by an organization carrying out the feasibility study in a document called Deposit Development Project. The documentation must be opinioned by the Commission for Mineral Reserves and then submitted for approval to the Ministry of the Environmental Protection, Natural Reserves and Forestry.

To ensure that economic reserves are reclassified as workable, the feasibility study must demonstrate that the following conditions are met:

- mining and processing costs for a given portion of a deposit or the average costs for the whole deposit, if it is to be mined simultaneously at different places, will not exceed the value of output;
- quality of the ore and products of its beneficiation will satisfy the consumer's requirements;
- geological and technical parameters of the deposit, i.e. width, structure and continuity, properties of wall rocks, gaseous conditions etc. will not contravene with the requirements of the selected mining method;
- exploitation will not contravene with the health and safety regulations in force;
- exploitation will not cause any unacceptable environmental changes and all changes resulting from exploitation and related activities will be acceptable or can be rehabilitated on completion of the exploitation.

Precision of resource estimates

The most fundamental weakness of all mineral resource classification systems is the uncertainty as to how precise and accurate resource estimates are. Since the accuracy can only be fully assessed through reconciliation with mine output records, i.e. after given blocks have been mined out, it cannot be defined during the deposit evaluation contrary to recent claims by some geostatisticians. It is therefore the question of precision that is more appropriate and should be addressed by every classification system. Whatever calculation method is used, the total error is a function of three types of errors, namely: (1) errors made in physical measuring of various parameters; (2) methodical errors inherent to the calculation method used; and (3) errors resulting from the assumption that global parameters of a deposit can be determined from spot measurements. Unfortunately, no calculation method can satisfactorily define the total error. The problem has been discussed by Gałkiewicz (1975), who suggested that the total error does not exceed 10% for estimates of reserves in category A, 30% for reserves in category B, 50% for reserves in category C_1, 70% for reserves in category C_2 and 90% for reserves in category D_1. According to Niec (1982), however, precision of estimates attained in practice has actually been higher: 15% for reserves in category B, 25% for reserves in category C_1 and 40% for reserves in category C_2. The writers have found an interesting example showing how prognostic reserves, estimated in a number of brown coal deposits in the west of Poland in 1981, compared with C_1 and C_2 reserves documented in the same deposits in 1991 (Table 4).

Table 5. *Correlation with the USGS and USBM 1976 classification system (after Smakowski 1978)*

USGS & USBM 1976 SYSTEM				TOTAL MINERAL RESOURCES						
				IDENTIFIED				UNDISCOVERED		
				DEMONSTRATED		INFER POS	HYP	SPECULATIVE		
				MEASURED PROVEN	IND PROB					
			POL	A	B	C_1	C_2	D_1	D_2	D_3	E
ECONOMIC	RECOVERABLE	WORKABLE		▓	▓	▓	▓				
	UNRECOVERABLE	UNWORKABLE									
SUBECONOMIC	PARA MARGINAL	UNECONOMIC									
	SUB MARGINAL										

Correlations with other resource and reserve classification systems

Correlations with other systems are difficult due to the relatively complex structure of the Comecon system. Smakowski (1978) has discussed this topic at some length and compared the Polish variant of the system to the American classification as presented by the US Geological Survey (1976). This is shown in Table 5. He concluded that although documented reserves A, B, C_1 and C_2 are broadly comparable to reserves in proven, probable and possible categories, the correlation is complicated by the fact that categories C_2 and D_1 straddle the fields of probable and possible reserves. Thus, although reserves in categories C_1 and C_2 are in large part comparable to probable reserves, some reserves of category C_2, which by definition are delineated on the basis of results obtained from isolated workings, drillholes and outcrops, cover the field of possible reserves. Furthermore, reserves documented in categories C_1 and C_2 often do not have precisely identified perimeters and require further investigations to enable the selection of appropriate methods of access, exploitation and processing. Smakowski (1978) therefore argues that, in some cases, these reserves may be more comparable to indicated resources rather than to probable reserves. A complication here is that some mineral deposits are too complex to be documented in categories higher than C_1, or even C_2, before a decision is made on their commercial development and because of that they are after all more comparable to probable reserves as defined by the USGS system (Popreňak 1993a). However, it is accepted that portions of documented reserves in category C_2 that are delineated with a lesser degree of confidence are similar to inferred resources. Prognostic reserves in category D_1 present similar difficulties. Small portions of these reserves correspond to inferred resources and some to hypothetical resources. There are no problems with reserves in categories A and B, which equate to proven reserves, and with prognostic reserves in category D_3 and theoretical reserves E, which together equate to speculative resources.

Correlation along the ordinate is self-explanatory provided it is realized that the more recent criteria of balance used in Poland and in some of the other countries of the former Comecon have generally been stricter than feasibility criteria used in the USA.

Put in the context of the IMM definitions of

Table 6. *Classification of mineral reserves and resources introduced in Czechoslovakia, now in force in the Czech Republic and Slovakia*

	RESERVES			RESOURCES
	DETECTED Z-1	PROBABLE Z-2	POSSIBLE Z-3	PROGNOSTIC
ECONOMIC: RECOVERABLE & NONRECOVERABLE				
UNECONOMIC				

resources and reserves, categories A and B may be taken as equivalent to measured mineral resources, whilst categories C_1 and C_2 are broadly comparable to indicated mineral resources. Prospective reserves straddle the fields covered by indicated resources and mineral potential. Industrial reserves are comparable to mineral reserves as defined by the IMM definitions.

Outlook

On 1 January 1992, Czechoslovakia officially abolished the Comecon-type mineral reserve classification system and introduced a new three-tier system. According to this new classification (Popreňak 1993b), reserves are classified as detected (Z-1), probable (Z-2) and inferred (Z-3) (Table 6). Undiscovered mineral deposits, the existence of which is assumed from showings, indications and by analogy with similar geological environments, are denominated as prognostic resources.

Detected reserves (Z-1) are those portions of mineral reserves that satisfy the following conditions:

- shape, dimensions, setting and internal structure are verified by exploratory workings;
- grade, technical characteristics and types of material are defined by small- or large-scale testing and space distribution of useful and deleterious constituents is proved;
- geological and geotechnical conditions determining the way the reserve can be exploited are proved.

Probable reserves (Z-2) satisfy the following conditions:

- shape, dimensions, setting and internal structure are interpreted from geochemical and geophysical data and confirmed by such amount of exploratory work which allows reasonable assumptions of continuity between observation points;
- grade, technical characteristics and spatial distribution of different types of material are determined from small-scale testing and spatial distribution of useful and deleterious constituents is known;
- geological and geotechnical conditions relevant to mining are interpreted from exploration work and inferred by analogy with similar deposits.

Possible reserves (Z-3) must satisfy the folowing conditions:

- shape, dimensions, setting and internal structure are interpreted from geological, geochemical and geophysical data and from outcrops and isolated drillholes and exploratory workings;
- grade and technical characteristics are determined from small-scale testing or by analogy with similar mineral deposits;
- distribution of useful and deleterious constituents is known only approximately;
- geological and technical conditions determining the way of reserve exploitation are derived from isolated data and by analogy with similar deposits.

Poland is considering the introduction of a similar classification system. The most favoured option is to combine categories A and B as reserves under exploitation, categories C_1 and C_2 as reserves before exploitation, categories D_1 and D_2 as possible reserves and categories D_3 and E as resources.

The authors express their gratitude to Ing J. Popreňak and Dr M. P. Martineau of Samax Ltd, who assisted with original data from Slovakia, to Mackay & Schnellmann for the permission to use its computer facilities and to W. G. Yuill for his valuable comments on various aspects of this paper.

References

DIATCHKOV, S. A. 1993. Principles of classification of reserves and resources in the Commonwealth of Independent States. *Elements*, 2, 6–10.

GALKIEWICZ, T., 1962. Zasoby przewidywane (perspektywiczne, prognostyczne). *Przeglad Geologiczny*, 6, 301–302.

—— 1975. Dokładnosc obliczen zasobów złóz kopalin stałych. *Rudy i Metale Niezelazne*, 20, 454–455.

KOZUBSKI, F. 1965. Zasady ustalania kryteriów bilansowosci złóz surowców mineralnych. *Przeglad Geologiczny*, 10, 422–426.

NIEC, M. 1982. *Geologia kopalniana*. Wydawnictwa Geologiczne, Warszawa.

POPREŇAK, J. 1993a. *Some comments to the map of reserves—Silver vein—Rožňava*. Unpublished.

—— 1993b. *Information on actual mineral resources classification system in Czechoslovakia and former classification system in Comecon countries*. Unpublished.

PRZENIOSLO, S. 1978. Prawidłowosci rozmieszczenia złóz i przesłanki poszukiwawcze. *Prace Instytutu Geologicznego*, 83, 312–317.

RUTOWSKI, T. 1967. Geologiczne kryteria bilansowosci zasobow złóz kopalin stałych. *Przeglad Geologiczny*, 1, 1–4.

SMAKOWSKI, T. 1978. Klasyfikacja zasobów słóz. *Przeglad Geologiczny*, 2, 125–129.

US GEOLOGICAL SURVEY. 1976. *The Unified Department of the Interior Classification Method*. Geological Survey Bulletin, 1450-A.

WANIELISTA, K. 1976. Uwagi o kryteriach bilansowosci złóz rud. *Rudy i Metale Niezelazne*, 21, 495–498.

—— & BUTRA, J. 1991. Kryteria bilansowosci złóz miedzi dla cienkich pokładów. *Rudy i Metale Niezelazne*, 36, 72–74.

ZOLTOWSKI, Z. 1964. *Prawo geologiczne*. Wydawnictwa Geologiczne, Warszawa.

Fault interpretation from coal exploration borehole data using SURPAC2 software

P. D. GRIBBLE

Mining Technology Consultants and Associates, 113 Roskear Road, Camborne, Cornwall TR14 8BY, UK

Abstract: The use of isopach or contour plans for geological surfaces as the basis for fault interpretation of coal data, even when a dense drilling pattern has been applied, is often unsatisfactory. These plans tend to 'smooth' or underemphasize the effects of faulting, making assessment of the likely fault pattern difficult.

Within the SURPAC2 Software system a series of modelling tools is available to assist with this interpretation process. An appropriately oriented grid is overlain on a digital terrain model (DTM) of the geological surface in question to regularize the data. The dip between data points is calculated, and a model of dip change created using the DTM method. A new contour model, which highlights trends in dip change, is produced, clearly showing likely fault traces and throw. This model may be sectioned to illustrate more graphically the disturbed and undisturbed areas. The fault model is then used as the basis for interpretation, to be compared with other models of the geology, and known local and regional faulting trends. Use of the macro facilities within SURPAC2 allows a series of surfaces to be rapidly analysed to build up a picture of likely fault continuity and pattern within the succession.

Once the fault pattern is established, it can then be used elsewhere in SURPAC2 for further modelling and evaluation. Application can be simple, such as the formation of boundaries of faulted areas for estimation of mineable coal resource. More complex modelling can be achieved by interaction between geological surfaces and the fault surfaces defined by the interpretation process. A complete model of the faults and their impingement on the coal and associated layers is formed, for further use within the SURPAC2 design tools.

Emphasis in the use of computing software today seems to be towards the 'need' for three-dimensional graphical applications and realizations. Whilst SURPAC2 possesses these capabilities, this paper sets out to show the straightforward possibilities for day-to-day application of the software. Here the need is to be able to update and model data rapidly, perhaps whilst in the midst of a drilling programme, as an aid to fault interpretation from the exploration borehole data. The aim is to maximize the usefulness of the data obtained from the drilling programme, and to provide clear guidance for continuation or extension of such a programme. All the processes described below are carried out entirely with the SURPAC2 software, many of them lending themselves to automatic processing using the macro system provided by the software.

Methods

Data storage

Borehole data in SURPAC2 is stored within a relational database which contains all the geological, quality (analytical) and survey data for any given hole, together with geophysical and similar data as required. The borehole data may be represented graphically, and changes to geological (or any other) data held in the database interactively modified. This is of great assistance during geological interpretation, particularly for seam or layered deposits.

For the purposes of this modelling application, the data are extracted on the basis of geological codes as stored in the database, previously defined by the user. Extraction may also be based on quality (i.e. value), area, or borehole type as suits the purpose. The main data structure used by the program is the 3D string modelling system, as originally described by Porter (1979) and Miller (1987), and data are extracted from the database in this form.

Initial modelling

The data for a given surface e.g. base of seam, are extracted from the database using the appropriate geological code, with the data set produced having the x, y and z coordinates for the base of the chosen lithological unit. The

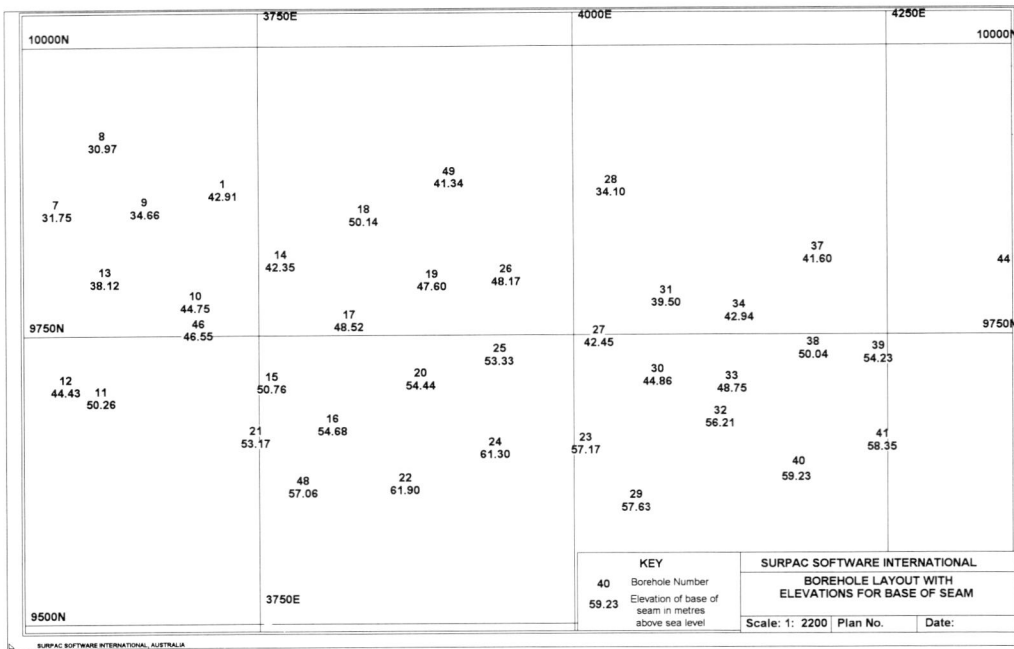

Fig. 1. Borehole layout with elevation values in metres above sea level for base of seam.

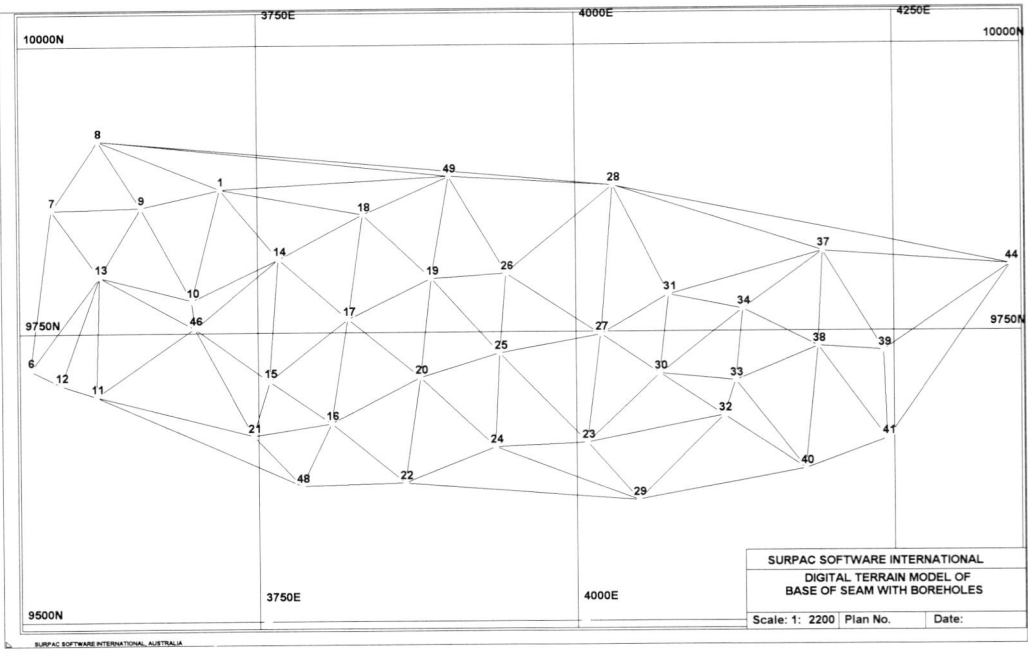

Fig. 2. Digital terrain model of raw data and borehole collars.

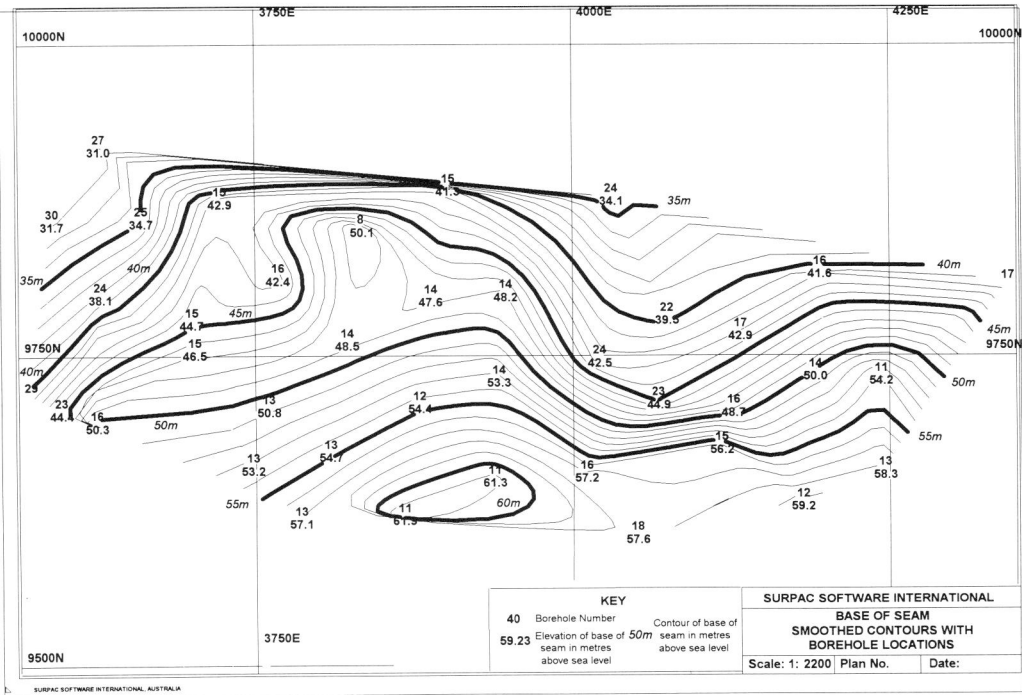

Fig. 3. Smoothed contours in metres above sea level for base of seam and borehole collars.

borehole locations and the z (elevation) values for the example database are shown in Fig. 1. This type of data is typically modelled as a contour or isopachyte plan. Such a plan is created by forming a digital terrain model (DTM), by the construction of triangles formed on sets of three data points. In this example, each point is the co-ordinate of the base of seam derived from the borehole data, and the resulting DTM is shown in Fig. 2. The method of calculation uses only those data points available, no mathematical interpolation or the original data points takes place. The 'digital surface' produced consists of a series of triangular facets. This DTM may be used for a variety of purposes, in this example, the creation of the isopachyte plan. The contour lines are calculated from the DTM, using methods similar to those used by hand to create footwall contour plans. The results, smoothed for presentation purposes, appear in Fig. 3.

The use of isopachyte maps or contour plan of a geological surface for the interpretation of faults, even when a dense drilling pattern has been completed, often produces unsatisfactory results. The plan shown in Fig. 3 tends to 'smooth' or underemphasize the effect of faulting, making assessment of the likely fault pattern difficult. In this example, the results are far from conclusive and do little to assist in the siting of future boreholes.

Further modelling

Further modelling is carried out using a series of the modelling tools available within SURPAC2 to assist with the interpretation process. This uses the following steps. Firstly, grids of points are overlaid on the DTM of the base of the seam, to regularize the data set. It should be noted that, as described below, this is used in an interpolative process, and does not make use of gridding algorithms. In order to analyse these data fully, both an east–west and a north–south biased grid are employed. This means that in the second step of the modelling process described below, calculations between points on the grid are either in an east–west or a north–south orientation. The different bias gives quite different results dependent on the orientation of the faulting pattern, as later demonstrated. The process of overlay interpolates the relevant

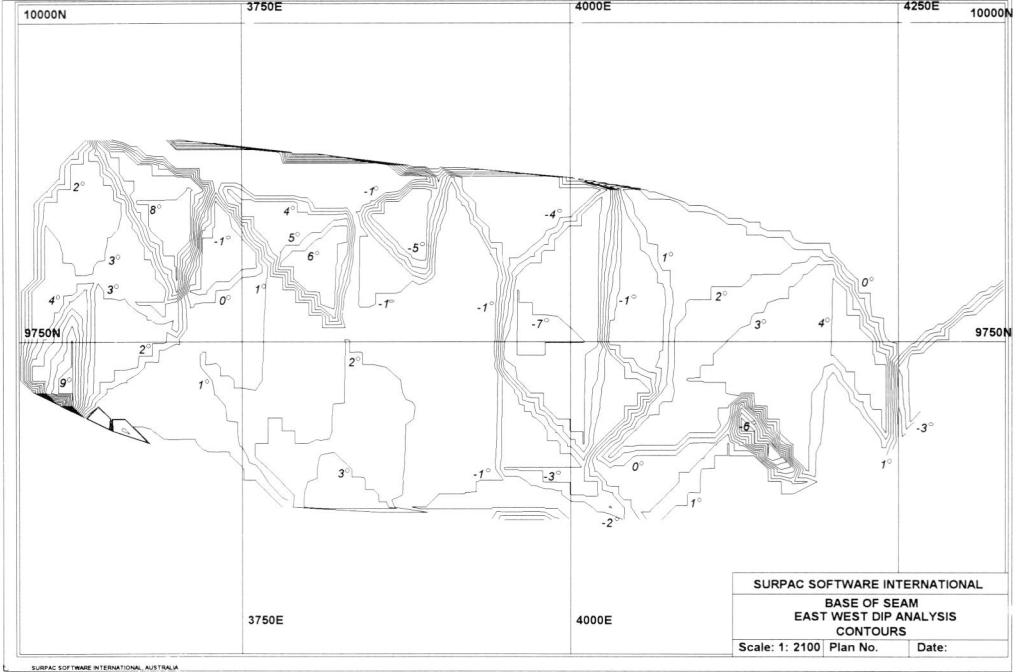

(a)

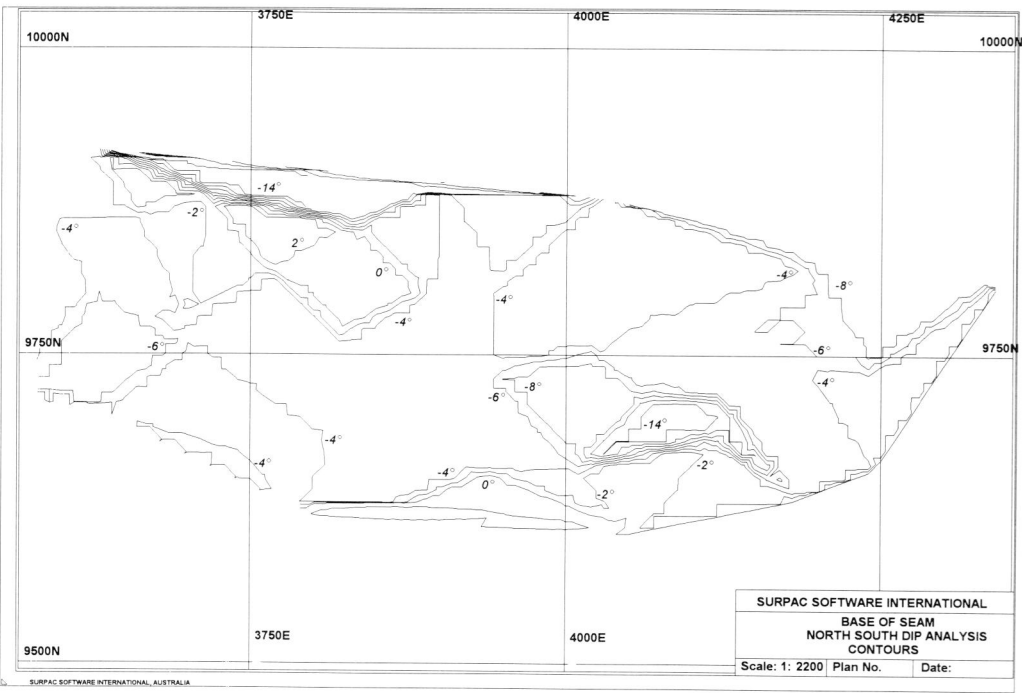

(b)
Fig. 4. Contours of dip from (**a**) the east–west dip model and (**b**) the north–south dip model.

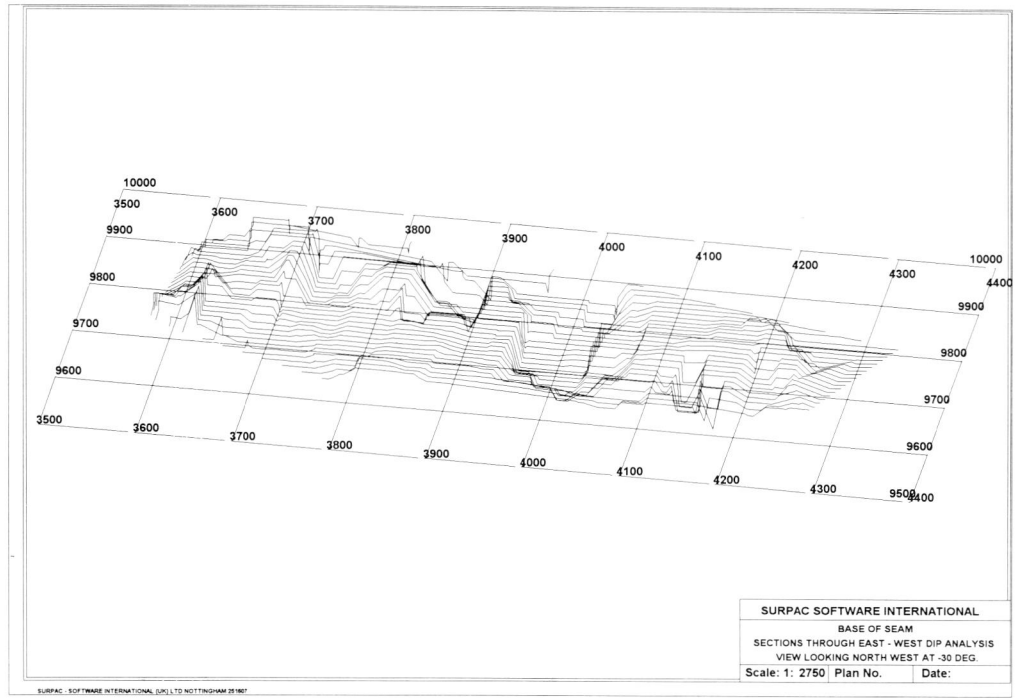

(a)

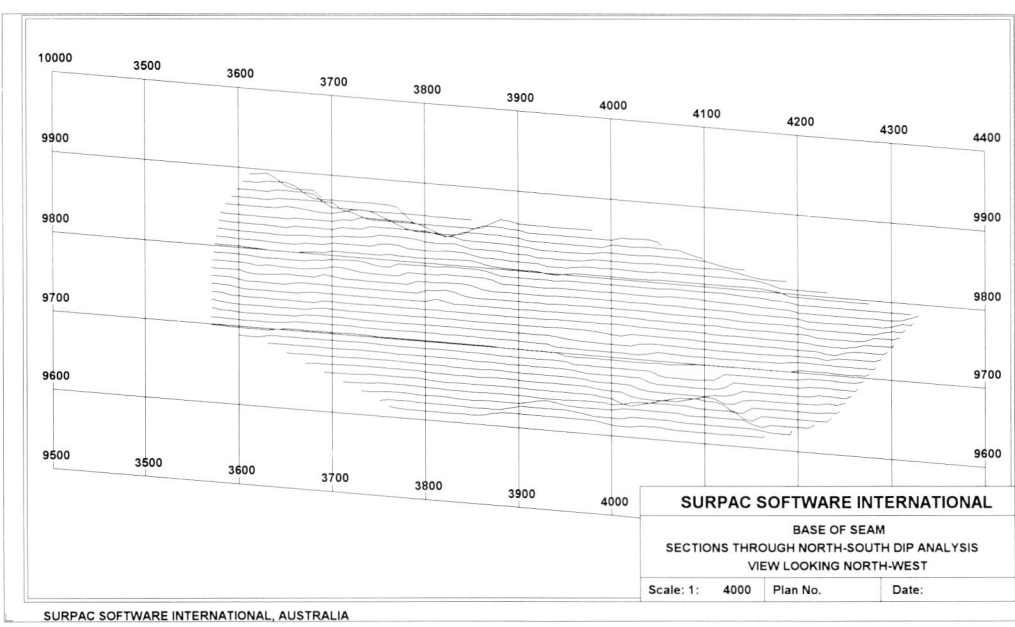

(b)
Fig. 5. 3D view of sections taken through (**a**) the east–west dip model and (**b**) the north–south dip model.

elevation value at each grid point according to its position within the triangle in which it falls. The purpose of the grid is to give regular data points whose elevation relates directly to the original data set, but which give further possibilities for modelling, including the method described here. Secondly, the dip (or slope if preferred) value between each of the data points on the grid is calculated for both east–west and north–south orientations, i.e. in the east–west example, the calculation of dip is made between one point and its neighbour to the east. The value obtained is plotted at the first of the two points. These values effectively show the change in dip across the model, and the values for change in dip are again modelled as a DTM, from which contours of changing dip are produced as shown in Fig. 4a and b.

The models of dip change show a pattern of plateaux, being the unfaulted areas separated by zones of rapid dip change, the possible location of faults. In this example Fig. 4a shows the disruption of the seam highlighted by the east–west analysis, whereas the north–south analysis, shown in Fig. 4b, fails to pick up the full extent of the probable faulting. Figure 4a and b thus illustrates the differences between the east–west and the north–south biasing of grid points, with the east–west analysis providing the better tool to assist with interpretation in this example. The differences between the two different analyses are further illustrated by producing cross sections from the two DTMs of dip change. The resulting three-dimensional views are shown in Fig. 5a and b, where once again, interpretation is much better served by the east–west analysis. Figures 4 and 5 demonstrate that the technique quickly provides a basis for further drilling and interpretation.

In practice, use of the macro facilities within the software allows for rapid analysis of a series of surfaces so that a picture of likely fault continuity and pattern within the succession can be built up. The final model can then be compared with known local faulting and regional trends to confirm validity.

Further applications

Once a fault pattern has been established, further tools within the software can be applied to the data.

Three-dimensional modelling

In a multi-seam deposit, the major faults will be quickly determined and their orientation and location finalized in the key layers. The modelling process can then be modified to show the pattern of faulting as it intersects with each seam or layer, without the need to carry out the entire process described above for each horizon. The method used in this instance is to model the fault planes themselves as DTMs. These DTM planes are then intersected with DTM models of the seam surfaces, and lines of intersection automatically produced. These can then be used as boundaries for other applications as described below.

Presentation

The three-dimensional model of seams and faults, formed using a combination of the methods described above, may be enhanced by use of colour rendering techniques. This provides a powerful aid to interpretation and presentation of the geology of the deposit under evaluation. In addition, cross sections showing boreholes, geology and faulting can be produced to complement the three-dimensional representations.

Resource estimation

The fault patterns created above can be used to define boundaries which are then used for volume and quality calculation within the model, typically in the undisrupted coal areas. Areas of influence around boreholes defining a resource estimation boundary can be modified using the limits determined by the faulting, using a simple shape intersection process. Volumes may then be calculated using DTM surfaces or sections derived from the DTMs, as preferred.

Mine design

In addition to assisting in fault modelling, the models of dip change may also be used to determine appropriate mining methods as a deposit changes along strike or dip. This would be achieved by defining regions, typically by isopachs, where the seam dip becomes unacceptable for a given method.

The fault boundaries patterns may also be used as an aid to pit design in two key areas. Firstly, in defining zones in which a constant face angle would be maintained during pit generation, and secondly, as an aid to the interactive design of bench outlines. The fault zones defined could also be used in the determination of underground development scheduling for longwall production.

Conclusions

The methods described here have been used effectively to define the faulting pattern in a well drilled area. Major fault zones were highlighted, as opposed to earlier interpretations which showed only a series of faults, thus drawing attention to the potential problem areas. Overall, routine application of the tools available within SURPAC2 to an ongoing drilling programme can provide guidance for further drilling, prior to utilization of more advanced computer modelling techniques.

References

MILLER, D. R. 1987. String and Block Modelling in Mine Planning. *In: Computers in Mine Planning.* ACADS Seminar, Melbourne, Australia.

PORTER, J. R. 1979. String Ground Model Surveys—A new role for the Engineering Surveyor. *In: South East Asian Survey Congress.*

INTMOV: a program for the interactive analysis of spatial data

W. HATTON

British Coal Opencast, Mansfield, Nottinghamshire, UK

Abstract: A set of programs have been developed to compute moving-windows statistics from spatial data, which are fully interactive, allowing the user to compute and display two plan views of the window statistics, the parent population histogram and the two histograms of the windows statistics all on one graphical display. Stationarity of a deposit can be checked instantly at differing window sizes with all the sample data or with partitioned subsets. Data can be read into the routine from a current exploration campaign or from working site-survey data, allowing comparisons between 'projected' and 'ground' truth. Screen and DXF file output of the local neighbourhood window statistics are plotted in a colour which corresponds to standard confidence bands of the parent or global distribution. Rapid identification of local anomalies is possible. Window statistics for all the neighbourhoods can be written to file at any subset number of the original sample number. Plots of the window means and standard deviations versus numbers of boreholes allow the exploration geologist to check stationarity in a retrospective way. The main aim in the development of these routines was to reduce the amount of user keystrokes needed to compute and display the maximum amount of neighbourhood information. The technique is demonstrated with examples from UK Coal Measures.

Exploratory data analysis is a key step in understanding the underlying variability in any spatial data. Any technique which uses spatial interpolation to describe and quantify a mineral deposit needs to be carefully chosen to suit the type of inherent variability and the type of mineral deposit. A thorough understanding of the spatial variability will help the investigator to choose the correct assumptions for the spatial model. A rapid detection of local anomalies is highly desirable in any set of modelling routines.

A technique which has been used in exploratory data analysis of spatial data is moving-windows statistics (Isaaks & Srivastava 1989; Murray & Baker 1991). Moving-windows statistics are simply the determination of any statistic within a local spatial neighbourhood across a deposit. The simplest window one can compute and display graphically is a rectangle. Local data are assigned to a particular window and the summary statistics are calculated. Plots of summary statistics of windows may reveal trends and anomalies in the data set. A very useful application for this technique is to check for stationarity in a deposit. The most widely used geostatistical estimation procedures use stationary random function models. The decision to view a particular sample data configuration as an outcome of a stationary random function model is strongly linked to the decision that these samples can be grouped together.

Murray & Baker's MWINDOW routine was applied successfully to British Coal Opencast (BCO) seam thickness and quality data, but needed its output re-directing for plotting to the US EPA public domain Geostatistical Environmental Assessment software (Englund & Sharp 1988). Each run of the program also needed user input of the deposit limits, source data file and the neighbourhood box size. Modelling many surfaces was a time-consuming exercise with plotting comparisons difficult. The time restriction was viewed as an extreme limit to the effectiveness of the exploratory technique. A routine called INTMOV was developed in Quick-BASIC 4.5, running on an IBM PS2 model 70, which would reduce the user input time by reading repetitive data input from a scratch file. It was also designed to speed plotting by combining the plotting routines into the same module.

The major aim of INTMOV was to maximize the exploratory analysis time by reducing user input and also increase the exploration time for the same user activity. There was also a requirement to check for stationarity throughout the life of a current exploration exercise and from case history data from sites as a 'postmortem' exercise. Studying plots of the window-statistics versus the numbers of boreholes drilled

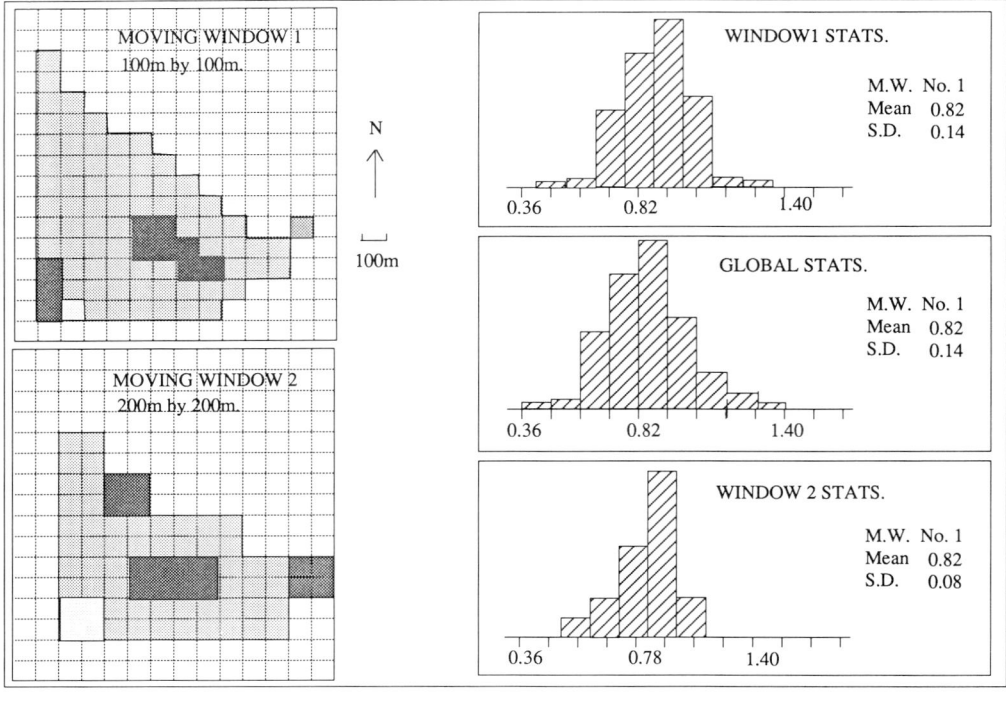

Fig. 1. Graphical output from INTMOV. An example of moving-windows statistics for coal seam thickness. Plan view 1 shows statistics calculated at 100 m by 100 m box sizes. Plan view 2 shows statistics calculated at 200 m by 200 m boxes. Histograms of plan view 1 statistics, global statistics and plan view 2 statistics are shown from top right to bottom right.

permitted BCO geologists to make more informed decisions on when to stop exploration programmes.

Programme design

It became obvious that an enhancement was needed to integrate the plotting of window statistics into the routine to reduce the user keystroke requirement and also improve the ergonomics of data input for the window computations. The reduction in keystrokes was easily achieved by reading the seam quality and thickness data from an ASCII file containing the input file name, coordinate limits and the neighbourhood box sizes for the current run. Maximizing the amount of moving-window output to the user for each run was more problematical, in that comparisons of differing window-sized computations needed to be made in plan on the same graphical display, along with the histogram output for the parent distribution and the two moving-window computations. This was achieved by writing all the graphical output to one graphics screen, subdividing 5 zones of output (Fig. 1). Two plan views can be plotted from the basal strip menu options MW1 and MW2 respectively, shown in the top left and bottom left of the screen output. The histogram for moving-windows computation 1 is displayed in the top right corner, with the parent or global population centrally below and the moving-windows statistic computation in the bottom right of the screen output.

Very rapid comparisons of differing window size computations can be carried out quickly. The colour coding of each neighbourhood is plotted in relation to standard confidence intervals in the parent or global distribution. For example if the window mean has a value greater than the global mean plus two standard deviations, then the local window box is painted red on the graphics screen. The colour legend for the local window colours is represented by grey-scale in Fig. 1. The actual screen output uses a colour coding system allowing the user to quickly spot any local statistics which are anomalous in relation to the global average of the deposit. Comparisons between output area 1 and 2 with differing neighbourhood sizes can be made efficiently, helping to ascertain critical distances in the auto-correlation. Rapid interactive graphical techniques have been applied to variography with good effect (Bradley & Haslett 1990) using dynamic data links from map to scatterplot views, rigorously testing variogram models quickly. In a more simplistic way INTMOV has achieved similar results by displaying several differing views of the same data.

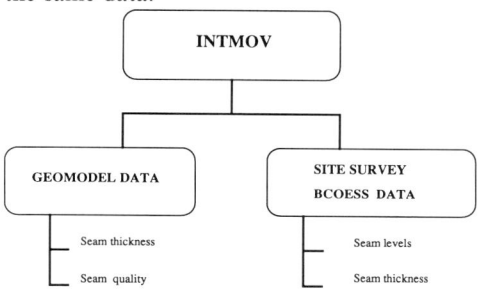

Fig. 2. INTOMV data structure. Imported data can come from two sources; exploration data and site survey data.

The remaining design feature which needed to be addressed was to include a method for retrospectively looking at site surveyed data from the British Coal Opencast Site Survey System (BCOESS) and comparing this with data imported from the British Coal exploration database Geomodel (Knight 1986). The key to these linkages had to be speed and ease of data transfer into INTMOV. The following section describes how this was achieved.

Intmov: structure, import and export data

Data import into INTMOV can be from two sources of data. Borehole exploration data can be extracted from BCO's Geomodel relational database and transferred into INTMOV via ASCII flat data files, similar in format to a GEOEAS DAT files. Seam thickness or analytical data is obtained in this manner. Working site data is held in the BCOESS system as AutoCAD drawing files. Worked seam thickness data sets can be exported from the drawing files into DXF files and converted into the same DAT file structure as the converted Geomodel data sets (Fig. 2).

Output of the processed moving-windows statistics for each plan view is automatically displayed to the graphics screen or optionally driven out to a DXF plot file, which then

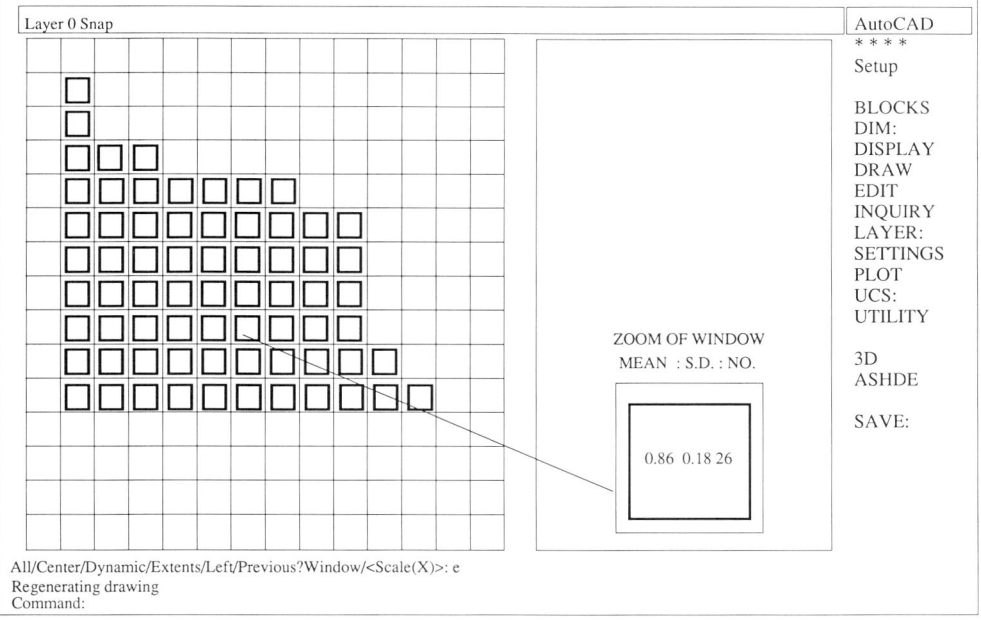

Fig. 3. Plotted output from INTMOV shown in AutoCAD. Plot of all window statistics for seam thickness. Plot shows zoomed view of window-statistics demonstrating window mean, standard deviation and the number of samples used in window calculation.

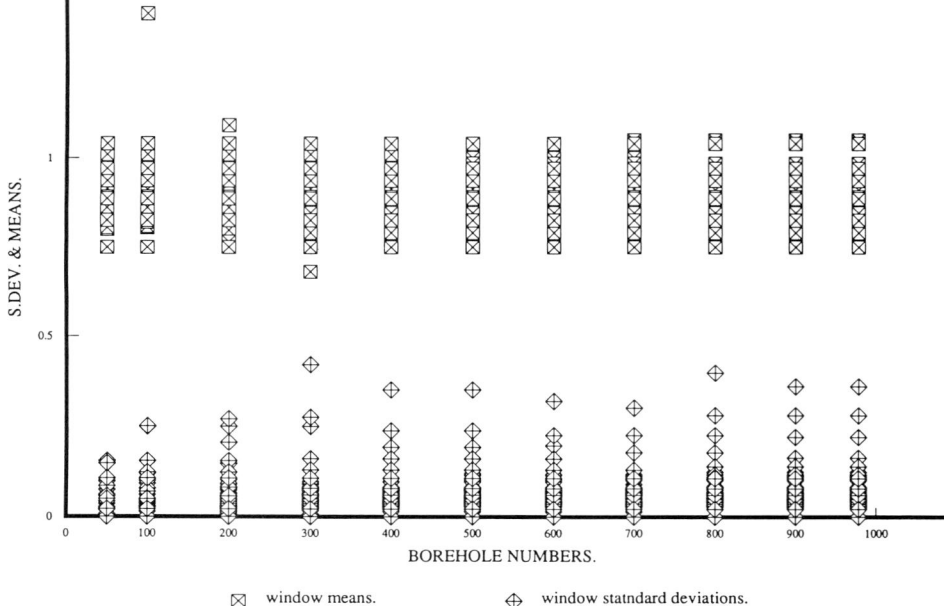

Fig. 4. Winplot output. Window means and standard deviations are plotted versus borehole numbers from retrospective subset sampling of the total borehole numbers.

is automatically picked up and processed by AutoCAD when the user terminates the INTMOV routines (Fig. 3).

Sequential views of the exploration and site data

The ability to look at an exploration programme at various key stages with increasing sample numbers is of prime importance to any geological sampler. Many geostatistical studies have concentrated on using various statistics and utility functions to ascertain when there has been sufficient sampling to describe with confidence the underlying variability of the deposit (Rendu 1970; Journel 1973; Brooker 1975; Scheck & Da-Rong Chou 1983; Aspine & Barnes 1989; Whateley 1991). In most of these studies the exploration programme has been critically reviewed by plotting some statistic versus the number of samples drilled, whether it is the estimation variance or some utility based primarily upon estimation variance does not matter. The underlying concept is to review and view the exploration data at regular time intervals. INTMOV was designed to allow the user to do this in two ways, firstly in a progressive manner and secondly retrospec-

tively using the SUBSET option. SUBSET allows the investigator to take the full data set as it exists and produce snapshots of the data and window-statistics at any sequentially ordered subset number. These subset values can be written to scratch file and subsequently plotted with a routine WINPLOT (Fig. 4). The window means and standard deviations can be viewed retrospectively through a drilling campaign very quickly, using case history data if required.

Case study: moving windows applied to UK coal seam thickness data

The routine was applied to many seam thickness and seam quality data sets from the UK Coal Measures. One example is documented here, using seam thickness data from East Pennine Coalfield.

The data sets were recorded and input into the Geomodel database, from which they were exported into the necessary format for Intmov processing. Moving-windows statistics were calculated for consecutive batches of 100 boreholes throughout the exploration. Each batch was incorporated into the previous sample cumulatively.

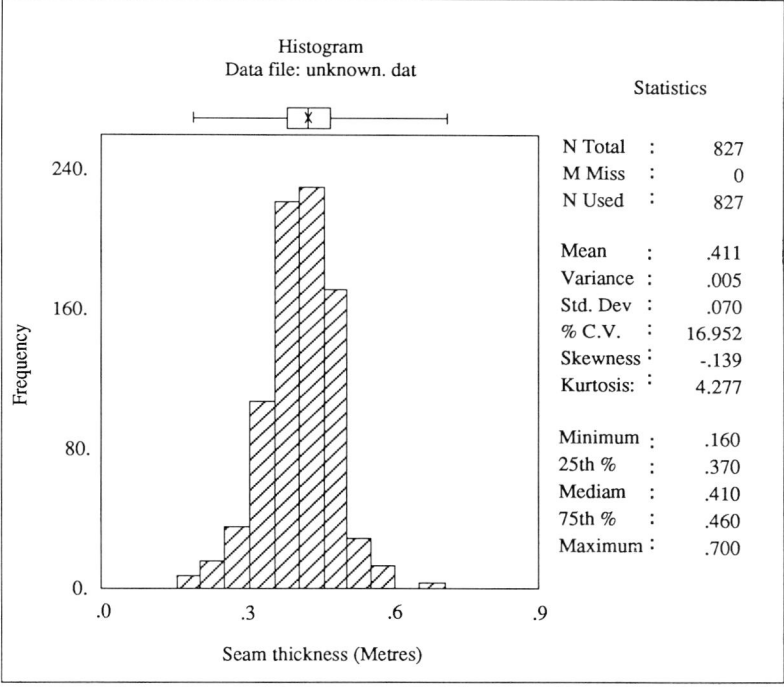

Fig. 5. Global seam thickness statistics and histogram.

Global seam statistics are displayed in Fig. 5. The 827 boreholes gave a mean thickness of 0.41 m and a standard deviation of 0.07 m. Window statistics were calculated for 100 m by 100 m blocks for each of the consecutive batches of 100 boreholes. The plot of the window statistics (Fig. 6) shows a wider spread in the first 200 boreholes, beyond which the window statistics are very uniform. This is very typical of very uniform coal seam deposits. Such unifor-

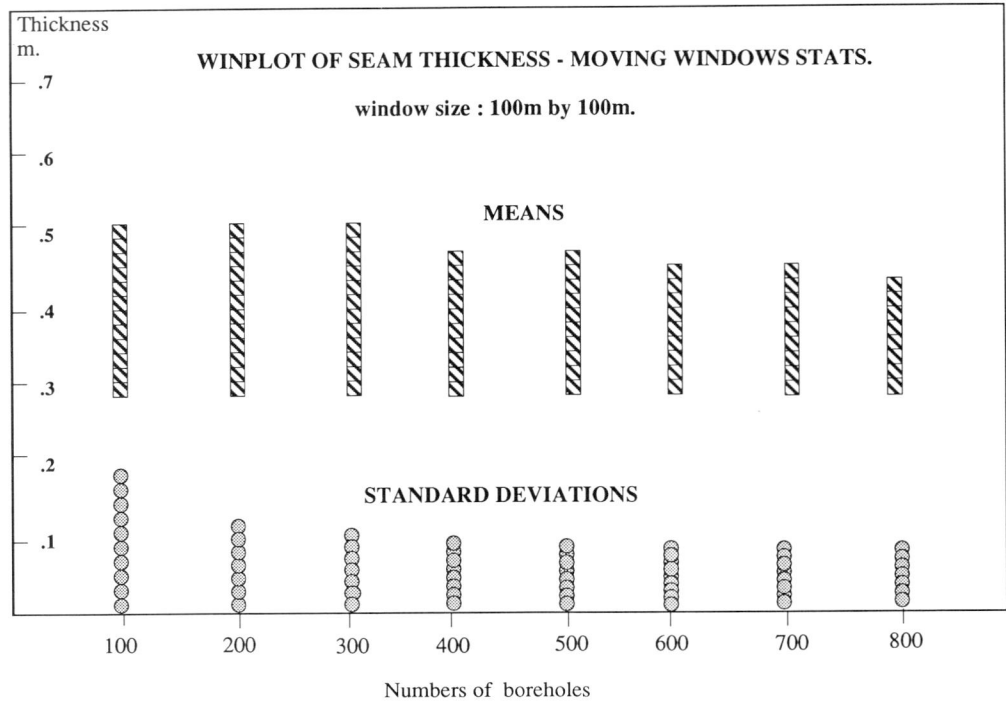

Fig. 6. Plot of the window statistics versus borehole numbers for moving windows statistics calculated at 100 m by 100 m boxes.

mity is normally described as stationarity. Comparison of the window statistics calculated for 100 m by 100 m and 200 m by 200 m blocks shows a high degree of similarity (Fig. 7). The Winplot results in Fig. 6 clearly demonstrate that there is very little extra understanding gained in the variability from the extra holes. Window means which lie greater than or less than 2 standard deviations from the global mean are very quickly identified at any stage of an exploration programme (Fig. 7). The comparison between window statistics calculated at 100 m by 100 m and 200 m by 200 m boxes is a very useful check on stationarity, especially if it is undertaken at regular stages of exploration.

This case history displays how the module can locate areas of local variability in an exploration programme and also be used as a performance indicator for an exploration campaign, using Winplot output.

Conclusions

The enhancement of Murray & Baker's (1991) MWINDOW routine in QuickBASIC 4.5 on an MS-DOS platform has resulted in the module INTMOV which is truly interactive, allowing rapid exploratory analysis of spatial data. Measurement of the mean and standard deviation for local neighbourhoods at differing window sizes can be plotted comparatively and quickly.

Enhancements to the modules could allow overlapping window computations to counteract sparsity of samples and more sophisticated window statistics could be computed. The module could easily cope with the latter enhancement, with the development of extra QuickBASIC libraries to add to the existing plotting shell.

The author wishes to thank British Coal Opencast for permission to publish this paper, their sponsorship, availability of data, time and patience. Special thanks go to the following individuals: K. M. Pickup, M. K. G. Whateley, J. L. Knight, R. G. D. Smith, H. Orme and my wife Felicity.

Trademarks: AutoCAD is a registered trademark of Autodesk Ltd. IBM is a registered trademark of International Business Machines Corporation. MS-DOS is a registered trademark of the Microsoft

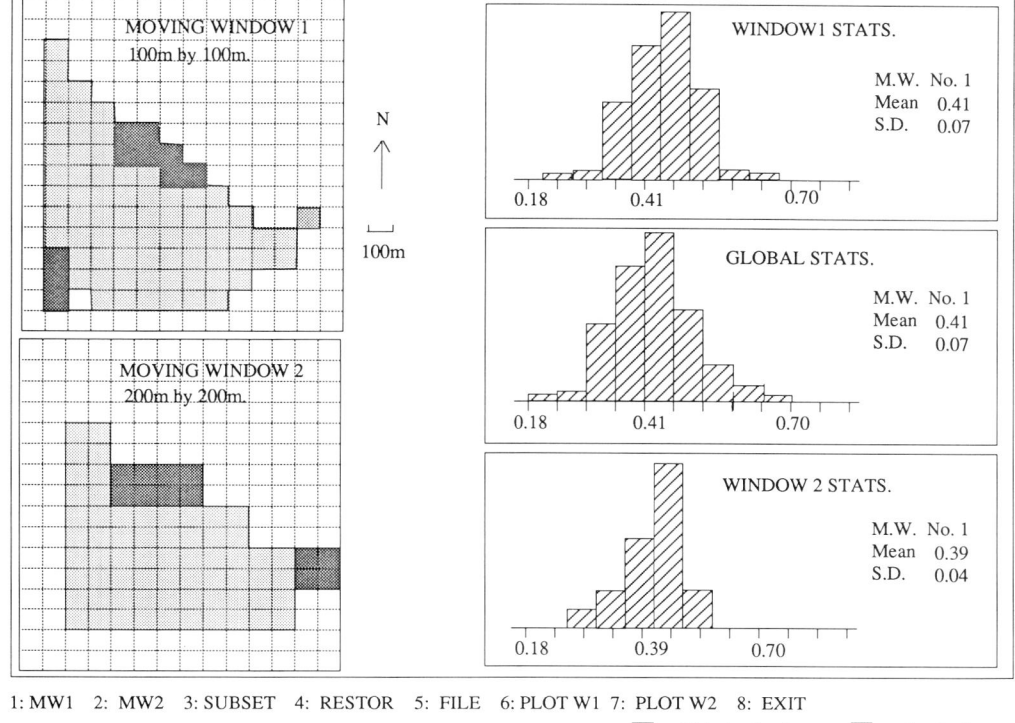

Fig. 7. INTMOV output for seam thickness. Plan view 1 shows statistics calculated at 100 m by 100 m box sizes. Plan view 2 shows statistics calculated at 200 m by 200 m boxes. Histograms of plan view 1 statistics, global statistics and plan view 2 statistics are shown from top right to bottom right.

Corporation. QuickBASIC is a registered trademark of the Microsoft Corporation.

References

ASPINE, D. & BARNES, R. J. 1989. Infill sampling design and the cost of classification errors. *Mathematical Geology*, **22**, 915–932.

BRADLEY, R. & HASLETT, J. 1990. Interactive graphics for the exploratory analysis of spatial data—the interactive variogram cloud. *Second CODATA conference on Geomathematics and Geostatistics*, Leeds.

BROOKER, P. I. 1975. Avoiding unnecessary drilling. *Proceedings of the Australasian IMM*, **253**, 21–23.

ISAAKS, E. H. & SRIVASTAVA, R. M. 1989. *Applied Geostatistics*. Oxford University Press.

JOURNEL, A. G. 1973. Geostatistics and sequential exploration. *Mining Engineering*, Oct., 44–48.

KNIGHT, J. L. 1986. Geomodel: a geological database for coal prospecting and site evaluation. *In: Computer Applications in Geotechnical Engineering*. Midlands Geotechnical Society Publications, 59–66.

MURRAY, M. R. & BAKER, D. E. 1991. MWINDOW: an interactive FORTRAN-77 program for calculating moving-windows statistics. *Computers and Geosciences*, **17**, 423–430.

RENDU, J. M. 1970. Some applications of geostatistics to decision-making in exploration. *In: Decision-making in the Mineral Industry*. CIMM, Special volumes, **12**, Montreal, 175–184.

SCHECK, D. E. & DA-RONG CHOU 1983. Optimum locations for exploratory drillholes. *International Journal of Mining Engineering*, **1**, 335–343.

ENGLUND, E. & SPARKS, A. 1988. *GEOEAS (Geostatistical Environmental Assessment Software) User's Guide*. US Environmental Protection Agency, EPA 600/4-88/033.

WHATELEY, M. K. G. 1991. Geostatistical determination of contour accuracy in evaluating coal seam parameters: an example from the Leicestershire Coalfield, England. *Bulletin de la Société Géologique de France*, **162**, 209–218.

Reserve assessment of a stratified deposit with special reference to opencast coal mining in Great Britain

C. P. NATHANAIL

Wimpey Environmental, Hargreaves Road, Groundwell, Swindon, Wiltshire SN2 5AZ, UK

Abstract. The assessment of reserves in a stratified deposit should define the areal extent of the mineral, the spatial variation of thickness and quality and the ratio of mineral to waste. The assessment of reserves at opencast coal mines in Great Britain also involves the estimation of the extent of areas where coal is absent or of reduced quality due to old mining activity, washout, alteration by igneous intrusions, lime burning and faulting. The most appropriate estimation technique will vary from site to site and, within an individual site, from seam to seam. Conventional non-spatial statistics can be used where the extent of areas of lost coal is not definable. Area-of-influence approaches such as Thiessen polygons are routinely used to estimate the spatial extent of old workings from point observations at boreholes. The use of indicator kriging may provide a refinement to this deterministic approach. Delaunay triangulations are used by popular mine design software for area and volume calculations. Kriging may also be used to model the variation of seam thickness and quality. Whatever method of estimation is used, poor quality input data will result in poor estimates. Metadata such as how, when and by whom information was collected should be used to decide whether or not a particular piece of information should be retained. Experience shows that each site must be treated as a unique entity with techniques and software being modified to suit the site circumstances. A black box 'expert system' approach would fail to identify site specific features and result in misleading estimates.

The assessment of reserves at opencast coal sites in Great Britain is conducted by the following organizations for the stated reasons:

(a) British Coal Opencast (BCO)—establish reserves;
(b) opencast contractors—price tenders for BCO sites;
(c) licensed operators—establish reserves at licensed sites.

This paper concentrates on the assessment of reserves which may be conducted in a limited time frame during the tendering period for a BCO site. Reserve assessments at opencast coal site involve four phases:

(a) data preparation;
(b) assessment of contractual and non-contractual quantities of coal;
(c) assessment of overburden quantities;
(d) preparation of methodology to enable forecasts of coal output.

The schedule of estimated quantities (SEQ) produced by British Coal Opencast defines the quantities of coal to be recovered from each contractual seam and details allowances for areas where coal is absent due to washout, burning, workings or faulting. The SEQ is produced in 'slow-time' and is, in effect, a bill of quantities. The quantities therein may reflect a conservative estimate in order to minimize the risk of claims or an optimistic estimate, perhaps to emphasize the viability of the site Coal may be also recovered from seams not included in the SEQ and an estimate of this non-contractual coal is needed.

Since the estimation exercise is being carried out by an organization other than the one responsible for the collection of the data, the quality of the data has to be assessed in order to determine which data elements to retain and which to discard. A discussion on data quality in geographical information systems (GIS), used to assist in the evaluation of data quality is followed by a description of the preparation of data and a discussion on the relative merits of different methods of interpolation.

Data quality in geographical information systems

De Freitas (1993) proposed the following

relationship as being of some use in designing ground characterization programmes:

(What we need to know − (what we already know) = (what we do *not* know).

To the above, Chrisman (1991), recognizing that not all information is equally reliable or useful, would add 'how do we know what we know'.

Chrisman (1983) defined data quality as *fitness for use*. Quality is a function of positional and attribute accuracy, precision, resolution, currency, logical consistency, completeness and lineage (Lyons *et al.* 1989; DCDSTF 1988; NCGIA 1990). The proposed US standard for the exchange of spatial information (DCDSTF 1988) recognized the above definition of data quality and required a quality report to provide the data user the basis upon which to judge suitability (Chrisman 1991). Researchers (see Goodchild & Gopal 1989) are aware of the need for the explicit incorporation of metadata, data about data (Rhind 1990) (Table 1), but the issues involved (Table 2) entail heavy computational requirements and have yet to be generally incorporated in commercial software.

Table 1. *Metadata: information about the information used in reserve assessments*

Pixel size
Age of data
Personnel involved
Agency of collection
Method of collection
Input equipment used
Precision of computations
Processing methods involved
Definitions of classes in different source documents

Table 2. *Quality-related issues in GIS (after Goodhild & Gopal 1989)*

a. Precision of GIS processing is effectively infinite.
b. All spatial data are of limited accuracy.
c. Precision of GIS processing exceeds accuracy of data.
d. In conventional map analysis, precision is usually adapted to accuracy.
e. Ability to change scale and combine data from various sources and scales in a GIS means that precision is usually NOT adapted to accuracy.
f. We have no adequate means to describe the accuracy of complex spatial objects.
g. Objective should be a measure of uncertainty on every GIS product.

Accuracy is defined as the closeness of results, computations or estimates to true values, or values accepted to be true (NCGIA 1990). *Precision* is defined as the number of decimal places or significant digits in a measurement (NCGIA 1990). GIS work at high precision, mostly much higher than the accuracy of the data.

The NCGIA (1990) advise that *positional accuracy* be tested with reference to an independent source of higher accuracy such as a larger scale map, the global positioning system (GPS) or raw survey data. They point out that evidence internal to the GIS, such as unclosed polygons and lines which overshoot or undershoot junctions, may also be used. Accuracy may be computed from knowledge of the errors introduced at different stages of data capture. If the sources of error combine independently, an estimate of overall accuracy is given by summing the squares of each component and taking the square root of the sum.

The NCGIA (1990) suggest that attribute accuracy must be analysed in different ways depending on the nature of the data. The scale of measurement employed (nominal, ordinal, cardinal) is one aspect of this nature. For *continuous cardinal attributes* accuracy may be expressed as a measurement error (e.g. elevation accurate to 1 m) or an estimation error. For *ordinal attributes* accuracy is a function of how appropriate, sufficiently detailed and well-defined the categories are. Attribute accuracy may be tested by preparing a *misclassification matrix* which compares the information stored in the database with ground truth (Guptil 1989; NCGIA 1990).

Logical consistency refers to the internal coherence of the data. For example, in a database of borehole information, the elevation of the top of a unit should be equal to the sum of the elevation of the base of the unit and the thickness of the unit. Furthermore, the elevation of the base of a unit should be equal to the elevation of the top of the underlying unit and the sum of all thicknesses should equal the borehole depth.

For *completeness* consistent data should be available over the whole area of interest (Lyons *et al.* 1989). This is of particular relevance in those raster GIS* which use a numerical code restricted to a range of 0–55 in place of a character attribute, such as lithology. The link between code and text is made in an attribute

*A *raster* GIS stores information in a series of grid or picture elements. *Vector* GIS store spatial information in terms of points, lines and polygons.

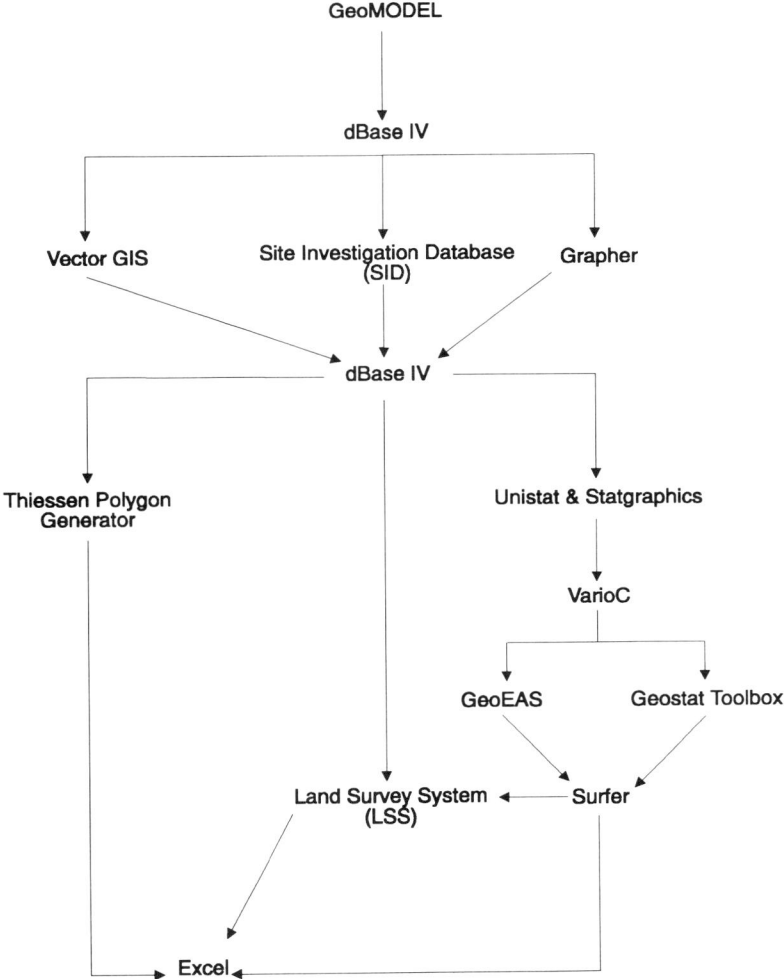

Fig. 1. Data flow during preparation and interpolation.

value table. All possible attribute values must be included in the table or, less elegantly, a catch-all class such as 'other' must be provided.

Lineage is a record of the data sources and of the operations which created the database and is often a strong indicator of accuracy. Rosenbaum (pers. comm.) discovered that strong directional trends in borehole data did not reflect palaeo-glacial movement but rather the three different drilling contractors who provided the data!

Resolution refers to the scale at which observations or measurements were made and therefore the scale at which, and purposes for which, it may be appropriate to use the data. In engineering geological mapping, rules have been set up to determine the content of maps at different scales (UNESCO 1976).

Lyons *et al.* (1989) define *currency* as the date of data collection. However, the rates at which the parameter and the accuracy of the measurement techniques are likely to change also influence the currency of a dataset. For example, lithology does not change very quickly, fracture intensity may increase due to stress relief effects on a recently exhumed rock mass and the accuracy and resolution of remote observation methods such as downhole geophysics has improved in recent years (MacAllum 1992).

Data preparation

Borehole schedule

The borehole schedule comprises digital information from cored, geophysically logged (geologged) and openhole boreholes supplied by British Coal Opencast to the contractor as ASCII datafiles from the GeoMODEL system. The data detail borehole locations and their strata intersections. The strata information comprises lithological descriptions, geometrical data, lineage and reliability indicators and, for most intersections of coal, a code indicating the seam name and leaf. The data form two separate sets. One gives information from openhole and cored boreholes and the other from geologged boreholes. In recent years the second data set has been a subjective assessment of the 'best' data from a particular borehole.

Data preparation potentially involves several different processes carried out using different software programs. The way in which data are moved between the various programs is illustrated in Fig. 1.

The first stage in deriving quantities of coal is to set up a database of borehole and strata information in dBASE IV. A list of the different seam codes used by BCO is produced by indexing on seam code (INDEX ON seam code TAG seamcode UNIQUE) and then consolidated to ensure that a given SEQ seam is only described by a single seam code.

In order to code as many of the uncoded coal intersections, strata information is then downloaded to the site investigation database (SID) program (Zytynski 1991) where a large number of cross sections are generated from the borehole data. These cross sections can be interpreted by the geologist and seam codes added to the dBASE database. Seam codings assigned by the geologist are stored in a separate field to those assigned by BCO; this always enables the originator of any given code to be identified at a later stage.

Continuous coal is occasionally split in the GeoMODEL data into separate records. In cases where this reflects variations in coal quality such intersections are left uncoded. Those intersections which could geologically be considered to be equivalent to coded intersections immediately above or below can be coded and taken into account in the estimate. Instances of inconsistent coding may also be discovered. These comprise continuous intersections of coal which have been assigned different seam codes in the GeoMODEL data. These intersections are re-coded.

Coal intersections which only contained traces of coal (bigram TR in the description field of the GeoMODEL data) are excluded from the assessment. Coal intersections described as shaly (bigram SY) or dirty (bigram DY) are usually coded by BCO and can be retained.

The seam coding exercise not only improves the coverage of contractual seams but also provides information on non-contractual coal which might be recovered during the working of the site.

Seam area boundaries

The limits within which coal is to be recovered are shown on the contract documents. The boundaries are digitized into ASCII files and loaded into the Land Survey System (LSS) mine design software (McCarthy & Taylor 1992) and other spatial processing programs. The digitized boundaries are compared with the areas given in the SEQ. Boundaries with gross discrepancies (>1%) are re-digitized. A conversion factor is then applied to digitized areas to bring them and any calculations based upon them into line with the SEQ areas.

Methods of interpolation

A variety of interpolation methods may be used depending on the nature of the data, the extent to which areas of washout can be identified and the time available.

Thiessen polygons

Thiessen polygons enclose the area that is closer to the central borehole than to any other borehole. They are based on the Delaunay triangulation, a unique method of forming a triangulation from a set of borehole locations (Davis 1986; Holliday & Nathanail 1991). The area of each polygon is calculated and multiplied by the seam thickness measured at the central borehole. A cookie-cutting (boundary-on-surface intersection) operation is carried out to ensure only coal within the seam area is considered. The volume of coal within the seam area is calculated by summing the volume of coal within each polygon or part of a polygon within the seam area. In some cases the seam area extends beyond the area covered by Thiessen polygons. The coal in this extra area is assessed by assuming a thickness equal to the average seam thickness.

The area allowance for old workings can be estimated by constructing Thiessen polygons *reference* around boreholes intersecting a parti-

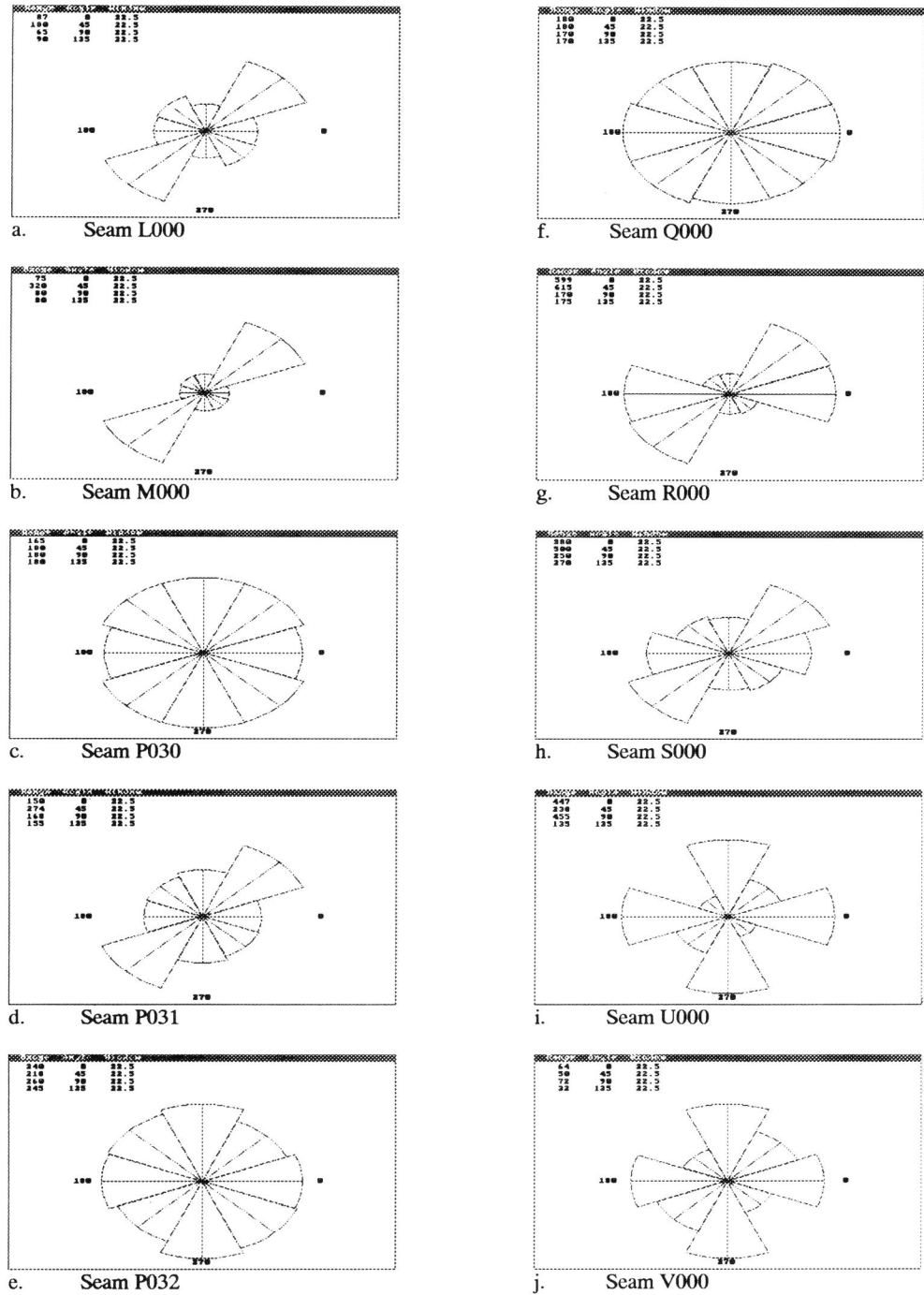

Fig. 2. Anisotropy rosettes from a sequence of seams. Seams are in stratigraphical order with L000 at the base and V000 at the top of the sequence.

cular seam and assigning to the boreholes a value of 1 if workings were encountered or 0 if no workings were encountered. A similar approach can be taken for burnt coal. Washouts can be assessed using Thiessen polygons or by assigning a thickness of zero metres to borehole intersections where the seam has been affected.

Delaunay triangulation

Delaunay triangulations of seam thickness are constructed using the LSS mine design software and the volume of coal calculated by reporting volumes through zones defined in a separate file containing the seam area boundaries. The procedure followed within LSS is:

(a) create models of seam thickness for each SEQ seam from 'load' files output from dBASE;
(b) create a model of borehole locations with zero level;
(c) create models of coal area boundaries;
(d) use the LSS *report volume to survey by zones* function to calculate the volume of coal.

Where the triangulation does not extend over the entire seam area, extra points have to be digitized with a seam thickness equal to the average seam thickness from all intersections of given seam. The extra points are located such that their influence on the triangulation is minimized. A minimum number of points is added. The points are added away from the seam boundary. The triangulation is then inspected to ensure it covers the entire contractual seam area. The digitized points are given a point feature code of PMBD to differentiate them from borehole data (code PMBH or PGBH; the initial 'P' is used by LSS to signify a point feature).

Average seam thickness

For those seams where the spatial distribution of parameters requiring an area allowance is not known, the average seam thickness for all intersections within the seam area multiplied by the area provides an estimate of coal volume. The area allowance for washouts is determined from the ratio of the number of boreholes which either do not encounter the seam when they ought to have done or encounter an abnormally reduced thickness of coal to the total number of boreholes intersecting a seam.

Geostatistics

The quantity of data in the borehole schedule usually means that there is a sufficient number of intersections of most seams to warrant a geostatistical study. Low cost software packages have been successfully used to krige coal seam thickness (Nathanail & Rosenbaum 1992).

In those cases where too few boreholes have intersected a particular seam to enable a variogram to be adequately modelled, the variogram for overlying seams with more intersections may sometimes be used. Of course this assumes that there are no geological discontinuities such as faults between the two seams and that the two seams can justifiably be treated as having the same spatial variability. Examples from a sequence of seams shows that variograms of adjacent seams are generally similar. However, seams farther apart in the stratigraphic succession have dissimilar variograms. Anisotropy rosettes of directional variograms from a sequence of seams from an opencast coal site in central England (Fig. 2) show similarity between several pairs of adjacent seams (e.g. L000 and M000 U000 and V000). However, the three leaves of seam P (P030, P031 and P032) show very different degrees of anisotropy.

Non-parametric geostatistics can be used to model the probability of encountering old workings. The datafile of '1's for boreholes intercepting workings and '0's for those not intercepting workings can be used to produce a variogram; the data may then be kriged. The results are contours ranging in value from zero to one. These may be interpreted as contours of probability of encountering workings within each panel or as estimates of the percentage of workings in each panel (Fig. 3).

Forecasting coal output

Once a contract has been awarded, it should be possible to incorporate the models of seam thickness and quality into the system to be used to store and process survey information during the working of the site. This requirement and the nature of the survey software should influence, but not determine, the method of interpolation used during the assessment.

The coal seam models can be calibrated during operation with actual coal recoveries to enable ever-improving coal output forecasts to be made. This calibration process can also be fed into assessments of other, especially neighbouring, sites.

Within LSS, the facility to report volumes by

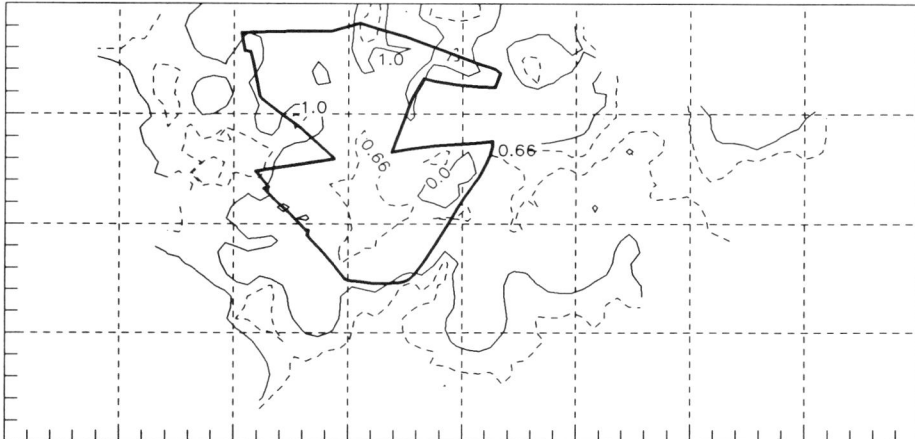

Fig. 3. Probability contours of encountering old workings (thick line delineates boundary of contractual seam area).

zones (see above) allows the volumes in regions coaled, regions outside the contract areas, regions remaining to be coaled and regions not coalable (e.g. due to mass movement), which are differentiated using various surface codes, to be separately reported in a single operation.

Conclusions

Reserve assessments in the stratified deposits of the Coal Measures based on the data supplied by British Coal Opencast to tenderers can be carried out in short time frames using a variety of techniques selected on the basis of the geological and contractual characteristics of each site. The quality of the data has to be established as part of the assessment, however short the time available. This can be achieved using principles established for data management in geographical information systems.

The execution of reserve assessments in a short time frame creates a need for imaginative and efficient use of several software programs in order to ensure reliable estimates. Such work is demanding and challenging and offers the geologist the opportunity to exercise skills of decision making which will have very large financial implications under less than ideal conditions. Tenders can be won or lost for comparatively small sums and high quality reliable reserve estimates within the required time frame can make the difference between a successful and unsuccessful tender.

References

CHRISMAN, N. R. 1983. The role of quality information in the long term functioning of a GIS. *Proceedings of Autocarto* 6 ASPRS, 2, 303–321.

—— 1991. The error component in spatial data *In:* MAGUIRE *et al.* (eds) *Geographical Information Systems: Principles and applications.* Longman Scientific & Technical, Harlow, 1, 165–174.

DAVIS, J. 1986. *Statistical analysis of geological data,* 2nd edition. John Wiley, New York.

DCDSTF 1988. The Proposed Standard for Digital Cartographic Data. *The American Cartographer,* 15, 9–140.

DE FREITAS, M. H. 1993. Discussion. *In:* CRIPPS, J. C. *et al.* (eds) *The engineering geology of weak rock.* Engineering Geology Special Publications 8, A A Balkema, Rotterdam, 493.

GOODCHILD, M. & GOPAL, S. 1989. *The accuracy of spatial databases.* Taylor & Francis, London.

GUPTIL, S. C. 1989. Inclusion of accuracy data in a feature based object-oriented data model. *In:* GOODCHILD, M. & GOPAL, S. (eds) *The accuracy of spatial databases.* Taylor & Francis, London, 91–97.

HOLLIDAY, J. & NATHANAIL, C. P. 1991. *THIESSEN. FOR, FORTRAN77 source code for Thiessen Polygon generation.* Wimpey Environmental Limited (unpublished).

LYONS, K. J., MOSS, O. F. & PERRETT, P. 1989. Geographic information systems. *In:* BALL, D. & BABBAGE, R. (eds) *Geographic Information Systems: Defence applications.* Brassey's Australia, Rushcutters Bay, Australia, 8–41.

MACALLUM, R. 1992. Geophysical logs and the search for opencast coal reserves. *In:* ANNELS, A. E. (ed.) *Case histories and methods in mineral resource evaluation.* Geological Society, London, Special Publications 63, 77–93.

McCarthy, S. & Taylor, B. 1992. *Land Survey System User Manual version 3.1.* Hall and Watts Systems, Birdlip, Gloucestershire.

Nathanail, C. P. & Rosenbaum, M. S. 1992. The use of low cost geostatistical software in reserve estimation. *In:* Annels, A. E. (ed.) *Case histories and methods in mineral resource evaluation.* Geological Society, London, Special Publications **63**, 169–177.

NCGIA 1990. *Core curriculum: Introduction to GIS.* National Centre for Geographic Information, Santa Barbara, California.

Rhind, D. 1990. The ubiquitous geographical information system. *Science and Public Affairs*, **5**, 57–66.

UNESCO 1976. *Engineering geological maps: A guide to their preparation.* UNESCO Press, Paris.

Zytinski, M. 1991 *SID geotechnical database system: User manual version 2.* MZ Associates, Carmarthen, Dyfed.

Phosphate rock: factors in economic and technical evaluation

A. J. G. NOTHOLT

Mineral Resource Consultancy, 12 Thornhill Road, Ickenham, Uxbridge, Middlesex UB10 8SF, UK

Abstract: Phosphate rock in the mineral industry refers to the marketable, usually beneficiated, product. Geologically, however, the term embraces a very wide variety of rock types, of both sedimentary and igneous origin, as well as their weathering derivatives. Phosphate deposits thus form in markedly different geological environments and ore characteristics are so varied that a combination of nearly identical factors applicable to their evaluation seldom exists. In spite of this diversity, the phosphate component in most deposits is a member of the apatite group represented by the two end members: fluorapatite, the main mineral in crustal igneous rocks, and carbonate fluorapatite or francolite, by far the most important phosphate component of commercial sedimentary deposits. The identification of the correct apatite composition is important in assessing the quality of marketable phosphate concentrates.

Evaluation usually begins with a detailed mineralogical, petrographical, and chemical study of a representative sample of prospective phosphate ore to determine the principal constituents in the ore and their mode of occurrence. The data obtained establish the grade (P_2O_5 content) of the rock and its chemical quality and, hence, the method and degree of beneficiation required to provide a commercially acceptable phosphate concentrate. Beneficiation technology is such that often the most difficult ore types (e.g. siliceous, dolomitic and Fe-Al-rich) can be upgraded satisfactorily, but the problems become more acute when impurities occur as substitutions in the apatite lattice. Quality rather than grade is often the deciding factor in determining the viability of a given deposit, in addition to other technical and economic factors.

Phosphate rock (PR) is well known as the basis of a major mineral industry of worldwide importance, its value residing in the fact that it is a vital source of phosphorus (P) in the manufacture of phosphate fertilizers and certain phosphate chemicals, notably detergents. Nowadays, most phosphate fertilizers are of the 'high-analysis' type, their manufacture being based on the acidulation of PR with phosphoric acid, which itself is produced by the reaction of PR with sulphuric acid. More than four-fifths of annual world PR production is used for this purpose, hence the overriding importance of phosphoric acid as an intermediate in the phosphate fertilizer industry.

More than 34 countries produce PR in various forms, world marketable production totalling some 140.4 Mt in 1992, of which about 21.9 Mt entered international export markets. Nearly four-fifths of world production is derived from deposits of sedimentary origin (phosphorites). Most of the remainder is obtained from various igneous rocks, notably from alkaline igneous intrusions, including carbonatites and their weathering derivatives, and from deposits produced by the weathering of phosphatic sedimentary limestones.

PR for practical purposes may be defined as naturally-occurring material containing one or more phosphate minerals and possessing chemical and physical characteristics that make it acceptable for commercial use as a source of phosphate (Notholt 1980). As such it is represented by an extremely wide variety of rock types of both sedimentary and igneous origin. Grade or phosphate content is normally expressed in terms of P_2O_5 determined by chemical analysis, although P_2O_5 does not occur in nature as such. The grade of products may also be expressed as % BPL or bone phosphate of lime, reminiscent of the early days of the world fertilizer industry when tricalcium phosphate, thought to be the chief component of animal bone, was the principal source of phosphate in fertilizer manufacture (% P_2O_5 × 2.1853 = % BPL; % BPL × 0.4576 = % P_2O_5). Commercial PR varies in grade from about 18% P_2O_5 (about 39.3% BPL) to 38.6% P_2O_5 (about 84.3% BPL). PR entering world trade markets averages 28.5% P_2O_5 (about 63% BPL) or more. As mined, PR ranges in grade from nearly 4% to about 39% P_2O_5.

Crustal rocks average about 0.1% P, equivalent to about 0.23% P_2O_5. Because of its great affinity for oxygen, the element P is not found in nature in a free or uncombined state combining readily in the form of orthophosphate, PO_4, with a number of cations, chiefly calcium. Many phosphate-bearing minerals are thus known in nature (Deer et al. 1962; McConnell 1973; Nriagu & Moore 1984) and these are among the most complex and varied in the entire mineral kingdom. However, by far the most important is the apatite (calcium phosphate) group, accounting for almost all of the known PR reserves and resources. Apatite is a widespread accessory mineral in many igneous and sedimentary rocks, in various residual deposits, including those produced by intensive weathering of sedimentary phosphatic limestones and of carbonatites, and as replacements of limestone coral rock. Apatites thus form in a wide variety of environments, giving rise to compositions that often differ markedly from the usually assumed theoretical composition.

Geological and graphical factors

A complex set of interrelated economic, technical, environmental and political/social factors determine the commercial potential of individual PR deposits (Cathcart 1968, 1980; Lehr & McClellan 1974). Unfortunately, these deposits vary so widely in terms of their overall geology, ore characteristics, and amenability to beneficiation, that nearly identical parameters are rarely, if ever, applicable. Thus, only generalizations may be offered regarding, for example, geological characteristics of the PR deposit, ore to waste and concentration ratios, ore quality and grade, recovery efficiency, transportation costs, location in respect of potential markets, and environmental constraints. Any of these variable factors can determine whether the deposit is a non-economic phosphate 'resource' or becomes an economic 'reserve'.

The most favourable mining conditions involve high-volume removal of both waste (overburden) and ore and a large tonnage output of PR, followed by the efficient beneficiation or upgrading of ore to specific marketable products (Emigh 1973; Lehr & McClellan 1974; Cathcart 1980; McClellan 1980a). Under ideal circumstances, deposits comprise one or more thick and reasonably level beds of high-grade ore of uniform texture and mineral composition, and a minimum of structural deformation combined with shallow overburden to allow easy open pit mining. However, these ideal conditions are seldom encountered in practice and various mining techniques have been developed to suit the nature of the deposit.

Most of the world production of PR comes from the open pit mining of sedimentary deposits somewhat less than 1 m to over 10 m thick, a characteristic feature being the use of large, capital intensive earth moving equipment epitomized by the electric walking dragline or the bucket-wheel excavator. These machines enable low grade but generally very extensive deposits to be worked economically by virtue of their high recovery efficiency and large tonnage throughput. The larger mines may have capacities of around 35 Mt per year or more of marketable product. However, other favourable parameters such as location close to sites or areas of consumption, high ore grades, abundant and cheap labour, or negligible overburden, can individually or collectively render very small tonnage operations profitable. To permit high-volume annual production rates over a reasonable period (at least 10 years) and at low-unit cost, reserves for a major new mine must be large, perhaps in the region of 50 Mt or more of recoverable ore (Everhart 1971 in Lehr & McClellan 1974). It is self-evident that reserves must also be sufficiently large to sustain and justify the cost of constructing processing plant.

Unconsolidated, flat-lying deposits, such as those of Tertiary age in the southeastern USA are among the easiest to mine. According to Cathcart et al. (1984), the total depth to which overburden and ore can be mined together is about 50 m and, because of the large size of the dragline buckets in use, the ore bed must be more than 1 m thick. As a rough guide, the practical upper limit of the ratio of overburden to phosphate ore, previously placed at about 3:1 (Cathcart 1968), is now around 5:1 (Cathcart 1991). In addition, the phosphate concentrate should contain more than 29% P_2O_5, less than 5% $Fe_2O_3 + Al_2O_3$, less than 1.5% MgO, the ratio of % CaO : % P_2O_5 being also less than 1.55:1. Mining of stratified deposits may also be by open pit methods where open geological structures exist and in areas where the PR beds are sufficiently close to the surface. Deposits in structurally complex geological settings are more expensive to mine and may even be uneconomic to exploit if drilling, blasting, ripping or other relatively expensive steps are required. In the folded and faulted Permian phosphate deposits of the Western US, for example, the ore bed must contain more than 18% P_2O_5, less than 3% $Fe_2O_3 + Al_2O_3$, less than 1.5% MgO, with a $CaO:P_2O_5$ ratio of less than 1.55:1 and the ore zone more than about 1.5 m thick. The minimum size for a new deposit is placed at 20 Mt (Cathcart et al. 1984).

Labour, mine equipment and energy costs are

now such that underground mining of PR is unlikely except in the most favourable circumstances. The only underground phosphate mine in North America, for example, is in Montana. Underground operations continue, nevertheless, to account for a significant proportion of output from mines, for example, in Russia, Morocco and Tunisia. Beds too deep for open-pit mining or steeply dipping, are uneconomic to work, as are sedimentary sequences consisting of PR beds intercalated with strata that are classed as waste or are of very low grade. There are few, if any, commercial mines operating under such adverse conditions, except perhaps where the operations are subsidized for political/social reasons.

As alternatives to conventional earthmoving techniques, mining of unconsolidated PR deposits by dredging methods have been employed commercially in North Carolina, while *in situ* slurrying with high pressure water injected through drill holes has been tested experimentally by the US Bureau of Mines in Florida and North Carolina (Savanick 1984, 1987). This method has potential notably in the extraction of relatively deep deposits of environmentally sensitive areas.

Because PR is of low-unit value, freight is usually a major and, on occasion, the overriding cost factor, determining the profitability of a deposit by virtue of its location in relation to domestic or international markets. Most commercial deposits competing successfully on world export markets are situated considerably less than 100 km from deep-water sea port terminals. Such coastal deposits may have progressively limited markets in the hinterland, however, if transport is by road or rail or, in more extreme cases, transport infrastructure is poor or non-existent. If transportation costs are high, it may be cheaper to import high-analysis fertilizers from other sources. The disadvantages of remote or unfavourable location can be offset, at least partially, if PR deposits are of high-grade, can be cheaply mined and require little or no beneficiation: a deposit which is of low grade, small and costly to mine is generally suitable only for local use in remote areas. A significantly less costly means of transporting PR overland may be by slurry pipeline. What is claimed to be the world's longest in the industry extends from a mine near Vernal, northeastern Utah, northwards to a chemical/fertilizer complex near Rock Springs, Wyoming, a total distance of 150 km. Its construction enabled output to be expanded from 726 000 t to 1.2 Mt per year of concentrate grading 31.5% P_2O_5 (Weber *et al.* 1987). Similarly, a 120 km pipeline in Minas Gerais, Brazil, extends from Tapira,

south of Araxá, to a terminal at Uberaba, NNW of São Paulo. Transportation problems can also be obviated by shipping value-added phosphate products or intermediates, both manufactured at or near the PR deposit, instead of merely PR concentrate.

Marketable PR generally contains more than 30% P_2O_5, and to achieve this grade most phosphate ores require beneficiation. Many techniques are available, including washing and screening, desliming, magnetic separation, flotation and calcination, and concentrations of these may be required to produce concentrates of an acceptable quality and grade (Table 1). The most sophisticated and expensive techniques involve flotation and thermal treatment. The beneficiation procedures adopted are determined mainly by ore characteristics and end-uses of the marketed product. In view of the extremely varied nature of PR, and particularly in the face of declining grades and quality, phosphate beneficiation has presented a considerable challenge to processing skill and resourcefulness (e.g. Walker 1990).

Table 1. *Range in chemical composition of selected commercial PR*

Source	Sedimentary %	Igneous %
BPL	63–82	76–83
P_2O_5	28.2–37.5	35.8–38.2
SiO_2	1.0–9.5	0.2–2.6
CaO	48.3–53.8	46.8–53.9
F	3.2–4.1	1.5–3.2
CO_2	1.7–7.5	0.2–5.4
Al_2O_3	0.1–1.4	0.1–0.5
Fe_2O_3	0.1–1.4	0.2–2.9
Na_2O	0.2–2.0	0.1–0.5
K_2O	0.01–0.4	0.02–0.44
MgO	0.02–0.8	0.06–1.5
Organic C	0.1–>0.5	0.03–0.09
Cl	0.01–0.23	0.006–0.022

Evaluation

The evaluation or 'characterization' of PR is a complicated procedure which involves thorough and careful analysis, together with continuous pilot-plant testwork, prior to its commercial use. Often complex textural and compositional characteristics may be revealed on examination, and these determine the most effective and economical means of beneficiating or processing PR. Pre-eminent in this field of investigation has been the Tennessee Valley Authority (Lehr 1980, 1984; Lehr & McClellan 1974; Lehr *et al.* 1967; McClellan & Lehr 1969; Smith & Lehr 1966) and, since its foundation in 1974, the International Fertilizer Development Center (McClellan 1980a, b; McClellan & Clayton 1980; McClellan & Gremillion 1980; McClellan & Hignett 1978; Roy & McClellan 1986; Van Kauwenbergh 1991; Van Kauwenbergh et al. 1990). This paper draws heavily on the results of the frequently innovative research carried out by these organizations.

Methods which have been most effective in characterizing potential phosphate ores include petrographic and electron microscopy, chemical analysis, X-ray powder diffraction and infrared spectroscopy (Lehr et al. 1967; McClellan & Gremillion 1980; Zevin et al. 1988). Complete evaluation is normally based on two or more of these techniques, XRD and infrared investigations generally providing rapid preliminary results in respect of widely varying samples. By these means ore characteristics can be identified and evaluated by first determining the composition of the apatite mineral present. The data obtained can then be combined with chemical analysis of the whole rock to evaluate the significance of other constituents, the primary objective being to determine how key impurities are distributed between the apatite and gangue-mineral fractions and, ultimately, the most appropriate beneficiation procedure. Theoretical P_2O_5 values obtained for apatites can be used to calculate the ratio of phosphate to gangue. In addition, the data can provide an indication of the chemical quality of the PR concentrates likely to be obtained.

Petrographic examination

Phosphate rock, whether of igneous or sedimentary origin, is not a very distinctive rock. Hand specimens can exhibit a wide variety of physical properties and in many cases the presence of phosphate can generally be determined only by geochemical field tests using or based on that originally developed by the US Geological Survey (Shapiro 1952). In most sedimentary PR deposits, apatite exhibits a variety of microstructures (Hewitt 1980) but it is usually micro- to cryptocrystalline in form and occurs in complex aggregates of crystallites. This is in marked contrast to the often medium to coarsely crystalline mode of occurrence of apatite in igneous and metamorphic rocks. Pelletal textures, frequently without internal structure, are common in major commercial deposits and have been described in detail (Mabie & Hess 1964; Trueman 1971; Cook 1972, 1976; Slansky 1986). Most pellets have a spherical or oval outline, frequently without internal structure. Apatite may also occur as angular to subrounded polished clasts, or form replacements in shells, coral, and faecal pellets, as well as in fossil vertebrate bones and teeth. Apatite particles are commonly stained by occlusions (endogangue) of finely divided Fe oxides and carbonaceous matter, and colourless occlusions of silt-size grains (e.g. quartz), clay minerals, opaline silica and, occasionally, fossil fragments. Major accessory minerals such as quartz, carbonates, feldspars, and heavy minerals (Table 2) are usually present as discrete grains. Non-pelletal varieties of PR also occur, the prevalent type consisting of authigenic microcrystalline mud for which the term 'microsphorite', a shortened version of microcrystalline phosphorite, analogous to 'micrite' in carbonate sediments, has been proposed (Riggs & Freas 1965). The sedimentary structures of microsphorite have been described in detail by Riggs (1979).

Table 2. *Generalized mineralogy of PR*

	Mineral Phases
Ore	
Phosphate (Ca, P)	Apatite (fluorapatite-francolite)
Gangue	
Phosphate	
Ca, Fe, Al	Crandallite, millisite
Fe, Al	Wavellite, variscite, strengite
Silica	Quartz, chalcedony, opal, cristobalite
Silicates	Clay minerals, micas, feldspars, pyroxenes, amphiboles
Carbonate	Calcite, dolomite, apatite
Evaporites	Halite, gypsum
Metallic impurities	
Fe	Magnetite, hematite, pyrite, goethite, limonite
Al	Clay minerals, micas
Mg	Dolomite, magnesite, pyroxenes
Ba	Barite
Ti	Rutile, ilmenite, anatase, perovskite
Organic matter	Indigenous compounds, beneficiation reagents

Because of its submicroscopic size, optical investigation of apatite in sedimentary PR gives, unfortunately, only limited data. Satisfactory identification has generally to rely on X-ray and chemical analysis (Smith & Lehr 1966; Lehr et al. 1967; McClellan & Gremillion 1980). Using large numbers of samples, statistical methods have been used to develop models which show a high degree of correlation between chemical composition and crystallographic properties, such as indices of refraction and a and c unit-cell dimensions (McClellan 1980b; McClellan & Van Kauwenbergh 1990). Microscopic study is a useful means of determining the index of refraction and its variation within grains.

Weathering and/or leaching is a variable that is often overlooked during evaluation, in spite of the fact that it can cause significant changes in physical and chemical properties of deposits being evaluated. The effects may be such that a deposit is rendered uneconomic, depending on the nature and extent of the alteration. Improvement of grade by weathering is a characteristic feature of many deposits; in the western US, mining of the Phosphoria Formation is restricted to weathered zones which often extend to depths greatly exceeding 200 m. The ore contains up to 8% P_2O_5 more than the deeper, unweathered rock. Leaching of carbonate cement in surface outcrops often results in a soft, porous rock that readily disintegrates upon handling, thus liberating the phosphate mineral. A particularly deleterious effect is the conversion of apatite to less desirable Fe and Al phosphates.

Chemical analysis

Extensive analytical data on both PR and apatites have been reported in the literature, the mineralogical and chemical properties of francolites having been studied in particular detail because of their overriding importance as a source of P (Lehr et al. 1967; McClellan & Lehr 1969; Gremillion & McClellan 1978; McClellan 1980a, b; McClellan & Van Kauwenbergh 1990). The composition of francolites can be described essentially in terms of only six components: P, Ca, Na, Mg, CO_2, and F, and, in addition can be represented quite satisfactorily by a compositional series having fluorapatite and carbonate fluorapatite as mineral end-members (Table 3). Apatites approaching fluorapatite in composition are the principal ore minerals in most igneous PR deposits; those occurring as carbonate fluorapatite, mainly within the varietal range assigned to francolite, i.e. containing more than 1% F and significant amounts of CO_2 (Sandell et al. 1939), are almost always the only apatites found in sedimentary PR. Francolite also occurs in igneous deposits but usually as a secondary mineral in supergene phases and residual/weathering enrichments. Such secondary apatites are similar to francolite in marine sedimentary rocks in being generally microcrystalline and containing some carbonate (Van Kauwenbergh 1991).

A relatively large number of isomorphic substitutions can take place in the francolite structure which can impose grade limits on some ores and also introduce impurities which cannot be removed by conventional processing methods. By far the most important is substitution of CO_3^{4} for PO_4^{3} which occurs essentially on a 1:1 basis (McClellan & Lehr 1969). Maximum substitution amounts to between 6 and 7% CO_2, with a corresponding decrease in the amount of P_2O_5 from the theoretical maximum of 42.2 in fluorapatite to about 34% P_2O_5 in the most highly substituted francolite (Table 3). This has important practical consequences since the degree of substitution determines the upper limit to which PR can be beneficiated. It is evident also that P_2O_5 values, while indicating the amount of phosphate mineral, are not necessarily directly related to the amount of gangue present, but rather to the degree of carbonate substitution in francolite.

Table 3. *Apatite parameters*

Fluorapatite←--------→Francolite
$Ca_{10}(PO_4)_6F_2$ $Ca_{10-a-b}Na_aMg_b(PO_4)_{6x}(CO_{3x})F_{0.4x}F_2$*

High temperature (Igneous & metamorphic) Limited occluded mineral matter	Low temperature (Sedimentary) Abundant occluded mineral matter
Composition	
	%	%
P_2O_5	42.2	34.0
CaO	55.6	55.10
CO_2	0.00	6.30
F	3.77	5.04
Na_2O	0.00	1.40
MgO	0.00	0.70
CaO:P_2O_5	1.318	1.621
F:P_2O_5	0.089	0.148

Substitutions	
Substituting ion	
Ca Na, Sr, Mn, Fe, K, U, Mg, REE	Ca Na, Mg
P C, S, Si, As, V, Cr, Al	P CO_3+F
F OH, Cl	F OH
O F, OH	

*Where a, b, and x = moles of Na, Mg, and Co_2^{3-} respectively.
Sources: McClellan (1980); McClellan & Lehr (1969).

X-ray analysis

X-ray powder diffraction (XRD) patterns of francolites are typically apatitic, slight shifts in peak positions and intensities indicating changes in cell parameters. Changes in unit-cell a dimensions with variations in carbonate content reported by Smith & Lehr (1966); Lehr (1967); and McClellan & Lehr (1969) indicate a linear relationship with francolite composition, values of a ranging from 9.320 to as much as 9.376 Å, and those of c from 6.877 to 6.900 Å.

Table 4. *Practical levels of impurities in PR*

Impurity	Source	% in PR	% in Phos. acid
Aluminium, Al_2O_3	Gangue	0.2–3	70–90
Iron, Fe_2O_3*	Gangue	0.1–2	60–90
Magnesium, MgO†	Gangue, apatite	0.2–0.6	n.a.
Fluorine, F*	Apatite	2–4	25–75
Silica, SiO_2	Gangue	1–10	5–40
Alkalis, Na_2O, K_2O	Gangue	n.a.	n.a
Chlorine, Cl	Gangue, apatite	0–0.05	100
Calcium, CaO†	Gangue, apatite	0.7–8.0	n.a
Organics, C	Gangue, reagents	0.1–1.5	15–70
Cadmium, Cd	Gangue	0.8–155 ppm	70
Uranium, U*	Apatite	35–400 ppm	75–80
Rare earths, REE*	Apatite	n.a.	n.a
Toxic elements (Se, As, Cr, V)	Gangue	n.a.	n.a.
Miscellaneous (Ti, Ba, etc.)	Gangue	n.a.	n.a.

* Recoverable
† May have nutrient value
n.a., not available.

Changes in composition from fluorapatite to francolite (and vice versa) correlate with changes in the *a* values, the lower the *a*-value the greater the amount of carbonate and other impurity substitution, i.e. the less altered the francolite (McClellan 1980b). Increases in *a*-values indicate the extent of alteration to produce forms of apatite approaching fluorapatite in composition (Van Kauwenbergh *et al.* 1990). Francolites that have been progressively weathered/altered have higher *a*-cell dimensions compared with less altered forms (Cathcart and Botinelly 1991). The effects of such post-depositional changes have been discussed also by Nathan (1984) and Nathan *et al.* (1990).

Infrared (ID) studies

The infrared absorption spectrum of francolite identifies both its composition and may, in addition, indicate the presence of extraneous phases and the degree of carbonate for phosphate substitution (Lehr *et al.* 1967; McClellan & Gremillion 1980). A characteristic CO_2 absorption doublet occurs at 1453 and 1420 cm^{-1}, this varying in amplitude according to the amount of CO_3^{2-} substitution. A CO_2 index can thus be derived which is based on the ratio of intensities of the C–O and P–O absorptions, the ratio being also directly proportional to the weight ratio $CO_3:PO_4$ and independent of the concentration of apatite. The CO_2 index and unit-cell *a* values for francolites show a linear relationship.

Impurities

The shift within the last 20 years or so to the manufacture of high-analysis phosphate fertilizers, based on the production of important intermediates such as phosphoric acid and, on a much more limited scale, superphosphoric acid, has resulted in a corresponding emphasis on rock quality rather than mere grade (Anon 1986; Kouloheris 1977; Lehr 1984). As a mineral raw material, PR has thus become predominantly process-specific, both physical and chemical characteristics (Table 4) playing a critical role in determining the suitability of the PR for a particular end-use. Such physical features as texture, hardness, porosity, particle size, crystallinity, and mode of occurrence (cement, coating, grains), as well as the composition of both phosphate and accessory minerals (Table 2), including the distribution of such elements as Fe, Al, Mg, Na, the heavy metals, and toxic elements, can individually or collectively determine the process to be adopted and the composition and character of the phosphoric acid produced. With declining grades mined, a higher proportion of acid-soluble mineral impurities is present.

Silica

In general, the presence of silica (SiO_2) as such is not considered as deleterious except in that it acts as a diluent and renders the rock more difficult to grind. Indeed, if sufficient quantities are present, silica aids the necessary removal of fluorine released from apatite during the preparation of 'wet-process' phosphoric acid (Robinson 1980). Most ores consisting mainly of quartz are usually amenable to economic beneficiation, usually involving flotation. Even quite low grade feeds can be treated. The method of beneficiating siliceous ores may vary considerably, depending on whether they contain clastic quartz, chalcedony or opaline silica in the free state (as exogangue) or are intimately mixed (as endogangue) with individual phosphate pellets. As endogangue, usually one grain of quartz occurs around which apatite is concentrically distributed; less frequently, the central nucleus comprises a grain of feldspar, a limestone fragment, or a dolomite rhomb. PR containing chalcedony and opaline silica tends to be hard and dense, and to require careful grinding to liberate the phosphate, often resulting in excessive phosphate losses. The latter is often true where there is a fine intermixture of quartz and apatite. Coarse, angular quartz grains are abrasive and may erode plant equipment. Silica may be present also as silicates in, for example, PR with high clay contents, i.e. high in Al_2O_3 and Fe_2O_3 (R_2O_3), and in many igneous phosphate deposits.

The beneficiation of low-grade siliceous phosphate ores is epitomized by the procedures adopted in central Florida, USA, where the famous land-pebble deposits have been worked since 1890. The apatite occurs as irregular rounded pellets that generally range from clay to sand size and many of the coarser clasts are compound, comprising three or more generations of pellets cemented by phosphate (Cathcart 1991). Traditionally, the 'double-float' or Crago (1940) flotation technique has been used to effect a separation of phosphate (carbonate apatite) from the gangue, this being chiefly in the form of silt to clay size quartz constituting about one-third of the ore, and clay minerals (Lawyer et al. 1975, 1982). More recently, what is claimed to be a unique process for selective fatty-acid flotation of phosphate minerals from siliceous ores has been developed (Anazia & Hanna 1990; Hanna & Anazia 1991, 1993) whereby Florida and North Carolina PR containing from 7.7% to 15.3% P_2O_5 and 62.3% to 77.2% siliceous gangue can be beneficiated to produce concentrates with 28.1% to 33.1% P_2O_5 and 4.1% to 9.9% SiO_2. These grades are comparable with those currently produced commercially.

Most deposits of igneous or metamorphic origin currently mined on a commercial scale are also highly siliceous. By far the most important of these are the remarkable apatite–nepheline deposits worked since 1930 near the south-western margin of the Khibiny intrusion near Kirovsk, in the Murmansk region of north-western Russia. A variety of ore types have been identified but usually the main minerals are apatite (10–80%), nepheline (15–20%), together with titanite, aegirine-augite, feldspar and titanaugite. Khibiny apatite–nepheline ores are highly crystalline and show average grades ranging from 6.64% to 31.43% P_2O_5 and 8.21% to 32.72% SiO_2. Despite a declining grade, to around 14% in 1990 for example, the grade of the flotation concentrate produced has been maintained at the high level of 39.5% P_2O_5 (86% BPL) (Notholt 1985). The SiO_2 content of the concentrate obtained averages only 1.73%.

Worked on a much smaller scale are steeply dipping Precambrian stromatolitic deposits near Maton in Rajasthan, India. These contain an average of 25.7% P_2O_5 and 38.1% SiO_2. Preliminary tests had shown that ore containing up to 40% SiO_2 is amenable to beneficiation by froth flotation, but attempts to reduce the silica content to below 10% resulted in a marked decline in phosphate recovery. A conventional flotation circuit is used to produce a PR concentrate containing 32% P_2O_5 but it has been reported (Anon 1986) that uncertainty still exists about the ability to produce concentrates with less than 10% SiO_2.

Iron and aluminium (R_2O_3)

Iron and aluminium are generally regarded as belonging to the same impurity category. They are particularly troublesome in wet-process phosphoric acid manufacture and PR high in Fe_2O_3 and Al_2O_3 (R_2O_3) is not normally regarded as usable, mainly because it cannot be beneficiated satisfactorily (Hignett et al. 1976). They also reduce the water solubility and overall quality of the fertilizers produced. Phosphate concentrates with up to 5–6% R_2O_3 may be tolerated for superphosphate production. As a guide, the $R_2O_3:P_2O_5$ ratio in PR should be around 0.10:1. Another problem results from the varied mineralogical forms in which either iron or aluminium, or both, are present. In sedimentary PR, scrubbing, desliming (to reduce Fe and Al as silicates, i.e. clays), flotation and high-intensity magnetic separation can all be effective on a commercial scale.

Non-phosphatic Fe/Al minerals. In many igneous and sedimentary deposits, iron may be present as magnetite, hematite, limonite, goethite, ilmenite or pyrites. Igneous PR deposits consisting essentially of magnetite–apatite mixtures can be readily beneficiated by magnetic separation. In

South Africa, for example, a dry beneficiation process based on the use of super-conducting magnetic separators was developed to treat pyroxenite ore averaging about 6–7% P_2O_5 formerly worked at Phalaborwa, northern Transvaal, and to recover phosphate from the slimes fraction of apatite tailings received from an adjacent copper operation (Roux 1985). In addition, very low ore grades can be accommodated. Pyroxenite, for example, averaging only 2.8% P_2O_5 from a deposit delineated at Loch Borralan, Scotland, was upgraded by separating the 180 μm fraction by dry, high-intensity, magnetic separation. Although recovery was poor, 58%, a product containing 20% P_2O_5 was achieved (Saavedra 1980 in Notholt et al. 1985). Beneficiation to the required product standards becomes more difficult or impossible, however, when Fe-minerals other than magnetite are present. For example, the brecciated and ferruginous deposit formerly worked near Glenover, South Africa, comprised mainly a capping of secondary apatite in a matrix of finely divided hematite and goethite, the ore averaging 32.8% P_2O_5 and 9.3% Fe_2O_3. Most of the hematite could have been removed by high-density magnetic separation, as at Phalaborwa, but the ore was upgraded only by screening the fine material and, as such, was unsuitable for phosphoric acid manufacture. However, Glenover rock proved well suited to the manufacture of elemental phosphorus and the entire output was used for that purpose.

In many magnetite–apatite deposits, magnetite is usually regarded as the primary product and there has been only limited recovery of apatite. Apatite is a characteristic mineral of the Kiruna iron ores worked for many years in northern Sweden, where the introduction of new beneficiation flowsheets and the use of new reagents proved effective in treating the fine-grained apatite-bearing ore. Apatite was recovered from non-magnetic tailings averaging 13% P_2O_5, following low-intensity wet magnetic separation for the production of dephosphorized iron ore pellet fines (Fagerberg & Sandgren 1980) A concentrate with 35.9% P_2O_5 was obtained by flotation. Production of apatite concentrate both at Kiruna and at Grängesberg in central Sweden has been variable, being dependent on iron ore production. Apatite production ceased in 1990. Most of the apatite in the Kiruna deposits is of the hydroxy-fluorapatite variety, as confirmed recently by Veiderma & Knubovets (1991). However, pure fluorapatite also occurs, as does chlorapatite with 1–2% Cl in some deposits (Frietsch 1974).

Both iron and aluminium may occur as complex silicates such as micas, feldspars, pyroxenes and various clay minerals. The clay fraction occurring as exogangue in sedimentary PR usually consists of silicate minerals such as attapulgite, montmorillonite, kaolinite and muscovite and its presence in large amounts leads to sludge problems in the storage of phosphoric acid if the R_2O_3 content exceeds 3 to 5%. Higher amounts make superphosphate fertilizers sticky and difficult to handle. Iron may also be released during acidulation from the oxide minerals goethite and hematite. Silicates such as orthoclase, microcline and plagioclase are much more stable under conventional acidulation treatment and are much less likely, therefore, to have any deleterious effects. Glauconite as separate grains in the endogangue is not common in higher grades of sedimentary PR, although mixtures of glauconite and apatite have been recorded.

Fe/Al phosphate minerals. Phosphate minerals, other than apatite, containing significant amounts of either aluminium or iron, and in some cases both, are to be found chiefly in tropical or subtropical regions where they form by intensive laterization of phosphatic bedrock. These minerals are generally complex, hydrous phosphates of Al and/or Fe whose structure is not well known. Their presence in various rocks can generally be established only by the use of chemical or X-ray diffraction methods. The supergene *crandallites* (including goyazite and gorceixite) and *wavellite* are the only minerals of this type to have been found in phosphate deposits of commercial or potential economic importance. When pure, crandallite contains 34.29% P_2O_5, 36.93% Al_2O_3, and 13.55% CaO, the CaO:P_2O_5 ratio being about 0.39:1; pure apatite, which contains 42.20% P_2O_5, 55.60% CaO, has a ratio of 1.32:1. Thus, total P_2O_5 values alone may give a false impression of the potential value of a PR deposit. If the CaO:P_2O_5 ratio is below 1.32, significant amounts of non-apatitic minerals may be present. Computer analysis can be useful in determining their distribution and extent if analytical work includes a calculation of CaO:P_2O_5 ratios. That crandallite and related secondary minerals are deleterious components in PR is emphasized by the fact that their complete removal cannot be achieved by conventional beneficiation procedures.

The only deposits currently worked on a commercial scale as a source of phosphate are those in western Senegal, on Thiès Plateau, where a lateritic zone developed over lower Tertiary PR beds, contains finely crystalline augelite, crandallite and a cryptocrystalline mineral, pallite, the last having apparently originated by the leaching of phosphatic montmorillonitic clays in the sedimentary succession. This mineral has been proved to be a variety of millisite (Capdecomme & Orliac 1967). The deposits, which range from 3 to 60 m in thickness, contain 28–30% P_2O_5, 8–11% CaO, 6–10% Fe_2O_3, and 27–32% Al_2O_3. The ore is calcined to produce a clinker which is then ground and marketed under the trade name of 'Phospal' for direct application to acid soils, usually as phosphate–potash mixtures. Phospal '34' (95% passing through a 0.16 mm screen) contains 34% P_2O_5 (minimum), 35% Al_2O_3, 10.4% CaO, and 11.5% Fe_2O_3

Carbonate

Carbonate exogangue is very common among sedimentary phosphate deposits, and probably well over two-thirds of the world's PR resources are of the carbonate-rich variety. Calcite predominates but dolomite is locally very abundant in some deposits. Calcite, dolomite and magnesite, as well as highly substituted apatites, cause foaming problems during the acidulation of PR. Moderate amounts of calcium carbonate (up to 4 or 5% as CO_2) are acceptable in superphosphate manufacture as a means of enhancing reactivity by making fertilizer porous and spongy. Drying is also facilitated.

The $CaO:P_2O_5$ weight ratio is one of the most important chemical factors affecting the economics of PR processing, since this determines acid consumption, each % CaO requiring some 17.5 kg H_2SO_4 per tonne of PR (Becker 1989). In phosphoric acid manufacture, it determines also the amount of waste calcium phosphate (phosphogypsum) which has to be filtered (Robinson 1980). A typical range for the $CaO:P_2O_5$ ratio is 1.4:1 to 1.6:1, irrespective of whether the PR is sedimentary or igneous in origin, beyond which excess quantities of acid are required which may ultimately render the entire process uneconomic.

The presence of magnesium (usually as dolomite) is never desirable since it creates high viscosity in phosphoric acid and is the principal source of insoluble phosphate precipitates present in ammonium phosphate liquid fertilizers. Apatites may contain up to 0.7% MgO in their structure and in such cases the magnesium content cannot be reduced by physical beneficiation. Higher values of Mg occur in PR when, for example, bentonitic clay or glauconite is present.

The beneficiation of carbonate-rich phosphate ores has been the subject of considerable research by the phosphate rock industry and others (Anon 1975, 1979, 1986, 1991; Anazia & Hanna 1990; Baumann & Snow 1980; Hanna & Anazia 1991; IMPHOS 1980; Lawver et al. 1975, 1982; Llewellyn et al. 1982; Mougdil & Vasudevan 1988; Parsonage 1986; Redeker 1984). However, the efficient separation of carbonate minerals from apatite in sedimentary PR ores continues to pose a challenging problem. Carbonate minerals and apatite have identical cations in the crystal lattice and very similar flotation properties, so that selective flotation, although technically feasible, has generally proved difficult, requiring careful control. None of the conventional flotation techniques has been able to reduce the dolomite content (as MgO) to below the 1–1.5% level normally required for phosphoric acid manufacture. Successful removal of dolomite from potential ores on a commercial scale, in the southern extension of the Florida phosphate field, for example, would greatly extend dwindling reserves.

Almost all commercial operations treating carbonate ores resort to high-temperature (800–900 °C) calcination followed by wet-chemical separation or slaking which aims to dissociate the carbonates present in both the exogangue and the apatite. It is a technique suitable for low-grade ores in which calcite is a major impurity, at the same time effectively eliminating traces of organic matter, although in some cases the process may be regarded as too expensive if the ore contains less than about 25% P_2O_5. In the western US, calcination plants were installed which removed about 40–50% of the calcium carbonate content, thereby raising the grade from 28% to 32% P_2O_5. In this way, it has proved possible to use western US ore more extensively for phosphoric acid manufacture, the ore having previously been regarded as too low in grade and high in organic matter (averaging 2%) to be suitable for this purpose. A notable development was the installation, in 1965, of the world's largest single phosphate kiln operating near Oron in the Negev Desert, Israel, capable of producing 600 000 t per year of calcined product.

In marked contrast, significant technological improvements have been achieved in respect of carbonate-rich ores of igneous origin. Apatite and calcite were first separated successfully on a commercial scale by flotation at Jacupiranga, São Paulo, Brazil, where carbonatite ore averages about 5% P_2O_5 (about 12% apatite), 71% calcite and 7% dolomite (Silvia & Andery 1972). The technique has been applied commercially to other Brazilian ores, including the weathered/leached deposits overlying the Araxá and Catalão complexes (Betz 1981). Similarly, a flotation technique developed in Finland (Kiukolla 1986; White 1984) at Siilinjärvi enables carbonatite ore grading only 3.8% P_2O_5 to be beneficiated to yield a 33% P_2O_5 concentrate. Well-developed crystallinity appears to be a favourable factor in beneficiating these types of ore.

Organic matter

Organic matter is almost always found as endogangue in phosphate pellets that have not been subjected to much secondary oxidation. It is usually of marine planktonic origin and composed essentially of humic acids. This constituent causes foaming problems in phosphoric acid processes, causing serious choking of filter cloths, as well as black discoloration of the products. In commercial practice, low-temperature calcination (450–500 °C) is required to remove organic matter in PR, a relatively low cost operation if maximum use is made of the fuel value of organic material. In North Carolina, for example, the world's largest producer of calcined PR, ore averaging about 15% P_2O_5 is passed through fluid bed calciners to remove water, oxidize the organic matter, and liberate carbon dioxide from flotation concentrates. Weight losses of 8% occur without loss of P_2O_5 at temperatures of approximately 810 °C (Hird 1985). The calcined product averages 72% BPL (33% P_2O_5).

Fluorine and chlorine

Fluorine is generally regarded as deleterious and normally should not exceed about 3 or 4%. The $F:P_2O_5$ is usually 0.09:0.13 in sedimentary ores and 0.04–0.06 in igneous deposits. PR concentrates high in F tend to give phosphoric acid also high in F, the content of which mainly determines the kinds and amount of insoluble precipitates that form in its ammoniation products. The concentration of F is particularly critical in liquid fertilizers.

The effect of soluble chlorine lies in its corrosive action on plant equipment rather than in chemical processing. Normally, a Cl content exceeding 0.1 to 0.2% cannot be tolerated because corrosion rates become too severe. Special steels have been developed, although only on a laboratory scale, which can cope with Cl contents of as much as 0.4% and even up to 4% (Becker 1989).

Chlorine in sedimentary PR is usually derived from salts (NaCl) found in most sediments, or from sea water used in beneficiation. In either case, the commercial practice is to reduce or eliminate the chlorine content simply by washing with fresh water. However, when present in apatite, i.e. as endogangue, thermal methods may be necessary to remove chloride. However, these are expensive procedures and not in commercial use at present.

Minor trace elements

PR may contain 16 or so trace elements, mainly as heavy metals and REE elements. Minor constituents such as Mn, Fe, and Cu, are beneficial as micronutrients. However, all eventually contribute to the precipitation of insoluble phosphate compounds. Other elements, including Cd, Pb, Cr, As, Hg, Se and V, are either toxic or potentially harmful in agricultural products. Of these, cadmium in fertilizers based on PR has received much publicity in recent years because of environmental implications. At present the only commercial process for Cd removal is high temperature calcination (1500 °C) and appears to be specific only to rock mined on Nauru in the Central Pacific (Anon 1986).

References

ANON 1975. *Seminar on Beneficiation of Lean Phosphates with Carbonate Gangue*, Cagliari, 23–24 April. Eleventh International Mineral Processing Congress, Cagliari.
—— 1979. *Proceedings of the Soviet–Swedish Symposium of the Beneficiation of Phosphate Rock*. Academy of Sciences of the Estonian SSSR, Tallinn.
—— 1986. Removing impurities to improve phosphoric acid quality: rock beneficiation. *Phosphorus & Potassium*, **146**, 27–31.
—— 1991. Selective flotation of dolomitic Florida phosphates, *Phosphorus & Potassium*, **172**, 23–27.
ANAZIA, I. J. & HANNA, J. 1990. Sequential separation of carbonate and siliceous gangue minerals during phosphate ore processing. In: HANNA, J. & ATTIA, Y. A. (eds) *Advances in Fine Particle Processing*. Elsevier, New York, 357–367.
BAUMANN, A. N. & SNOW, R. E. 1980. Processing techiques for separating MgO impurities from phosphate products. In: *Proceedings of the 2nd International Congress on Phosphate Compounds*, Boston, Massachusetts. Institut Mondial du Phosphat, Paris, 269–280.
BECKER, P. 1989. *Phosphates and Phosphoric Acid*. Marcel Dekker, New York.
BETZ, E. W. 1981. Beneficiation of Brazilian phosphates. In: LASKOWSKI, J. (ed.) *Mineral Processing*. Proceedings of the Thirteenth International Mineral Processing Congress, Warsaw, June 4–9, 1979, Pt. 8, 1846–1877.

CAPDECOMME, L. & ORLIAC, M. 1967. Sur les caractères chimiques et thermiques des phosphates alumineux de la région de Thiès (Sénégal). In: Colloque International sur les Phosphates Minéraux Solides. **2**, Structures et Propriétés des Phosphates. Société Chimique de France, Toulouse, 45–55.

CATHCART, J. B. 1968. Marine phosphorite deposits—economic considerations. In: Proceedings of the Seminar on Sources of Mineral Raw Materials for the Fertilizer Industry in Asia and the Far East. ECAFE Mineral Resources Development Series, **32**. United Nations, New York, 295–300.

—— 1980. The phosphate industry of the United States. In: KHASAWNEH, F. E., SAMPLE, E. C. & KAMPRATH, E. J. (eds) The Role of Phosphorus in Agriculture. American Society of Agronomy, Inc., Crop Science Society of America, Inc. & Soil Science Society of America, Inc., Madison, Wisconsin, 19–42.

—— 1991. Phosphate deposits of the United States—discovery, development; economic geology and outlook for the future. In: GLUSKOTER, H. J., RICE, D. D. & TAYLOR, R. B. (eds) Geology of North America, P-2, Economic Geology, Geological Society of America, Boulder, Colorado, 153–163.

—— & BOTINELLY, T. 1991. Mineralogy and chemistry of samples from a drill hole in the southern extension of the land-pebble district, Florida. US Geological Survey Bulletin 1978.

——, SHELDON, R. P. & GULBRANDSEN, R. A. 1984. Phosphate-rock resources of the United States. US Geological Survey Circular 888.

COOK, P. J. 1972. Petrology and geochemistry of the phosphate deposits of northwest Queensland. Economic Geology, **67**, 1193–1213.

—— 1976. Sedimentary phosphate deposits. In: WOLF, K. H. (ed.) Handbook of Stratabound and Stratiform Ore Deposits, **7**, Elsevier Scientific, Amsterdam, 505–535.

CRAGO, A. 1940. Process of concentrating phosphate minerals. US Patent 2,293,640.

DEER, W. A., HOWIE, R. A. & ZUSSMAN, J. 1962. Rock-forming Minerals, **5**, Non-Silicates, Longman, Green & Co. Ltd, London, 323–338.

EMIGH, G. D. 1973. Economic phosphate deposits. In: GRIFFITH, E., BEETON, J., SPENCER, J. M. & MITCHELL, D. T. (eds) Environmental Phosphorus Handbook. John Wiley, New York, 97–116.

FAGERBERG, B. & SANDGREN, P.-M. 1980. Apatite production from iron ore tailings. Mining Magazine, London, **143**, 439, 441, 443.

FRIETSCH, R. 1974. The occurrence and composition of apatite with special reference to iron ores and rocks in northern Sweden. Sveriges Geologiska Undersökning Afhandlingar, Series D.

GREMILLION, L. R. & McCLELLAN, G. H. 1978. The importance of chemical and mineralogical data in evaluating apatitic phosphate ores. SME-AIME Fall Meeting, Lake Buena Vista, Florida, Preprint no. **78-B-308**.

HANNA, J. & ANAZIA, I. J. 1991. A new process for the selective fatty acid flotation of phosphate from carbonate and siliceous ores. In: XVIIIth International Mineral Processing Congress, Dresden, FRG, 23–28 September. Preprints, **IV**, 263–275.

—— & —— 1993. New process for direct phosphate flotation from US siliceous ores. Mining Engineering, **45**, 184–188.

HEWITT, R. A. 1980. Microstructural contrasts between some sedimentary francolites. Journal of the Geological Society, London, **137**, 661–667.

HIGNETT, T. P., DOLL, E. C., LIVINGSTONE, O. W. & RAISTRICK, B. 1976. Utilization of difficult phosphate ores. In: ISMA Technical Conference, The Hague (Netherlands). Elsevier Scientific Publishing Company, Amsterdam, 273–288.

HIRD, J. M. 1985. The emergence of North Carolina's phosphate district as an international supplier. In: MOORE, A. I. (ed.) Proceedings of the International Conference 'Fertilizer 85', London, 10–13 February, 1985, **2**, The British Sulphur Corporation Limited, London, 385–396.

IMPHOS 1980. Proceedings of the 2nd International Congress on Phosphate Compounds, Boston, Massachusetts. Institut Mondial du Phosphat, Paris, 203–280.

KIUKOLLA, K. 1986. Mining and beneficiation practice at Siilinjärvi, Finland. Transactions of the Institution of Mining and Metallurgy (Section A: Mining Industry), **76**, A143–150.

KOULOHERIS, A. P. 1977. Solving problems in chemical processing of low quality phosphate rock. Engineering and Mining Journal, **178**, 104–108.

LAWVER, J. E., McCLINTOCK, W. O. & SNOW, R. E. 1975. Beneficiation of phosphate rock. The state of the art review. Mineral Science Engineering, **10**, 278–294.

——, SNOW, R. E., WIEGEL, R. L. & HWANG, C. L. 1982. Phosphate reserves enhancement by beneficiation. Mining Congress Journal, **68**, 27–31.

LEHR, J. R. 1967. Variations in composition of phosphate ores and related reactivity. Proceedings of the 17th Annual Meeting, Fertilizer Industry Round Table, Washington, DC, USA.

—— 1980. Phosphate raw materials and fertilizers: Part 1—a look ahead. In: KHASAWNEH, F. E., SAMPLE, E. C. & KAMPRATH, E. J. (eds) The Role of Phosphorus in Agriculture. American Society of Agronomy, Inc., Crop Science Society of America, Inc. & Soil Science Society of America, Inc., Madison, Wisconsin, 81–120.

—— 1984. Impact of phosphate rock quality on fertilizer market uses. In: Phosphates—what prospects for growth?, Orlando, Florida. Metal Bulletin Inc., New York, 273–314.

—— & McCLELLAN, G. H. 1974. Phosphate rocks: important factors in their economic and technical evaluation. In: CENTO Symposium on Mining Beneficiation of Fertilizer Minerals, Istanbul, November 19–24, 1973. Central Treaty Organization, Ankara, 194–242.

——, ——, SMITH, J. P. & FRAZIER, J. W. 1967. Characterization of apatites in commercial phosphate rocks. In: Colloque International sur les Phosphates Minéraux Solides. **2**, Structures et

Propriétés des Phosphates. Société Chimique de France, Toulouse, 29–44.

LLEWELLYN, T. O., DAVIS, B. E., SULLIVAN, G. V. & HANSEN, J. P. 1982. *Beneficiation of high-magnesium phosphate from southern Florida.* US Bureau of Mines Report of Investigations 8609.

MABIE, C. P. & HESS, H. P. 1964. *Petrographic study and classification of Western phosphate ores.* US Bureau of Mines Report of Investigations 6468.

MCCLELLAN, G. H. 1980a. Quality factors of phosphate raw materials. *In:* SHELDON, R. P. & BURNETT, W. C. (eds) *Fertilizer Mineral Potential in Asia and the Pacific.* East–West Resource Systems Institute, Honolulu, 259–377.

—— 1980b. Mineralogy of carbonate fluorapatites. *Journal of the Geological Society, London,* **137**, 675–681.

—— & GREMILLION, L. R. 1980. Evaluation of phosphatic raw materials. *In:* KHASAWNEH, F. E., SAMPLE, E. C. & KAMPRATH, E. J. (eds) *The Role of Phosphorus in Agriculture.* American Society of Agronomy, Inc., Crop Science Society of America, Inc. & Soil Science Society of America, Inc., Madison, Wisconsin, 43–80.

—— & CLAYTON, W. R. 1980. Francolite: the commercial phosphate mineral. *In: Proceedings of the 2nd International Congress on Phosphate Compounds, Boston, Massachusetts.* Institut Mondial du Phosphat, Paris, 131–143.

—— & HIGNETT, T. P. 1978. Some economic and technical factors affecting use of phosphate raw material. *In: Phosphorus in the Environment: its Chemistry and Biochemistry.* Elsevier, London, 49–64.

—— & LEHR, J. R. 1969. Crystal chemistry investigation of natural apatites. *American Mineralogist,* **23**, 1374–1391.

—— & VAN KAUWENBERGH, S. J. 1990. Mineralogy of sedimentary apatites. *In:* NOTHOLT, A. J. G. & JARVIS, I. (eds) *Phosphorite Research and Development.* Geological Society, London, Special Publications, **52**, 23–31

—— &—— 1991. Mineralogical and chemical variation of francolites with geological time. *Journal of the Geological Society, London,* **148**, 809–812.

MCCONNELL, D. 1973. *Apatite: its Crystal Chemistry, Mineralogy, Utilization, and Geologic and Biologic Occurrences.* Springer-Verlag, Vienna, New York.

MOUDGIL, B. M. & VASUDEVAN, T. V. 1988. Beneficiation of phosphate ores containing carbonate and silica gangue. *Mineral and Metallurgical Processing,* **5**, 120–124.

NATHAN, Y. 1984. Mineralogy and chemistry of phosphorites. *In:* NRIAGU, J. O. & MOORE, P. B. (eds) *Phosphate Minerals.* Springer-Verlag, Berlin, 275–291.

——, SOUDRY, D. & AVIGOUR, A. 1990. Geological significance of carbonate substitution in apatites: Israeli phosphorites as an example. *In:* NOTHOLT, A. J. G. & JARVIS, I. (eds) *Phosphorite Research and Development.* Geological Society, London, Special Publications, **52**, 179–191.

NOTHOLT, A. J. G. 1980. Economic phosphatic sediments. *Journal of the Geological Society, London,* **137**, 793–805.

—— 1985. Igneous phosphate rock: a growing source of P_2O_5 in phosphate fertilizer manufacture. *In:* MOORE, A. I. (ed.) *Proceedings of the International Conference 'Fertilizer 85',* London, 10–13 February, 1985, **2**, The British Sulphur Corporation Limited, London, 403–409.

——, HIGHLEY, D. E. & HARDING, R. R. 1985. Investigation of phosphate (apatite) potential of Loch Borralan igneous complex, northwest Highlands, Scotland. *Transactions of the Institution of Mining and Metallurgy (Section B: Applied Earth Science),* **94**, B58–B65.

NRIAGU, J. O. & MOORE, P. B. (eds) 1984. *Phosphate Minerals.* Springer-Verlag, Berlin.

PARSONAGE, P. 1986. Treatment of carbonate–phosphate rock by selective magnetic coating. *Transactions of the Institution of Mining and Metallurgy (Section A: Mining Industry),* **76**, A154–A158.

REDEKER, I. H. 1984. Resources to reserves by phosphate processing. *In: Phosphates—what prospects for growth?,* Orlando, Florida. Metal Bulletin Inc., New York, 315–337.

RIGGS, S. R. 1979. Petrology of the Tertiary phosphorite system of Florida. *Economic Geology,* **74**, 195–220.

—— & FREAS, D. H. 1965. *Stratigraphy and sedimentation of phosphorite in the Central Florida Phosphate District.* AIME Preprint, 65-H-84.

ROBINSON, N. 1980. Phosphoric acid technology. *In:* KHASAWNEH, F. E., SAMPLE, E. C. & KAMPRATH, E. J. (eds) *The Role of Phosphorus in Agriculture.* American Society of Agronomy, Inc., Crop Science Society of America, Inc. & Soil Science Society of America, Inc., Madison, Wisconsin, 151–193.

ROUX, E. H. 1985. Mining and beneficiation of Phalaborwa phosphate rock. Prospects for technological developments. *In: Raw materials in South Africa,* Part III, *Igneous phosphate rock.* ISMA paper A/83/58c.

ROY, A. H. & MCCLELLAN, G. H. 1986. Processing phosphate ores into fertilizers. *In:* UZO MOKWUNYE & VLEK, P. L. G. (eds) *Management of Nitrogen and Phosphorus Fertilizers in Sub-Saharan Africa.* Martinus Nijhoff Publishers, Dordrecht, 225–252.

SANDELL, E. B., HEY, M. H. & MCCONNELL, D. 1939. The composition of francolite. *Mineralogical Magazine,* **25**, 395–401.

SAVANICK, G. A. 1984. New developments in hydraulic borehole mining of phosphates. *In: Phosphates— what prospects for growth?,* Orlando, Florida. Metal Bulletin Inc., New York, 166–193.

—— 1987. *Borehole (slurry) mining of coal, uraniferous sandstone, oil sands, and phosphate ore.* US Bureau of Mines Report of Investigations **9101**.

SHAPIRO, L. 1952. Simple field method for the determination of phosphate in phosphate rocks. *American Mineralogist,* **37**, 341–342.

SILVIA, A. F. & ANDERY, P. A. 1972. Mining and beneficiation of apatite rock at the Jacupiranga mine, Brazil. *Phosphorus & Potassium*, 57, 37–40.

SLANSKY, M. 1986. *Geology of Sedimentary Phosphates*. North Oxford Academic Publishers, London.

SMITH, J. P. & LEHR, J. R. 1966. An X-ray investigation of carbonate apatites. *Agricultural and Food Chemistry*, 14, 342–349.

TRUEMAN, N. A. 1971. A petrological study of some sedimentary phosphorite deposits. *Bulletin of the Australian Mineral Development Laboratories*, 11, 1–71.

VAN KAUWENBERGH, S. J. 1991. Overview of phosphate deposits in East and Southeast Africa. *Fertilizer Research*, 30, 127–150.

——, CATHCART, J. B. & MCCLELLAN, G. H. 1990. *Mineralogy and alteration of the phosphate deposits of Florida*. US Geological Survey Bulletin 1914.

VEIDERMA, M. & KNUBOVETS, R. 1991. Kiruna apatite. *Scandinavian Journal of Metallurgy*, 20, 329–330.

WALKER, G. S. 1990. Processing options for low grade phosphate ores. *In:* GRIFFITHS, J. B. (ed.) *Processing developments*. Industrial Minerals, 271 (Supplement), 16–22.

WEBER, J. R., HADDENHAM, R. D. & BAILEY, J. H. 1987. Vernal Phosphate completes expansions and slurry pipeline without missing any concentrate shipments. *Mining Engineering*, 39, 14–20.

WHITE, L. 1984. R & D are keys to low-grade apatite mining at Kemira Oy. *Engineering & Mining Journal*, 185, 34–42.

ZEVIN, L., LACH, S., LEVY, Y. & PREGERSON, B. 1988. Mineralogical aspects of beneficiation of phosphates. *International Journal of Mineral Processing*, 24, 235–245.

Evaluation of grade estimation techniques

T. M. BELL[1] & M. K. G. WHATELEY[2]

[1] *C.L.M. Ltd (a member of the C.P. Holdings group of companies Ltd.) Nottingham Road, Giltbrook, Nottingham, UK*
[2] *Geology Department, University of Leicester, Leicester LE1 7RH, UK*

Abstract: A sample set was simulated from an exhaustive data set of silver values derived from blast holes of the 5220 bench. The sample set was used to estimate the ditribution of grade for the 5220 bench using a variety of estimating techniques. The estimating processes were controlled by the knowledge of the statistical and spatial distribution of the exhaustive data set. This approach was used to test the accuracy and precision of the different estimation techniques. The results from the simpler estimation methods (linear interpolation and triangulation) compared most closely to the blast hole samples. The more complex techniques (inverse distance and kriging), using weighting factors, were found to exaggerate the variance and smooth the interquartile range.

The dilemma that any mining geologist or engineer faces is 'How do I estimate a mean value that accurately reflects the quality of material that I am trying to estimate? What fluctuations about the mean value can I expect within a given environment?' There are many estimation techniques available but which one will work? Two aspects to grade estimation include the calculation of a global estimate for determining the total quantity of material and calculating local estimates to aid mine planning and scheduling. Local estimation will require more detail and is often a precursor to calculating a global estimate. Calculating a local estimate requires the spatial nature of the data values to be correctly defined, and in a mining environment is dependent upon geological control. Given that the spatial distribution is intrinsically linked to the statistical distribution, statistical inferences can be used to control processes of grade estimation. The data derived from an estimation process should always honour and reflect the statistical variation and spatial distribution of the sample data set. Techniques that characteristically smooth the data may provide a good global estimate but deliver invalid results as regard spatial distribution and local estimation. Mining ore as waste and vice-versa is always to be avoided. It is, therefore, the aim of this paper to evaluate the accuracy and precision of methods of grade estimation in a typical mining environment of high variability of grade. It is always important to have a thorough statistical, spatial and geological understanding of the sample data set prior to deciding which estimation technique is to be used.

A slice has been taken through an ore body and an exhaustive sample set precisely defines the characteristics of that slice. From knowing the actual statistical and spatial distribution of assay values we attempt to redefine the slice from a sub set of the data using this prior knowledge. It is therefore possible to test the effectiveness of different estimating techniques and also illustrate how a sub set of samples can impoverish the results. Typically, attempting to create detail from limited information is always dependent upon sample pattern and size. In an environment of high variance with high value outliers, where spatial structure of the distribution of values is highly organized, estimation processes need to adopt a localized approach.

The sample data set is from the Trinity Silver Mine situated within Pershing County, NW Nevada (Fig. 1), located on the NW flanks of the Trinity Range, 25 km NNW of Lovelock. The mine was an open pit, heap leach silver mining operation extracting rhyolite hosted, disseminated, hydrothermal, silver oxide mineralization. The geology and mineralization is described by Johnson (1977), Ashleman (1988) and Bell (1989). The exhaustive data set used for this evaluation was taken from a single bench of the open pit. The 5220 level data set contained 1386 blast hole samples (Fig. 2). No spatial bias or clustering effects are evident as the sampling pattern was at regular 4.5 m centres. The location of 51 boreholes, from the exploratory drilling programme, that intercepted the 5220

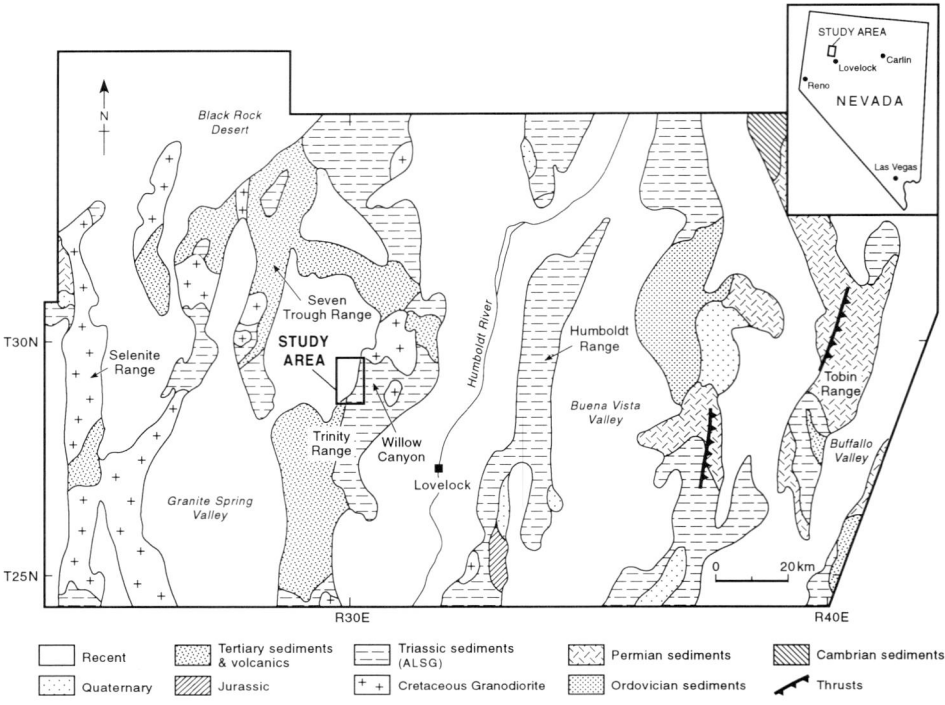

Fig. 1. Map of Nevada showing the location of the Trinity Silver Mine, Pershing County.

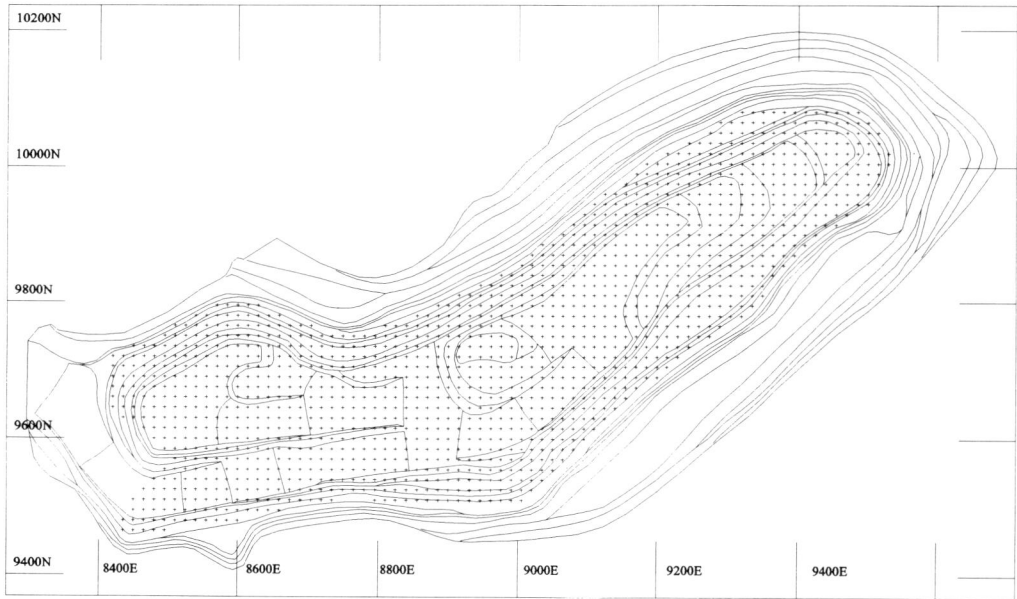

Fig. 2. Outline of the worked open pit and pattern of blast hole samples for the 5220 bench.

level were used for simulation. Table 1 outlines the set procedures followed for the purpose of this evaluation. This evaluation is based upon a single case study and therefore reflects the characteristics of this specific deposit. The conclusions may be of assistance in evaluating other deposits.

Table 1. *Procedures adopted for evaluating methods of grade estimation*

Global statistics
  Descriptive
  Histogram
  Probability plots
Population analysis
  Derived from global statistics
  Descriptive statistics
Spatial distribution
  Block model
  Contour maps
  Indicator maps
  Moving windows
  Semi-variograms
Link populations to spatial context
Borehole simulation
  Triangulated model
  Linear interpolation of grades
Grade estimation
  Triangles
  Linear interpolation
  Inverse distance
  Ordinary kriging
Estimation of errors
  Error maps
  Mean squared error
  Ore envelopes and spatial integrity

Statistical assessment of the exhaustive data set

The purpose of a statistical assessment of the data set is to derive assumptions about the accuracy and precision of subsequent grade estimates. This requires an understanding of the spread and character of the data. The univariate statistics, histogram and probability plot provide us with insight into the character of the statistical distribution of data values. The mean is of limited use if expected fluctuations in data values about the mean cannot be predicted within a given environment. The confidence of the global or local mean, as described by the variance, provides a measure of repeatability in time and space. The calculation of a mean and its confidence limits is frustrated by the implication that variables from a geological environment commonly consist of many quite small values and a few large ones. Deviations from symmetrical distributions and the linearity of a normal or log normal probability plot confound the modelling process and question the assumptions of applying a specific type of distribution to the data (Isaaks & Srivastava 1989). Changes in the characteristics of the cumulative frequencies of a data set over different intervals need always to be explored and checked against the possibility of multiple populations. It is evident from statistical information (Table 2, Fig. 3) that the silver values ($oz\,t^{-1}$) for the 5220 level neither conform to a Gaussian nor Log Normal distribution. If such distributions were modelled over or under estimation, respectively, is likely to occur.

The important features of the distribution are captured by the univariate statistics (Table 2) and illustrate immediately the problem of grade estimation for the 5220 bench. That is a pronounced peak below the mean and a large spread of data toward extreme high values (Fig. 3). The summary statistics provide measures of location, spread and shape. The mean Ag content of the bench is recorded as $1.4\,oz\,t^{-1}$, which is an accurate estimate for regular spaced data for the calculation of the total silver quantity of the bench. However, other estimates of central tendency that are more resistant to outliers (median, trimean, biweight etc.) invoke caution with respect to the precision of the mean (Fig. 4). The high variance describes the high variability of the data set and explains the poor precision of the mean as a local or point estimate. A strong positive skew with a long tail of high values to the right is evident from the high positive skewness ($+4.37$). The degree of asymmetry is also supported by the coefficient of variance (2.06).

The frequency histogram and probability plot (Fig. 3) illustrate how the data are proportioned. It is important to note that even though the data values range from 0 to $32.8\,oz\,t^{-1}$ Ag, only 23% are >1.3, 7.4% are >5 and 2.7% are $>10\,oz\,t^{-1}$. The upper quartile indicates that 75% of the entire data set lie between 0 and $1.14\,oz\,t^{-1}$ Ag, and the mean which is higher does not then reflect the majority of the data resulting in local or point overestimation. It is obvious therefore that the positive skew has a disproportionate influence upon grade estimation. Such a conclusion can be derived if the coefficient of variance exceeds 2 and therefore any local estimate within the confines of the 5220 bench could generate a mean that is considerably different from the global mean.

A skewed distribution of this nature with high value outliers needs to be evaluated for potential modality and departures from a continuous

Table 2. *Statistics calculated for the 5220 bench, including total and sub-populations*

	Total	Ore > 1.3 oz t^{-1}	0–0.4 oz t^{-1}	0.4–0.8 oz t^{-1}	0.8–1.3 oz t^{-1}	1.3–5 oz t^{-1}	5–10 oz t^{-1}	> 10 oz t^{-1}
No.	1386	323	730	217	116	221	65	37
Min.	0	1.303	0	0.401	0.802	1.303	5.133	10.115
Max.	32.8	32.8	0.398	0.797	1.292	4/997	9.906	32.8
25th%	0.169	1.913	0.1	0.4475	0.895	1.617	5.8028	11.1287
75th%	1.141	5.9505	0.257	0.6438	1.11	3.2543	7.9138	17.3173
Mean	1.402	4.8679	0.1823	0.5502	1.0202	2.5414	7.0679	14.8994
Median	0.357	3.167	0.1795	0.53	1.0345	2.326	6.702	13.357
Biweight	0.4464	3.6173	0.1798	0.5455	1.0193	2.4674	6.9665	14.0346
Trimean	0.506	3.5745	0.179	0.5383	1.0197	2.3845	6.783	13.8265
Sichel t	1.3291	4.7268	0.2027	0.5501	1.0205	2.5392	7.0659	14.8483
Variance	8.3798	20.0419	0.01	0.0126	0.0189	1.0266	2.0817	23.898
Std dev.	2.8948	4.4768	0.0999	0.1125	0.1374	1.0132	1.4428	4.8886
MAD	0.323	1.8193	0.0798	0.0985	0.115	0.8294	1.1585	3.0156
Skewness	4.3656	2.3424	0.2284	0.4601	0.1095	0.7067	0.4979	1.6538
Kurtosis	28.4261	10.1776	2.1985	1.9570	1.9726	2.4729	2.2040	6.1592
Coef. var.	2.0647	0.9197	0.5478	0.2044	0.1347	0.3987	0.2041	0.3281

distribution. Detailed exposure of the histogram and probability plot can highlight transitions from one distribution to another. A marked change in frequency from one class to another, or deviations of the cumulative frequency polygon from a straight line provides the necessary statistical support for population splitting. Often positively skewed data reflect the mixing of sub-populations or grade classes. Any such conclusion and attempt to model Gaussian distributions by population splitting requires statistical, spatial and geological support. Processes of overprinting, enrichment and depletion can provide geological support for the existence of sub-populations. The samples from the 5220 bench were split into five grade categories. Statistical support is evident for breaks in the distribution at 0.4, 0.8 and 1.3 oz t^{-1}; however, arbitrary breaks at 5 and 10 oz t^{-1} were chosen to split the tail of the positive skew and would require spatial or geological support. The statistics (Table 2) and measures of central tendency (Fig. 4) suggest that Gaussian distributions were approximated. The variance and skewness decreased for each class, compared to the global statistics, restricting expected fluctuations which should result in more reliable grade estimates. The value of such an exercise comes if the sub-populations prove to be spatially coherent.

Spatial description of the exhaustive data set

A coded block model and contours of silver grade of the 5220 bench (Fig. 5a & b) illustrates clear structure to the data. No random spread is evident and grade classes define specific structures. The bulk of the mineralized zone to the east forms an elongated linear structure with grade envelopes or shells surrounding high grade cores. Each shell describes a specific transition from one grade class to another. Alternatively discrete ore lobes or zones can be defined to the west and contain values specific to a single grade class. Sub-populations can be located within specific lobes or envelopes the boundaries of which are clearly defined and represent areas of high local variance. A 3D profile (Fig. 6) through the mineralized zones illustrates the transition from rapidly changing grade values to areas of stable grade. Peaks in the grade profile (Fig. 6) or sharp changes in grade are the prime factor in causing poor grade estimates and provide a case for spatially separating data for the purpose of calculating grade estimates.

The indicator maps (Fig. 7) support the structural integrity of the sub-populations illustrating the localized clustering of specific grade values. The 1.3 oz t^{-1} threshold defines the NE trend of the data whilst the 5 and 10 oz t^{-1} thresholds represent clusters of high values that are continuous over shorter distances. There is a degree of scatter to the low values less than 0.4 oz t^{-1}. The semi-variogram parameters (Fig. 9) quantitatively verify these observations. Such clusters can be separated and evaluated as single classes and grade estimates calculated for given zones. Such a method would reduce the overall variance within each zone and provide a more reliable set of local grade estimates for the

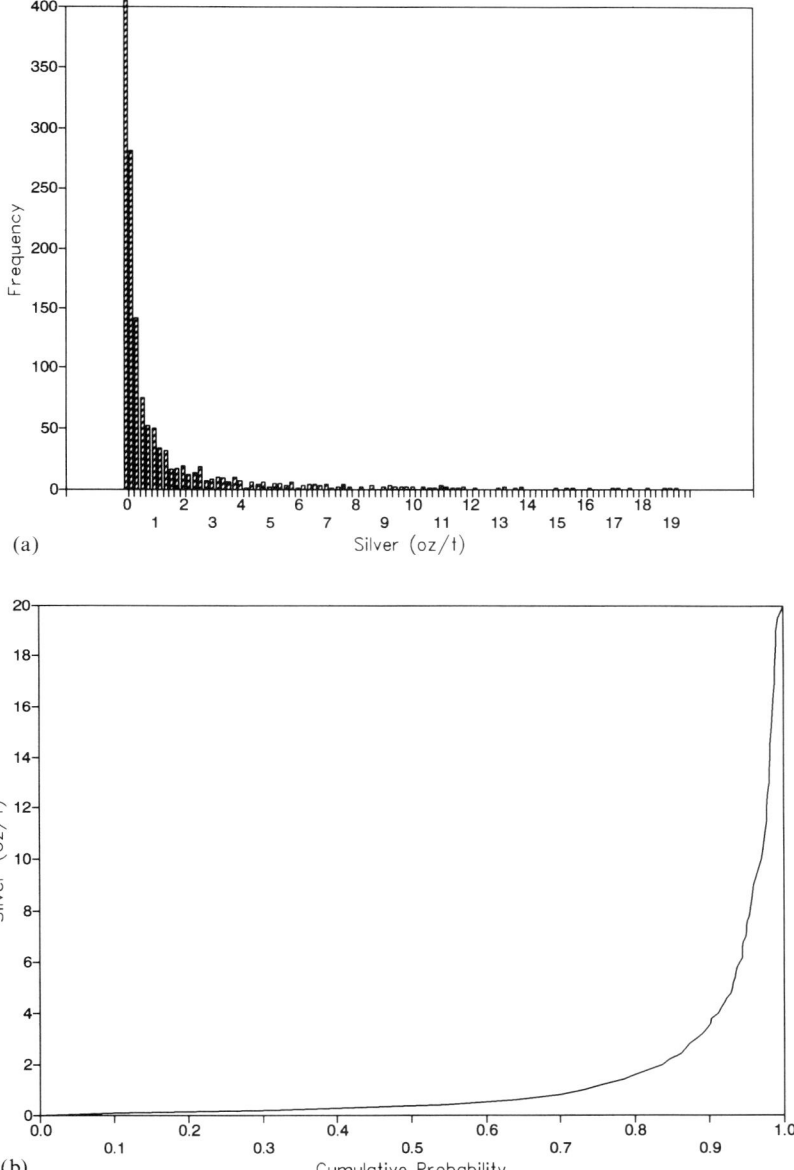

Fig. 3. Histogram (**a**) and normal probability plot (**b**) of the blast hole silver grades.

purpose of mine planning.

By applying moving window statistics and calculating the mean and variance for a given overlapping window (30 m^2) areas of high local variance can be defined. The contours (Fig. 8) show that the average silver values and variability change locally across the area. In general the change in variability reflects the change in the mean. The strong relationship between the mean and variability is referred to as the proportional effect which implies that the variability is predictable. Where that relationship is destroyed confidence in the mean is reduced. Contours of standard deviation high-

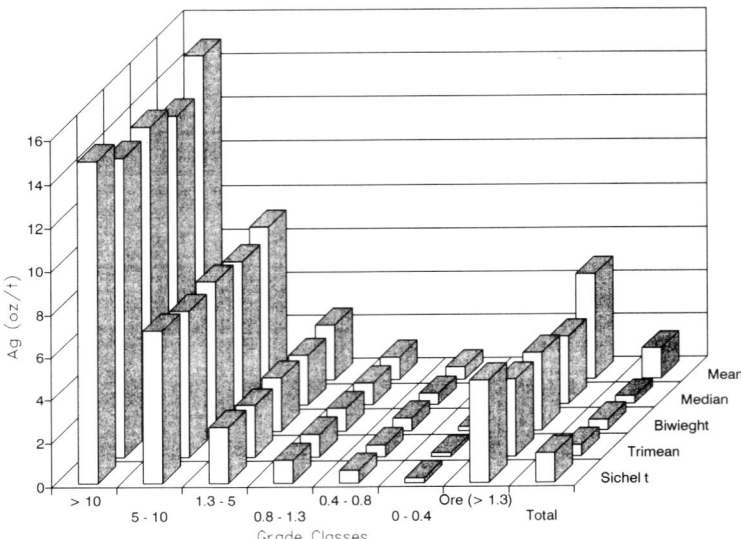

Fig. 4. Measures of central tendency calculated for the exhaustive data set and specified grade classes in oz t^{-1}.

light the areas of high variance and poor precision of the mean, which tend to coincide with the transition zones from one grade class to another or the mixing of two populations. Profiles of the window statistics (Fig. 8) illustrate that where the standard deviation exceeds the mean the grade estimate at that point would be an overestimate due to the influence of unrelated higher values. This provides a case for ensuring that samples from one zone are not used to estimate values within another zone.

The semi-variogram is a tool which can quantify the spatial continuity of the data (Whateley 1992). The indicator maps and window statistics show that there is a restricted distance over which samples are comparable, and that values from one zone/population can not be used to calculate grade estimates in other zones/populations. From omni-directional and directional semi-variograms (Fig. 9) distances of spatial continuity can be defined and an ellipse of anisotropy constructed. The anisotropy can be used to constrain the processes of grade estimation. The semi-variograms show that the east–west range reflects the discrete ore lobes (semi-major axis 130 m), the northeast range is the orientation of the main linear ore shells (major axis 330 m) and the northwest range the sharp boundaries to the ore shells (minor axis 80 m). Two ellipses of anisotropy can be defined and reflect the two distinct structural zones within the ore body.

Borehole simulation

Using different estimating techniques a series of re-defined block models were calculated from a sub-set of the original blast hole data. The subset represents 51 boreholes that intercepted the 5220 bench (Fig. 10). Assay values were simulated by simple linear interpolation from a computer generated triangulated model of the original blast hole samples. From the statistical and spatial assessment of the blast hole samples the following points were considered during the iterative process of grade estimation. The data are highly skewed resulting in a bias toward high values by the mean. The data contain subpopulations that are statistically and spatially supported, exhibiting well defined trends and discrete grade zones. The spatial structure is quantifiable and the anisotropy used to restrain the search distance during estimation. Without considering the influence of outliers, spatial continuity of the data and statistical breaks in the distribution most estimating techniques are likely to result in severe local overestimation.

The statistics of the simulated bore hole assays reasonably mirror the statistical distribution of the exhaustive data set (blast hole samples). The skewness (2.38) and coefficient of variance (1.74) are not as pronounced, and the mean is higher (2.29 oz t^{-1}). The drilling results reflect the high variability of data (variance = 15.9) and illustrate the potential problem of local overestimation and confidence in the mean. The sample data set

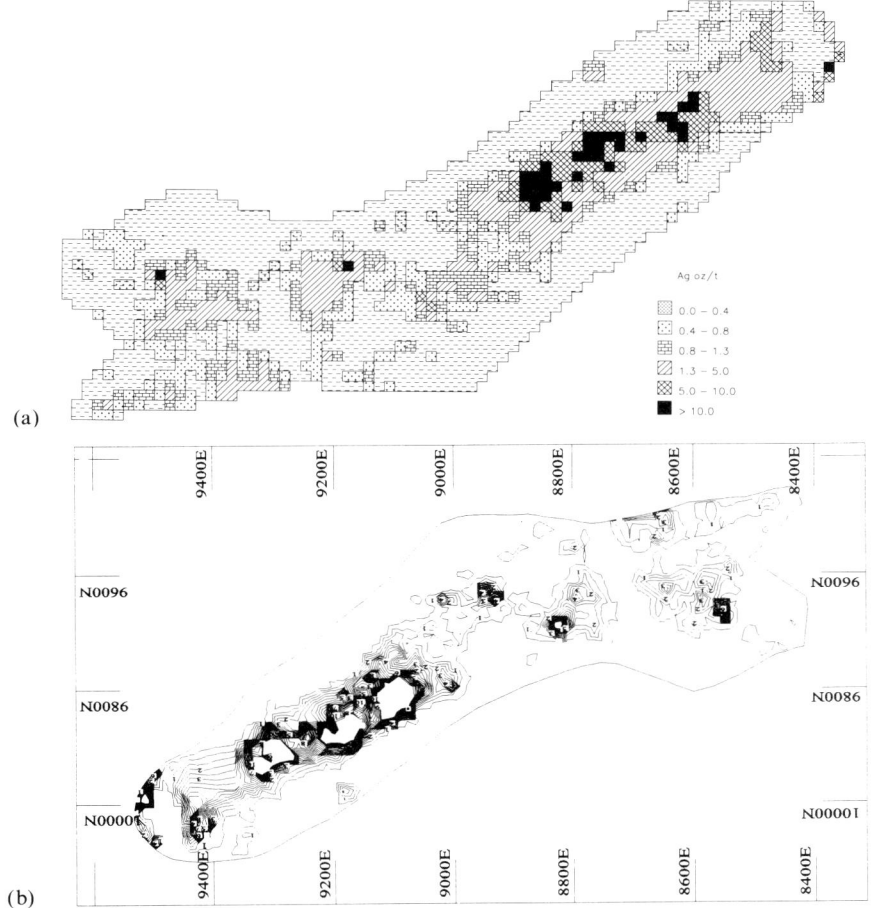

Fig. 5. The spatial distribution of silver grades on the 5220 bench; (**a**) block model coded according to grade classes; (**b**) contours of silver grade at $0.5\,oz\,t^{-1}$ contour interval.

shows no spatial clustering and conforms to a fairly regular pattern (Fig. 10), although a high proportion of drill holes are located within the eastern ore zone.

Is it possible from a sample set that is biased toward high values to generate a simulated block model that reflects the original blast hole samples without spatially or statistically splitting the sample set? To test this concept four different processes of grade estimation were used to create assay block models for the 5220 bench, namely inverse distance, kriging, linear interpolation and triangulation. Each method was restrained by search distances. The effect of isotropy, extracted from the semi-variograms of the blast hole samples, was applied only to the inverse distance method. The power of the inverse distance method was also varied but found to have negligible effect and is therefore not included.

Grade estimation and statistical comparison

The sample set is considered to be statistically representative of the exhaustive data, although it is a small sample size (4% of the total). It is unlikely that the sample set will model the spatial variation in values accurately, especially given the sharp boundaries that exist between grade zones. How accurate will estimated block models be in providing a true picture of the grade distribution on the 5220 bench? To evaluate the results (Tables 3 & 4) of the different grade estimation strategies four criteria were compared and assessed for replications.

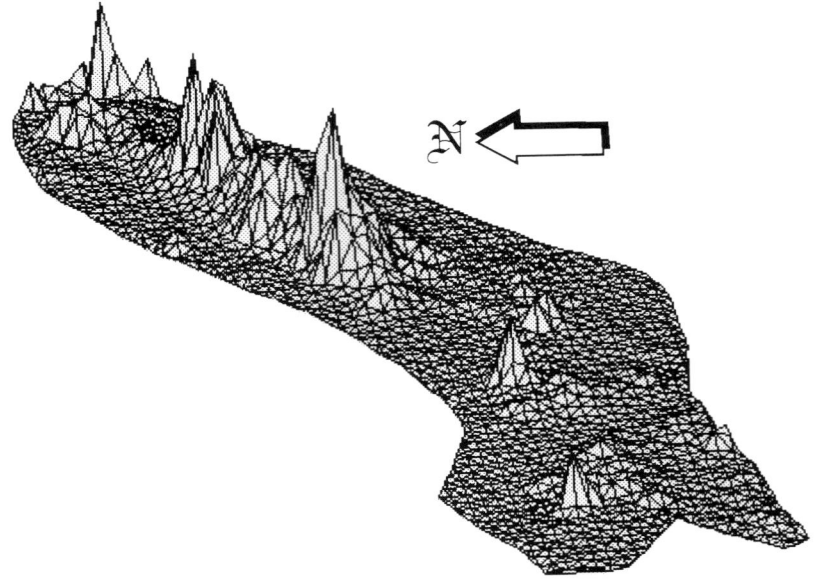

Fig. 6. Three-dimensional model of the blast hole silver grades showing the location of pronounced peaks in the distribution of grade values.

(1) the global arithmetic mean;
(2) statistical distribution based on the variance and skew;
(3) spatial distribution of values;
(4) definition of the ore zone (> 1.3 oz t^{-1}).

The arithmetic mean used as a global estimate of silver grade for the 5220 bench in every case is overestimated (Fig. 11a) with the results of linear interpolation and triangles less pronounced. However, the kriged and inverse distance means compare favourably with the mean of the drilling results. The statistical distributions measured by the variance and skewness vary considerably. A positive skew is always maintained and the variance fluctuates above and below the true variance. Inverse distance with a search distance based on the semi-major axis (120 m) of anisotropy appears to mirror the variability of the blast hole samples. Increasing the search distance to the maximum axis clearly destroys the statistical distribution, a function of excess smoothing and comparing unlike values. All the estimating techniques have transformed the distribution to the right, introducing a greater bias toward higher grade values indicated by a shift in the median value by a factor of two. This is again less pronounced for the linear interpolation and triangles method. This would imply that the later two methods have reduced the influence of high grade outliers producing a more restricted distribution.

The spatial distribution of grade values (Fig. 12) provides further evidence that the different estimating techniques have enhanced the inter quartile range of values i.e. between 0.8 oz t^{-1} and 5 oz t^{-1}. The structure of the data is destroyed by inverse distance (Fig. 12 a–c) and kriging (Fig. 12d); the discrete ore zones are lost and the high grade cores are exaggerated. This is most pronounced with the kriged model and increasing the search distance (Fig. 12c) to the maximum (330 m) obliterates the high grade cores causing a broad spread of values between 1.3 oz t^{-1} and 5 oz t^{-1}. Both linear interpolation (Fig. 12e) and triangulation (Fig. 12f) maintain the grade shells, high grade cores and discrete grade zones, although the sharp boundary between the 0.8 oz t^{-1} and 1.3 oz t^{-1} grade class is lost. Overall the different estimating techniques tend to be biased toward different grade classes. All the methods maintain the NE trending elongated linear nature of the grade zoning illustrating the influence of the ellipse of anisotropy, but to the extent that the discrete ore zones to the west are lost. The geometry of the ore grade envelope (> 1.3 oz t^{-1}) appears to be

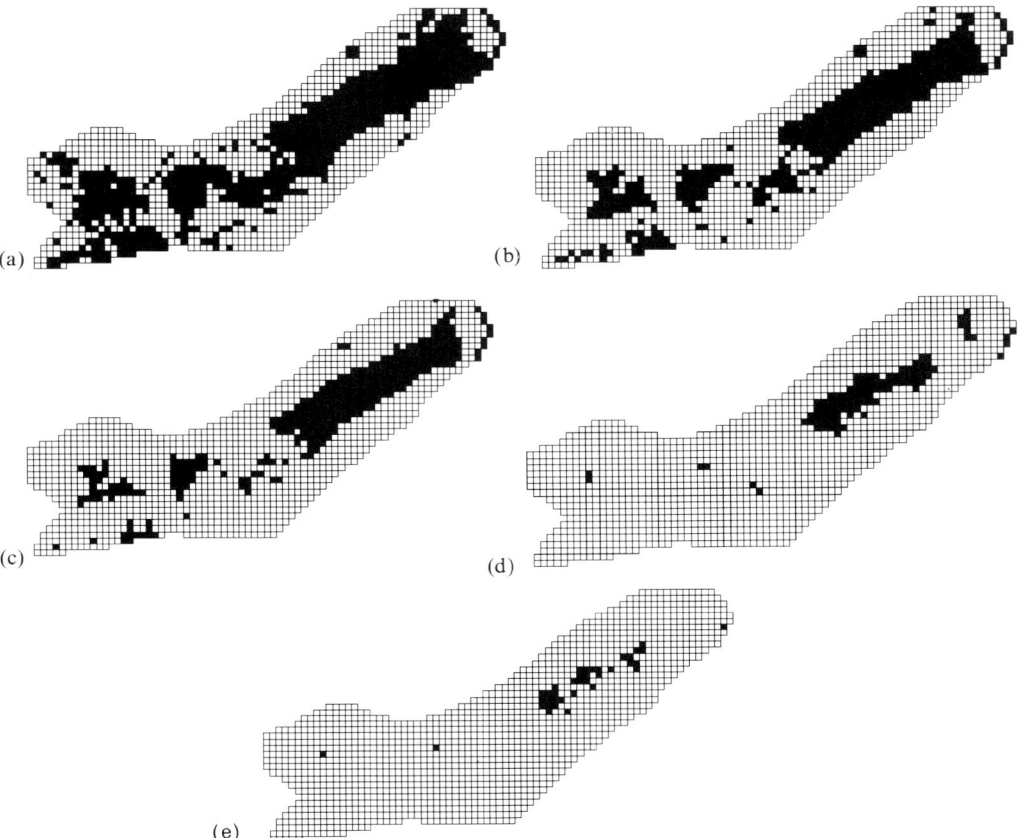

Fig. 7. Indicator maps of Ag oz t^{-1} at specified grade boundaries; (a) 0.4, (b) 0.8, (c) 1.3, (d) 5.0, (e) 10.0.

reasonably maintained although the percentage of blocks defined as ore has increased, mainly within the main eastern ore body. A high proportion of waste blocks will be flagged as ore and to the west most of the ore would be mined as waste. Both linear interpolation and triangles provide the best results in terms of ore waste definition.

The actual mean grade of the ore zone (Fig. 11b) is significantly overestimated by kriging and inverse distance, and underestimated when the search distance is increased. Triangles provide the most accurate estimate (4.62 oz t^{-1}) compared with the actual (4.87 oz t^{-1}). The positive skew of all the models is significantly reduced as high grade outliers become less significant. Kriging and inverse distance produce high variances similar to the blast hole samples, and all methods except triangles exaggerate the upper range of grade values indicated by the higher median and upper quartile. The low variance of the triangle model indicates that the extremes of the distribution have been curtailed, thus the high grade cores although maintained are less pronounced.

Estimation of error

The degree of error associated with each estimation strategy can be calculated by direct comparison of each estimated block with the actual mined block. Absolute errors can be calculated to give a ± Ag difference. By calculating the square of the actual error the relative deviation from the actual can be ascertained. The measures of central tendency (Table 5) for all estimations (mean and median) imply that the majority of estimated blocks are only slightly overestimated. The negative skew on the other hand implies a degree of underestimation particularly for the linear interpolated and triangulated models. The true picture is best illustrated by the histogram (Fig. 13) and variance of absolute errors. The high variance

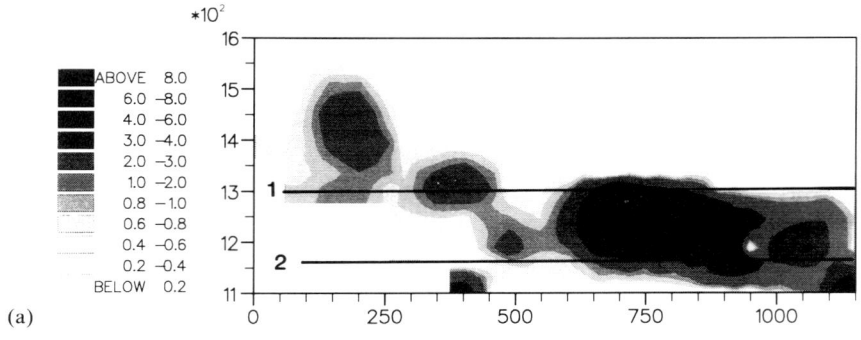

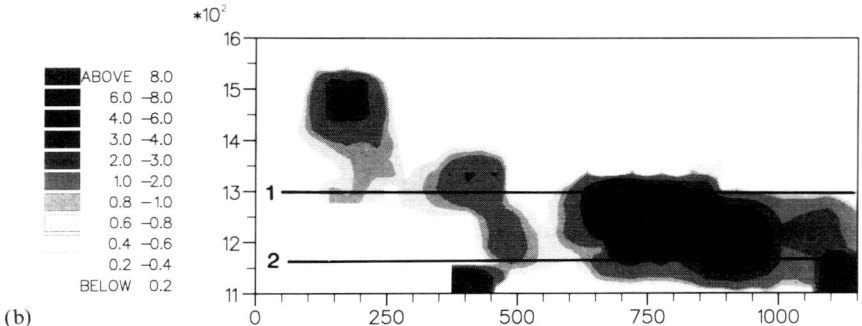

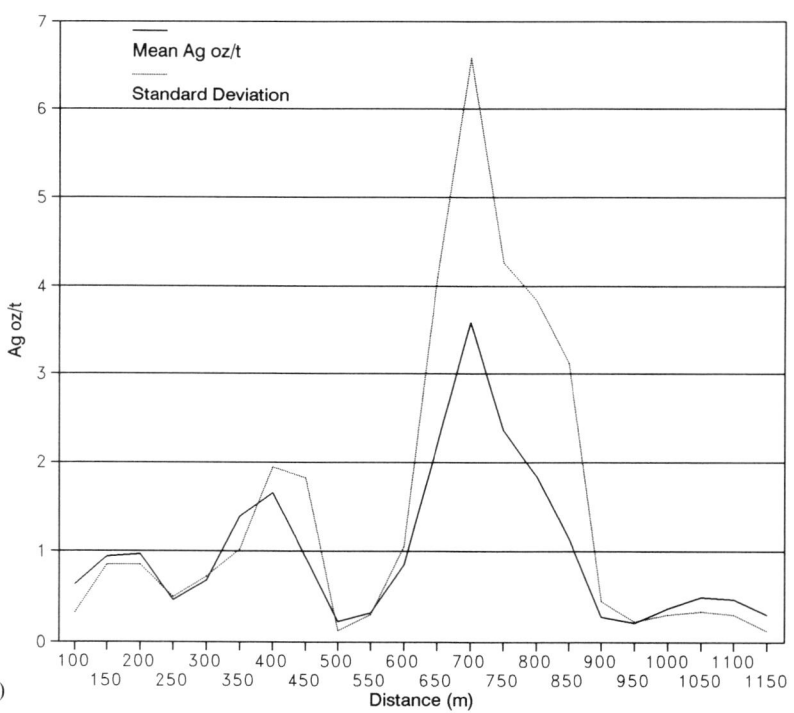

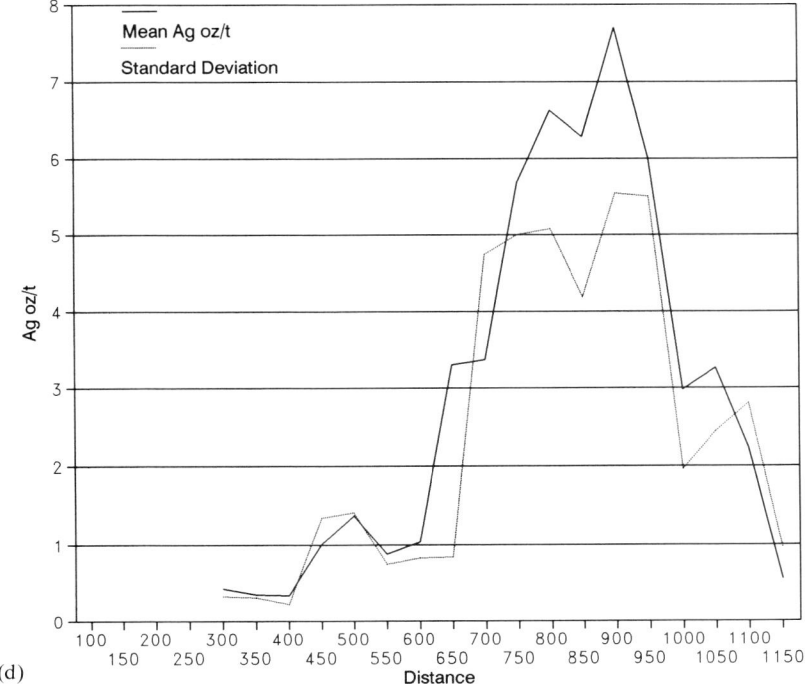

Fig. 8. Window statistics based on transformed grid oriented east–west with 30 m overlapping moving window. (**a**) Contours of mean Ag oz t^{-1}, (**b**) contours of standard deviation, (**c**) and (**d**) profiles 1 and 2 respectively of the window mean and standard deviation showing the proportional effect and highlighting areas of expected high error.

of the kriged and inverse distance models and the high proportion of extreme errors illustrate that both techniques result in a proportion of blocks with high magnitudes of error, compared with models derived from linear interpolation and triangles. The error maps (Fig. 14) provide further support with inverse distance (Fig. 14a–c) and kriging (Fig. 14d) creating almost equal proportions of over and underestimated blocks, causing a balancing of errors, but the magnitude of overestimation being greater. Increasing the search distance (Fig. 14 b–c) creates more overestimation whilst the linear interpolated (Fig. 14e) and triangulated (Fig. 14f) models overestimate low values and underestimate high values but the magnitude is small. The statistics of the squared errors and the error fluctuation graphs support the above premise (Fig. 15).

Conclusions

A sample from a given population, provided that it is not biased, will generally reflect the statistical nature of the population. Estimates derived from a sample data set should always maintain the statistical characteristics of that set of samples. The calculation of a point or block model from a sub set of the blast hole samples (e.g. drilling results) will always result in a certain degree of overestimation because of the high variability and strong positive skew to the original data. Methods that reduce the magnitude of error tend to be those that are simple (linear interpolation and triangles). Such methods were found to reduce the variance only to the extent of minimizing the influence of high grade outliers, which was acceptable. The more complex techniques (kriging and inverse distance) exaggerated the variance by inappropriately weighting extreme values but at the same time smoothing the inter quartile range of the data. Global estimates are best achieved from methods (linear interpolation and triangles) that gave equal or reduced weighting to high grade outliers and honoured the spatial distribution of the data points. Correct local grade estimates are clearly dependent upon variability and spatial distribution, quantitatively captured by the anisotropy of assay values. Anisotropy applied to estimating is essential in maintaining statistical and spatial structures. Local estimates for mine planning and scheduling can be drastically

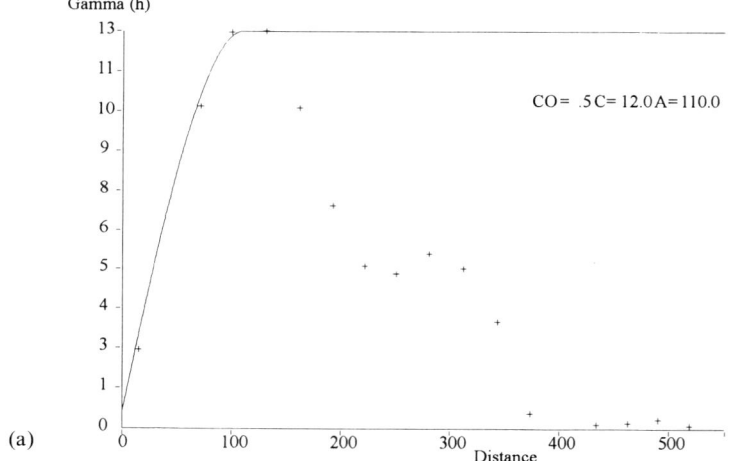

(a)

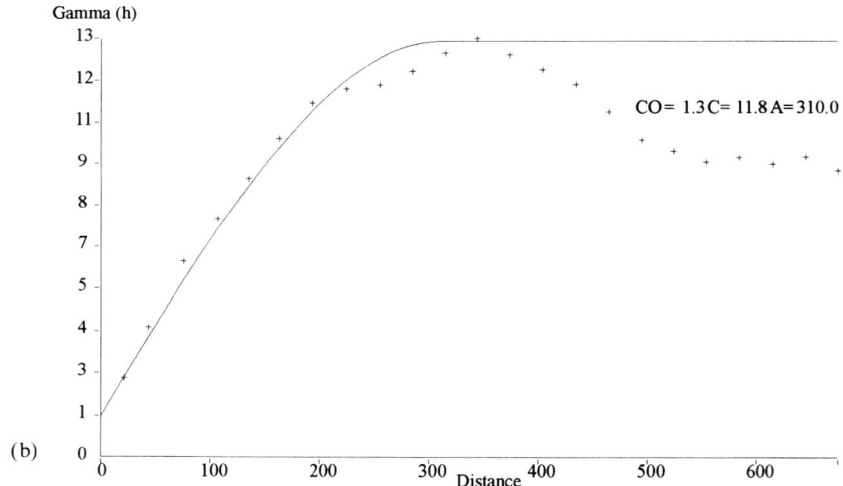

(b)

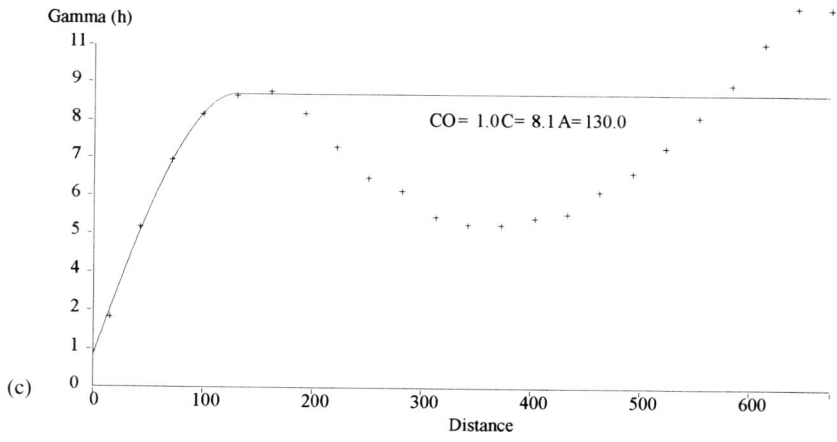

(c)

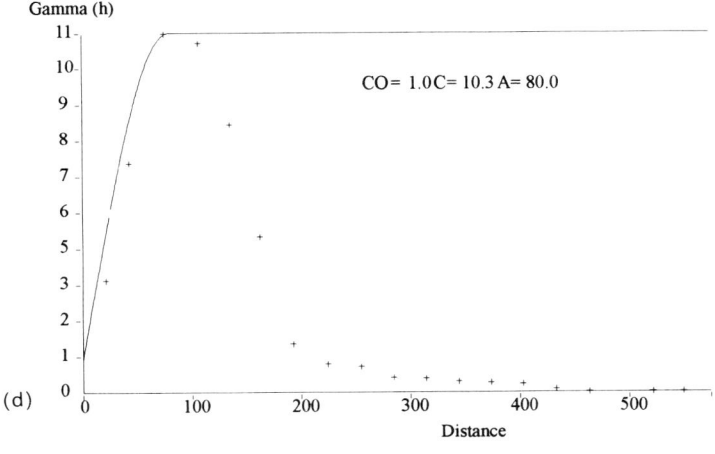

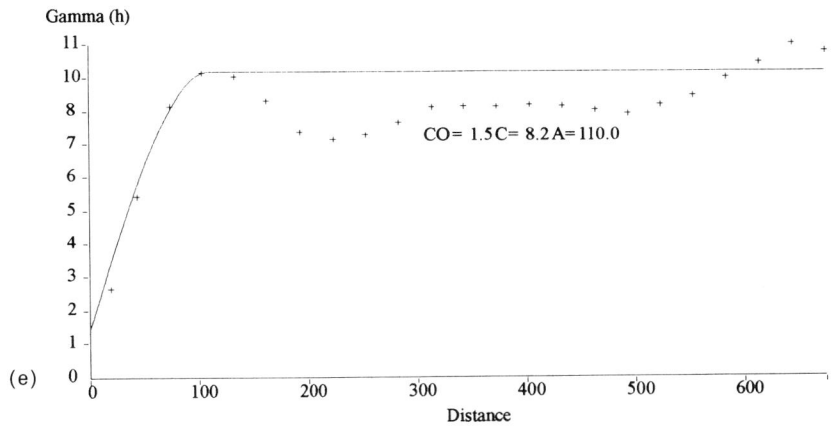

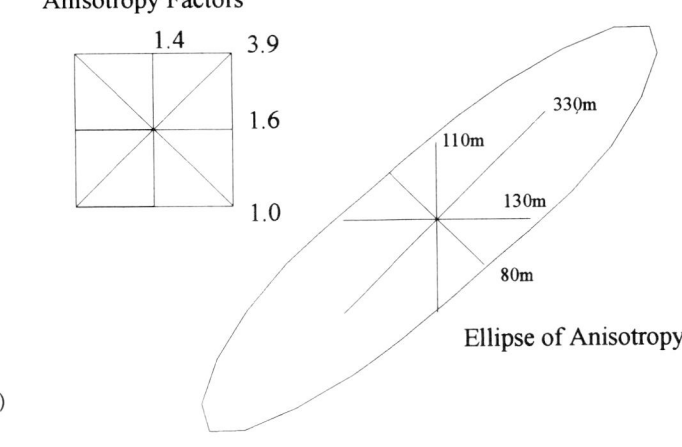

Fig. 9. Spherical modelled semi-variograms for the silver assays of the 5220 bench. (**a**) N–S direction, (**b**) NE–SW direction, (**c**) E–W direction, (**d**) SE–NW direction, (**e**) omnidirectional, (**f**) ellipse of anisotropy and anisotropy factors.

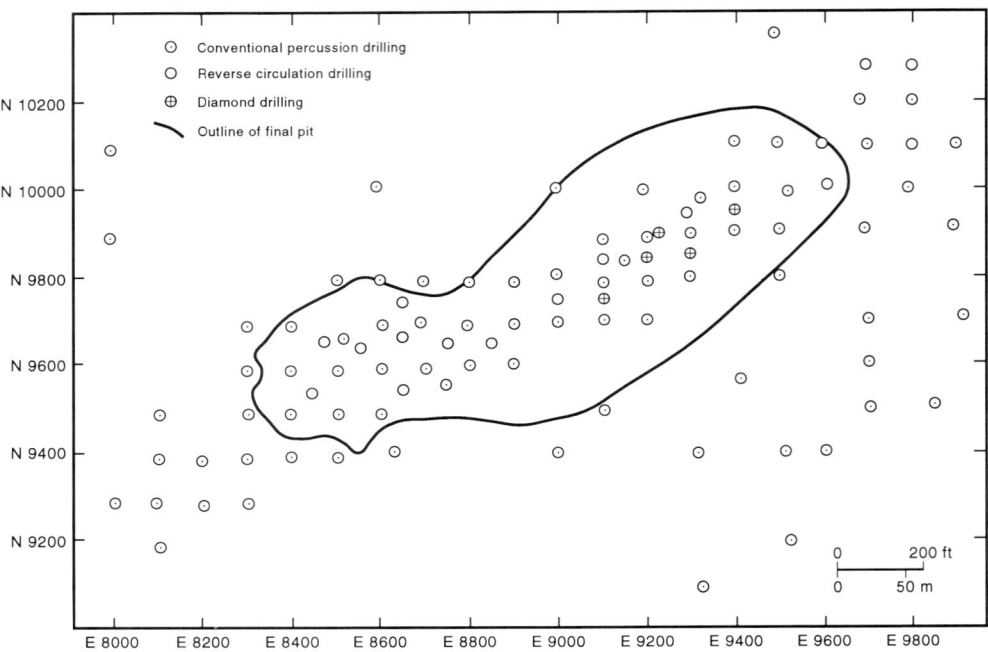

Fig. 10. Location of drill holes and intercepts with the 5220 bench. Grades were simulated by linear interpolation from a triangulation model of the blast hole samples.

Table 3. *Statistics for different grade estimation strategies calculated for the 5220 bench (total data set)*

	Inverse distance	Inverse distance (120 m search)	Inverse distance (330 m search)	Kriged	Linear interpolation	Triangles
No.	974	1207	1388	964	1386	1386
Min.	0.001	0.001	0.015	0.001	0.001	0.012
Max	18.62	16.31	10.82	18.62	17.39	16.04
$25^{th}\%$	0.27	0.28	0.47	0.27	0.29	0.34
$75^{th}\%$	1.52	1.44	2.94	1.52	1.29	1.46
Mean	2.10	1.85	1.97	2.12	1.61	1.61
Median	0.63	0.64	0.82	0.66	0.59	0.63
Biweight	0.63	0.60	1.43	0.64	0.58	0.64
Trimean	0.76	0.75	1.26	0.78	0.69	0.76
Variance	13.00	8.29	4.59	13.07	6.38	5.36
Std dev.	3.61	2.88	2.14	3.61	2.53	2.31
MAD	0.40	0.36	1.01	0.40	0.36	0.37
Skewness	2.49	2.22	1.56	2.48	2.61	2.33
Kurtosis	9.01	7.32	4.76	8.96	10.14	8.41
Coef. var.	1.72	1.56	1.09	1.71	1.57	1.44

Table 4. *Statistics for the ore population (>1.3 oz t^{-1}) from different grade estimation strategies*

	Inverse distance	Inverse distance (120 m search)	Inverse distance (330 m search)	Kriged	Linear interpolation	Triangles
No.	266	324	617	264	344	371
25th%	2.28	2.28	2.21	2.28	2.22	2.32
75th%	9.48	8.39	4.84	9.48	7.11	6.47
Mean	6.42	5.54	3.78	6.45	4.99	4.62
Median	5.94	5.03	3.17	5.93	4.30	3.96
Variance	21.63	11.83	4.34	21.66	10.21	7.43
Std dev.	4.65	3.44	2.08	4.65	3.19	2.72
Skewness	0.79	0.57	0.99	0.79	0.99	0.90

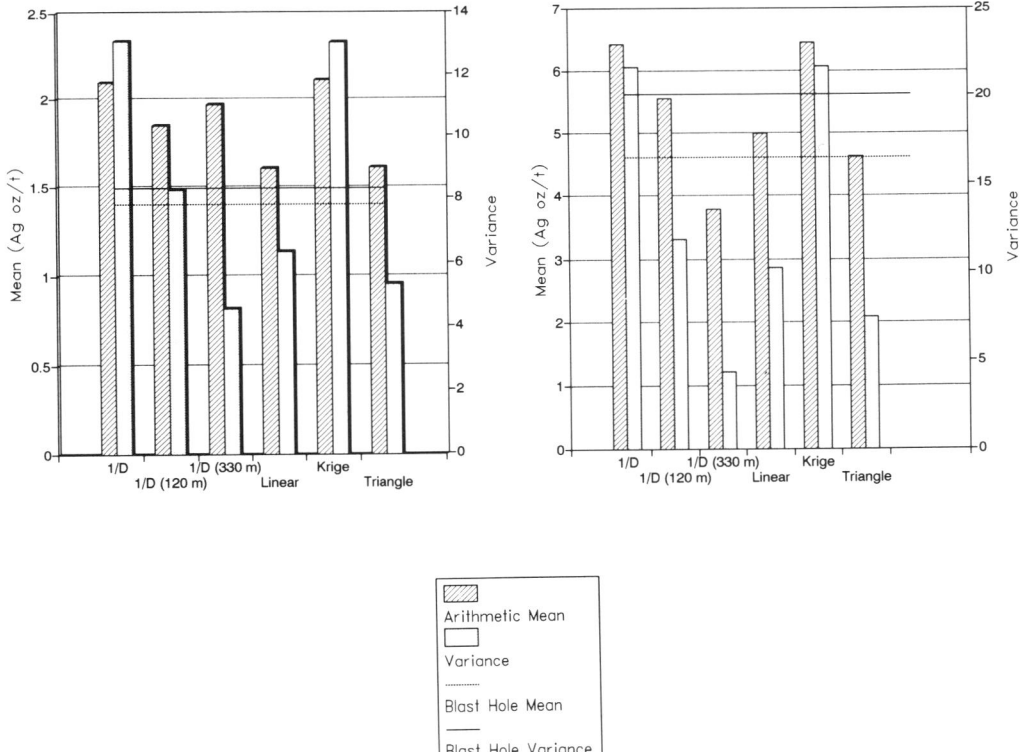

Fig. 11. Calculated mean grade estimates and variance for different estimation processes. (**a**) Global population, (**b**) ore population >1.3 oz t^{-1}.

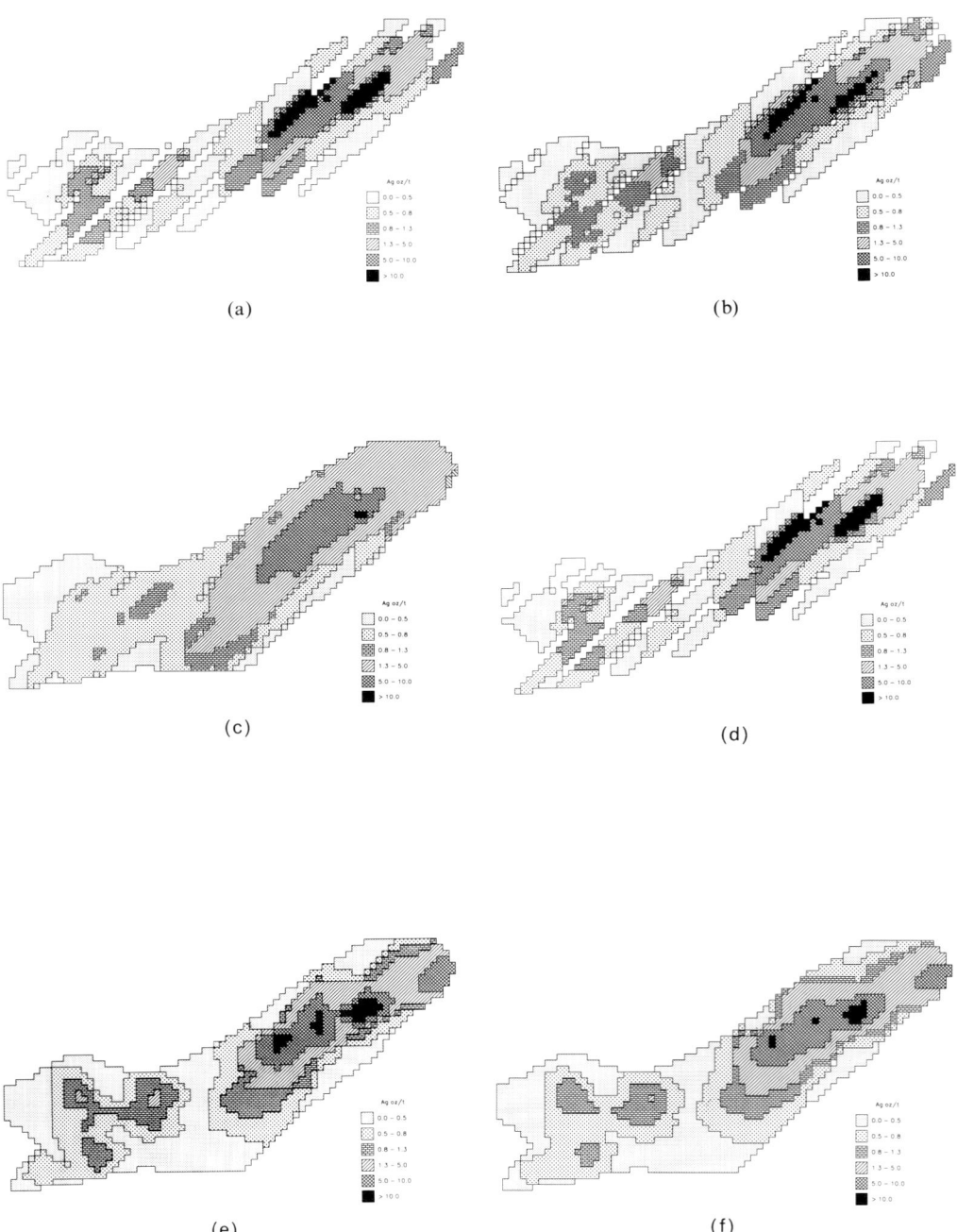

Fig. 12. Block models of estimated grade values. (**a**) Inverse distance with an 80 m maximum search distance, (**b**) inverse distance with a 130 m maximum search distance, (**c**) inverse distance with an 330 m maximum search distance, (**d**) kriged model, (**e**) simple linear interpolation, (**f**) triangulation model.

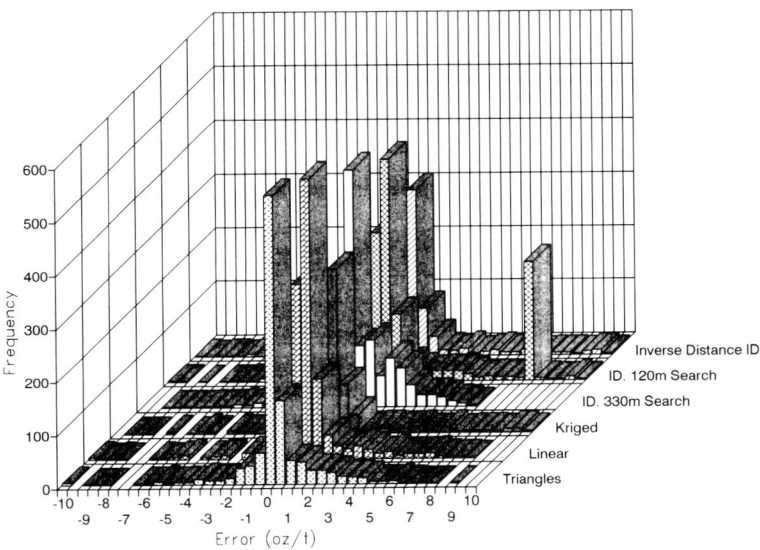

Fig. 13. Histogram of absolute errors for the different estimation processes.

Table 5. *Statistics of absolute errors*

	Inverse distance	Inverse distance (120 m search)	Inverse distance (330 m search)	Kriged	Linear interpolation	Triangles
$25^{th}\%$	−0.23	−0.20	−0.001	−0.29	−0.16	−0.13
$75^{th}\%$	0.48	0.54	1.49	0.48	0.49	0.55
Mean	0.31	0.29	0.57	0.30	0.20	0.21
Median	0.06	0.12	0.35	0.06	0.12	0.16
Variance	7.81	5.23	4.78	7.78	3.69	3.53
Std dev.	2.80	2.29	2.19	2.79	1.92	1.88
Skewness	0.74	−0.72	−2.94	0.75	−1.62	−2.55

Table 6. *Statistics of sqaured errors*

	Inverse distance	Inverse distance (120 m search)	Inverse distance (330 m search)	Kriged	Linear interpolation	Triangles
$25^{th}\%$	0.02	0.02	0.06	0.02	0.02	0.02
$75^{th}\%$	1.32	1.29	4.23	1.32	0.81	0.96
Mean	7.90	5.30	5.09	7.86	3.73	3.57
Median	0.16	0.14	0.42	0.16	0.12	0.13
Std dev.	29.12	21.16	22.78	29.06	17.51	18.54
Skewness	6.85	10.41	17.42	6.87	15.97	18.00

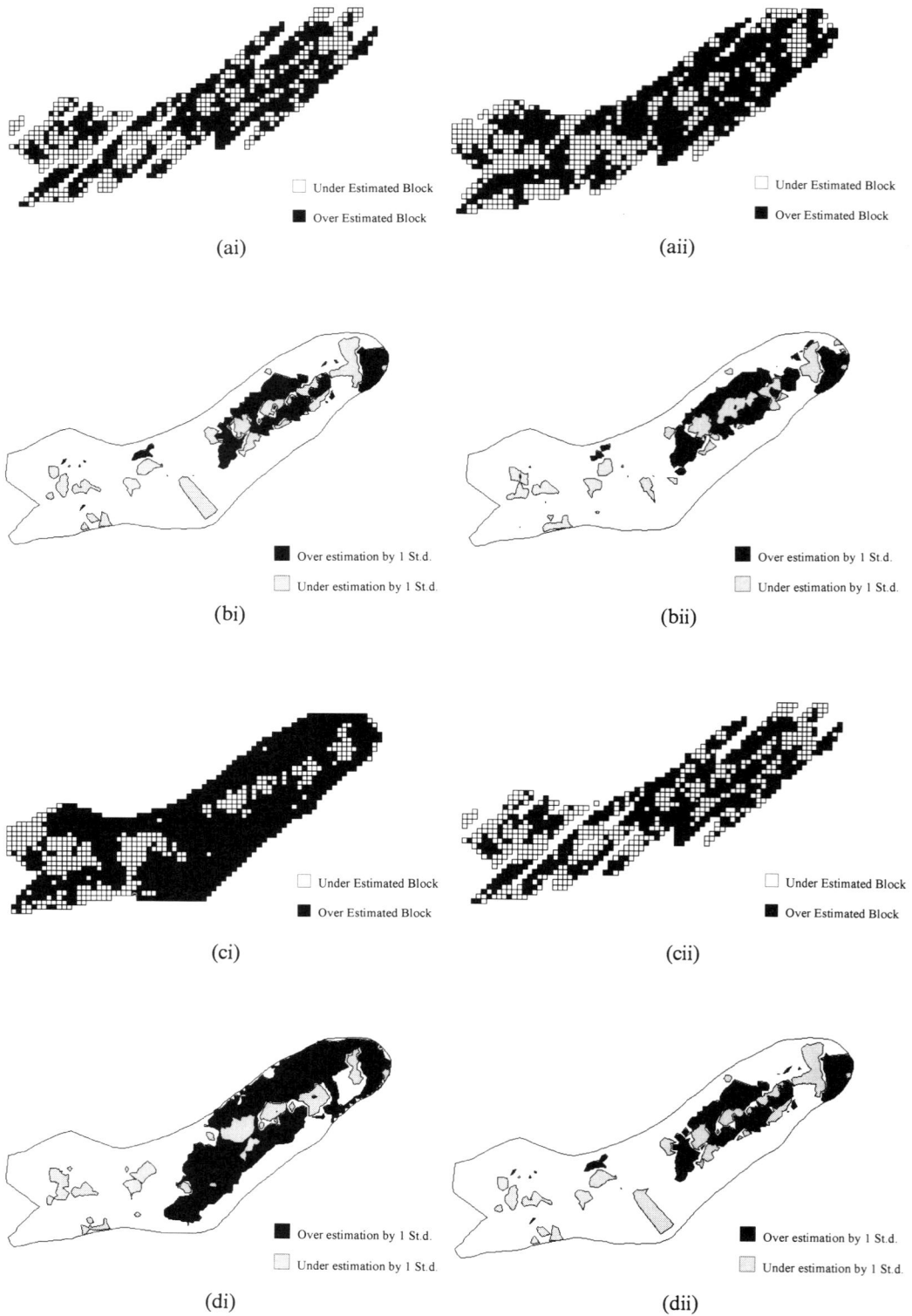

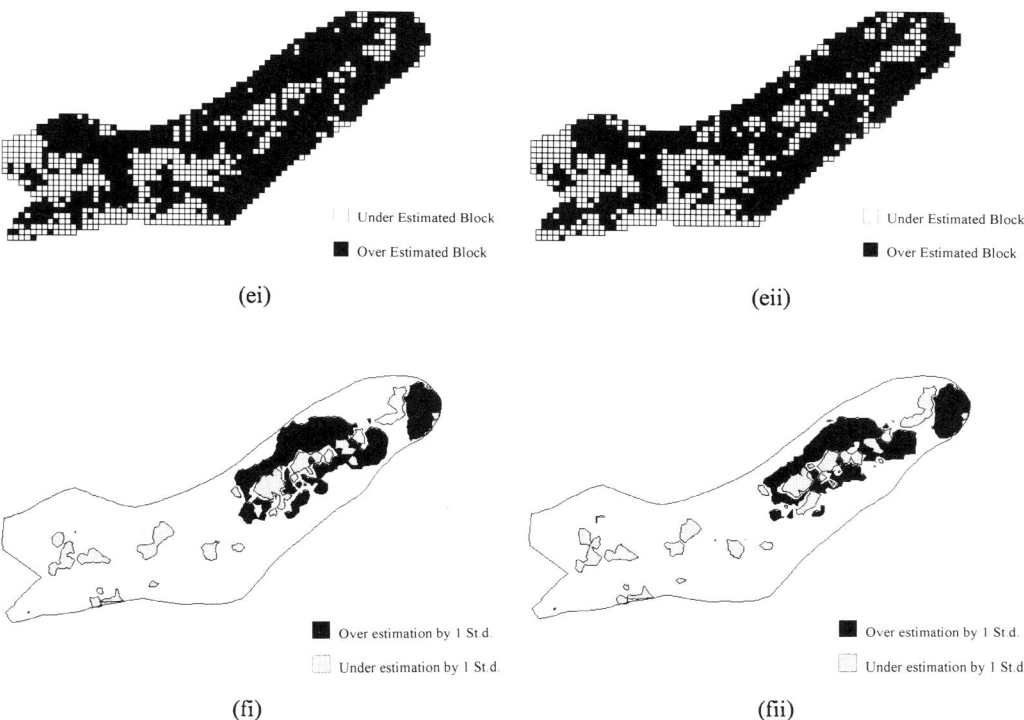

Fig. 14. Error maps showing (*i*) blocks that are over or under estimated, (*ii*) zones of high error classed by exceeding 1 standard deviation of the mean absolute error. (**a**) Inverse distance with an 80 m maximum search distance, (**b**) inverse distance with a 130 m maximum search distance, (**c**) inverse distance with a 330 m maximum search distance, (**d**) kriged model, (**e**) simple linear interpolation, (**f**) triangulation model.

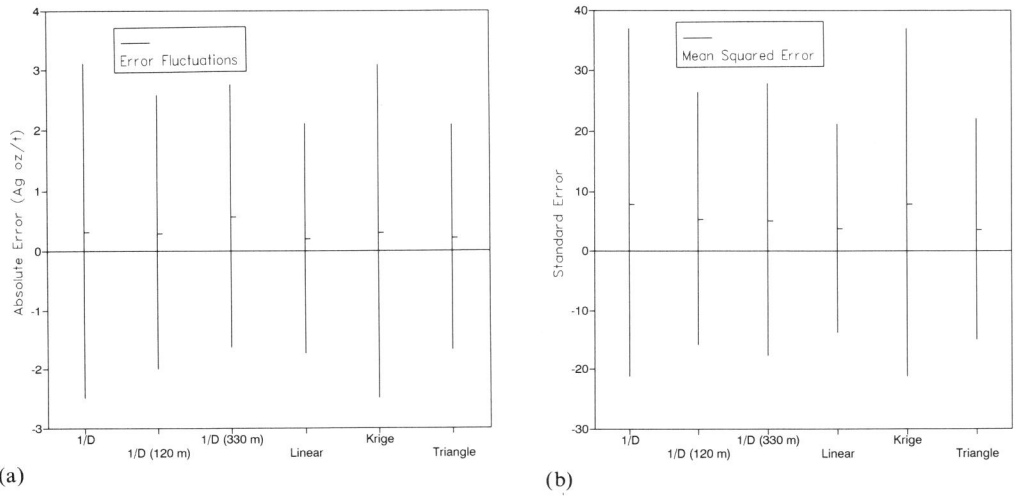

Fig. 15. Graphs representing the calculated mean (tick) and expected fluctuations (vertical bar) about the mean of (**a**) absolute errors, (**b**) standard error.

wrong if the spatial peculiarities are destroyed, although the global estimate can be correct. Yet the calculation of a given ore envelope at a stated cut off will be imprecisely located and subsequent global estimate of ore grade material will be incorrect if spatial variability and structure are not honoured.

References

ASHLEMAN, J. C. 1988. *The Trinity Silver Deposit, Pershing County, Nevada.* Unpublished Field Guide.

BELL, T. 1989. *Ore reconciliation and statistical evaluation of the Trinity Silver Mine, Pershing County, Nevada.* MSc Dissertation, University of Leicester.

ISAAKS, E. H. & SRIVASTAVA, R. M. 1989. *An Introduction to Applied Geostatistics.* Oxford University Press.

JOHNSON, M. G. 1977. *Geology and mineral deposits of Pershing County, Nevada.* Nevada Bureau of Mines and Geology. Bulletin 80.

WHATELEY, M. K. G. 1992. The evaluation of coal borehole data for reserve estimation and mine design. *In:* ANNELS, A. E. (ed.) *Case Histories and Methods Mineral Resource Evaluation.* Geological Society, London, Special Publications 63, 95–106.

Optimal open pit design: sensitivity to estimated block values

P. A. DOWD

Department of Mining and Mineral Engineering, University of Leeds, Leeds LS2 9JT, UK

Abstract: This paper discusses two important, and often overlooked, aspects of optimum open pit design: the information and support effects. Pit designs are ultimately based on estimated block grades and, as such, their optimality can only ever be approximate; as the errors of estimation of the block grades increase the pit design will deviate more and more from true optimality. Estimation errors depend on the amount of data available (information effect) and on the size of the block to be estimated (support effect). There is a lower limit to the block size used in grade and revenue block models for pit design and this lower limit is determined by the drilling grid. The use of blocks which are significantly smaller than the drilling grid will produce block models which will yield erroneous pit designs. Geostatistical simulation is used to quantify these effects for a particular case and to demonstrate the most effective way of dealing with small selective mining units at the planning stage.

Optimal open pit design is essentially a computer based implementation of an algorithm which is applied to a three-dimensional block model of an orebody.

Almost all optimal open pit design algorithms, with the exception of elementary methods applied to some stratiform deposits, are applied to a regular, fixed, three-dimensional block model of the orebody. The orebody is subdivided into regular blocks and a value is estimated for each block. This value is almost always the net (undiscounted) revenue that would be obtained by mining and treating the block and selling its contents. Some methods, such as parameterization (Francois-Bongarçon & Guibal 1982; Francois-Bongarçon & Marechal 1976; Matheron 1975a,b,c), use grade values in the block model. Stuart (1992) proposes an irregular three-dimensional model in which the orebody is represented by a series of arbitrary geometrical solids. Whilst such a model is a useful way of representing highly irregular and complex-shaped stratiform deposits it is doubtful whether sophisticated computer algorithms are really necessary for the design of optimal pits in such cases.

Whatever block values are used in optimal pit design they are based on estimated grade values and the reliability of these estimates depends on the amount of data available and the variability of the mineralization.

The purpose of this paper is to assess the effects of sparse data and of block size on the design of open pits. By using a geostatistically simulated orebody it is possible to design pits on the basis of different block sizes each with grades estimated from different drilling grids and compare these estimated 'optimal pits' with the true optimal pit based on the true grades of the smallest selective mining unit.

Block size

Sensitivity to block size is perhaps one of the most misunderstood concepts in optimal open pit design. Much has been written about the choice of block size used in block models for optimal pit design (Whittle 1989; Cai 1992) but the discussions seem to revolve around the ability of different block sizes to describe the geometry of the pit and the orebody. There is an obvious advantage to be gained in the decreased computing time which comes from a larger block size; the disadvantage of the larger block is the loss of definition of grade (and hence revenue) variations within the orebody. However, at the pit design stage, such small scale definition of the orebody is usually illusory simply because it cannot be inferred from the relatively sparse data available at the time.

The overwhelming restriction on block size is the amount of data available to estimate the grades of blocks. The numerical grade, and hence revenue, values assigned to blocks are values that have been estimated from the available sample data and they necessarily have an error associated with them. In general, for a given amount of data, the smaller the block size the greater is the error of estimation of its grade and therefore the greater the unreliability of the revenue block model used in pit optimization. The consequences of ignoring this discrepancy

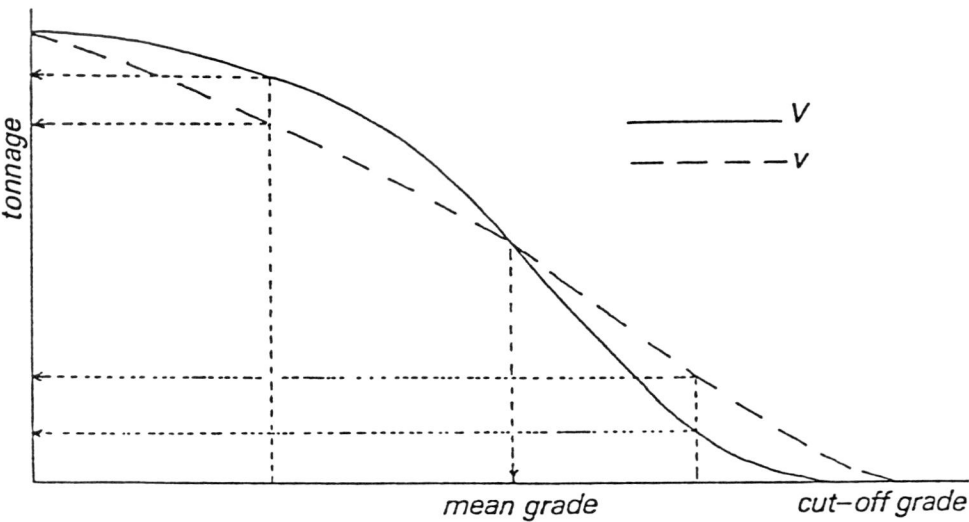

Fig. 1. The influence of support on recovery illustrated for two volumes v and V where $V > v$.

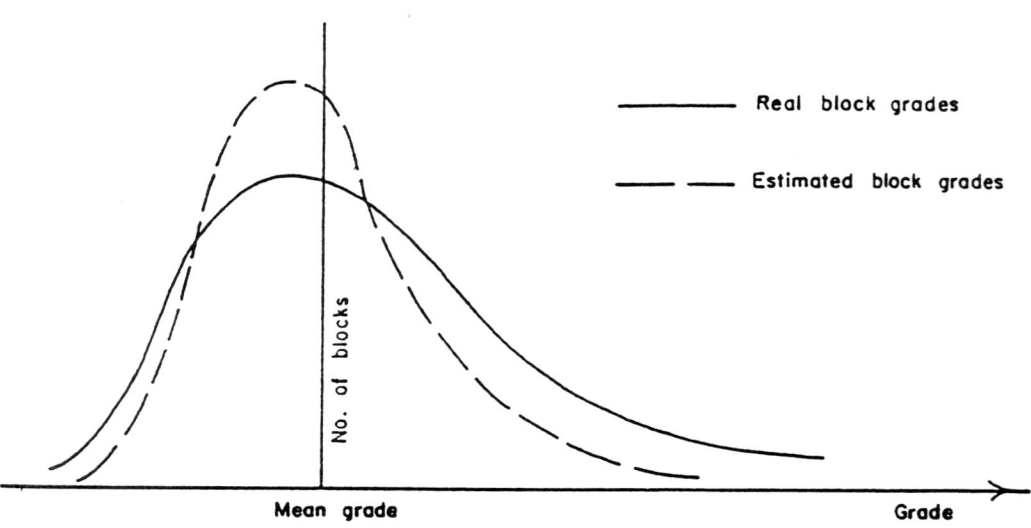

Fig. 2. Effect of information on recoverability: conversion of histograms to grade/tonnage curves would give similar effects to those shown in Fig. 1.

between true and estimated block grade (and revenue) values can be disastrous (David *et al.* 1974; Dowd & David 1976).

As a general rule of thumb the horizontal block dimensions should be limited to the size of the drilling grid; blocks with dimensions which are significantly smaller than this cannot be estimated with sufficient accuracy to provide a reliable block grade/revenue model on which to design the pit. An alternative approach (David *et al.* 1974; Dowd & David 1976) is to optimize the pit design on the basis of conditionally

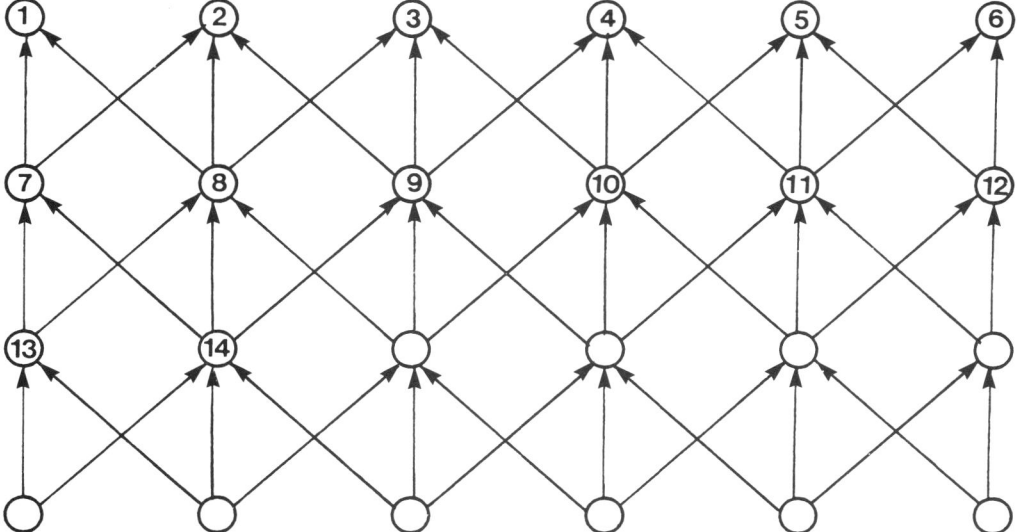

Fig. 3. Directed graph representing a two-dimensional model. Nodes represent blocks and arcs define mining constraints.

simulated grades of smaller blocks. The small scale block model of the orebody can be simulated a number of times and an optimal pit can be designed for each simulation thereby providing a probabilistic answer to a problem which should, more realistically, be posed as probabilistic.

The limitations on block dimensions will also limit the number of blocks in the orebody model. For example, a 1 km × 1 km area drilled on a 30 m × 15 m grid to a depth of 500 m is effectively limited to a 110 000 block orebody model for blocks with a vertical dimension of 10 m and a 220 000 block orebody model for blocks of 5 m height. On the basis of the information available it is illusory and highly misleading to subdivide the model into smaller blocks to account for supposed variations in grade and geology. It may be argued that, during mining, ore and waste are discriminated on the basis of much smaller selective mining units (e.g. planning blocks may be 30 m × 15 m × 10 m and selective mining units may be 5 m × 5 m × 5 m). If this is so then the correct approach is either via simulation or by estimating recoverable reserves for selective mining units within each planning block by well-documented geostatistical methods. In the latter case there are two estimates for each block, tonnage above cut-off grade and mean grade of tonnage above cut-off grade, which are then used in the calculation of net monetary value of the planning block.

Support effect and information effect

In geostatistical applications the volume on which grades are measured and/or estimated is referred to as a *support*. All grades are values averaged over a support. The volume–variance relationship expresses the most significant effect of measuring grades on different supports: the variance of a set of grades is inversely proportional to the volumes on which they have been measured. The major consequence of this relationship for mining applications is the effect it has on recoverability as illustrated in Fig. 1.

A block grade model for selective mining units (e.g. 5 m × 5 m × 5 m) will more accurately reflect mining recoverability than will a block grade model for large planning blocks (e.g. 30 m × 15 m × 10 m). In addition, the former will yield a more accurate revenue block model which will in turn yield a pit design which will be closer to the true optimum.

However, true grades are not known prior to mining and the grade models used for evaluation and design are *estimated* block grade models. The more information available the closer the estimated block grade models will be to the corresponding true block grade models. The major difference between block grade models based on different amounts of information (e.g. different drilling grids) is the variance of the estimated grade values as shown in Fig. 2. The consequence of this *information effect* in mining applications is a change in recoverability.

It is precisely these two effects, information and support, which cause the problems in assessing supposedly *optimal* open pit designs. All 'optimal' open pit designs are based on estimated values of a given support and, as such, their optimality is always open to question. One way of assessing the impact of the information and support effects on the design of open pits is via geostatistical simulation of orebodies (Journel & Huijbregts 1976). An orebody can be simulated to provide 'true' block grade models for various block sizes; the grades of these same blocks can be estimated from different drilling/sampling grids to provide corresponding estimated block grade models for each block size. Optimal open pits can then be designed for each true and estimated block grade model and the results can be compared.

The Lerchs–Grossmann algorithm

The Lerchs–Grossmann algorithm (Lerchs & Grossmann 1965; Dowd & Onur 1992) for optimal open pit design has spawned one of the most active areas of mineral industry operational research.

The Lerchs–Grossmann algorithm converts the three-dimensional grid of blocks in the orebody model into a directed graph. Each block in the grid is represented by a vertex which is assigned a mass equal to the net revenue value of the corresponding block. The vertices are connected by arcs in such a way that the connections leading from a particular vertex to the surface define the set of vertices (blocks) which must be removed if that vertex (block) is to be mined. A simple two-dimensional (vertical) example is shown in Fig. 3.

Vertices connected by an arc pointing away from a vertex are termed successors of that vertex, i.e. the vertex y is a successor of the vertex x if there exists an arc directed from x to y. The set of all successors of x is denoted Γx. For example, in Fig. 3. $\Gamma x_9 = \{x_2, x_3, x_4\}$. A closure of a directed graph, which consists of a set of vertices X, is a set of vertices $Y \subset X$ such that if $x \in Y$ then $\Gamma x \in Y$. For example, in Fig. 3, $Y = \{x_1, x_2, x_3, x_4, x_5, x_8, x_9, x_{10}\}$ is a closure of the directed graph. The value of a closure is the sum of the masses of the vertices in the closure. Each closure defines a possible pit; the closure with the maximum value defines the optimal pit.

Geostatistical simulation

Although geostatistical simulation, in the form of the turning bands method, was introduced some 20 years ago it has not fulfilled its early promise as a powerful tool in the mining industry. There are two principal reasons for this: firstly there has been some confusion on the part of many end-users (and some geostatisticians) as to the meaning and significance of simulated models and secondly, there are some shortcomings in the turning bands method which, although recognized early on, have been slow to be acknowledged and rectified.

Sequential methods, proposed by Journel & Alabert (1989, 1990), are an application of Bayes' theorem based on Devroye (1986). The n dependent events A_i, $i = 1, \ldots n$ can be sequentially simulated using the expression:

$$P(A_1, A_2, \ldots A_n) = P(A_n \mid A_1, \ldots A_{n-1}) \times$$

$$P(A_{n-1} \mid A_1, \ldots A_{n-2}) \ldots P(A_2 \mid A_1) \times P(A_1)$$

The technique requires the inference of the successive $n - 1$ conditional probability distributions. This can be achieved in either of two ways. The first is by means of a Gaussian transform and the second is to infer the distribution directly by the use of indicators. Sequential Gaussian simulation has been used in this study.

Sequential Gaussian simulation. The sequential gaussian simulation algorithm consists of the following steps:

(1) transform all conditioning data to standard Gaussian values;
(2) calculate and model the semivariogram of the transformed conditioning data;
(3) define a random path through all n grid points on which values are to be simulated;
(4) at each simulation grid point krige a value from all other values (conditioning and simulated);
(5) the kriged value and the associated kriging variance are the parameters of the conditional gaussian distribution at the given grid point given the conditioning data and all previously simulated values; draw a value random from this distribution and add it to the set of simulated values;
(6) return to step 4 until values have been simulated at all grid points;
(7) take the inverse transform of the gaussian conditionally simulated values.

The major advantages of this method are:

- the conditioning is an integral part of the simulation and does not have to be performed as a separate step;

- anisotropies are handled automatically;
- it can be applied for any covariance function;
- an efficient kriging algorithm (using a moving neighbourhood search) is all that is required for implementation.

The only apparent drawbacks of the method are the perceived disadvantage of using the intermediary gaussian distribution and a question about the degree of variability between successive simulations (Dowd 1992). The method can, of course, only be used for *conditional* simulations, i.e. the simulation requires a *seed* of a set of data values.

A case study

The orebody

The orebody used for this study is a tectonically controlled gold deposit with a lognormal distribution of gold grades defined on 1 m drill core samples. For the purposes of this study the grades of the 1 m drill core samples were composited into 5 m sample values by taking arithmetic averages; the compositing process effectively removed an initial, short range (approximately 4 m) variogram structure in the cross-dip (north–south) direction and the nested structure of the variogram of 1 m gold grades was replaced by the single structure (plus nugget variance) of the grades of 5 m gold grades. The geostatistical characteristics of the mineralization are summarized in Table 1.

Table 1. *Geostatistical characteristics of the gold mineralization studied*

Mean grade		1.68 g/t
Variance		122.5 (g/t)^2
Coefficient of variation		6.59
Experimental variogram		
C_o =	28.6(g/t)^2	
C_1 =	87.4(g/t)^2	
ranges:	east–west (strike)	50 m
	north–south (across dip)	10 m
	vertical (down dip)	30 m
Experimental variogram of logarithms:		
C_o =	0.80(ln g/t)^2	
C_1 =	2.73(ln g/t)^2	
ranges:	east–west (strike)	50 m
	north–south (across dip)	10 m
	vertical (down dip)	30 m

In this orebody the standard drilling grid for evaluation purposes is 30 m (E–W) × 15 m (N–S) and the selective mining unit is 5 m × 5 m × 5 m; the block size used for planning purposes is 30 m × 15 m × 10 m.

The sequential gaussian method was used to simulate grade values of 1 m drill core samples on a 1 m (E–W) × 1 m (N–S) × 5 m (vertical) grid over a total volume of 600 m (E–W) × 400 m (N–S) × 60 m (vertical) giving a total of 2 880 000 simulated values

The grades of 30 m (E–W) × 15 m (N–S) × 10 m blocks and of 5 m × 5 m × 5 m blocks were simulated by taking the arithmetic average of the values of the simulated grades within them; these average values are taken as the 'true' grades of the blocks. The grades of the 30 m × 15 m × 10 m blocks and of the 5 m × 5 m × 5 m blocks were estimated by ordinary kriging, using the simulated grades of samples on a 30 m × 15 m drilling grid.

One way of including the effects of the selective mining units on the pit design is to estimate the recoverable reserves for each 30 m × 15 m × 10 m block. In this study the *lognormal shortcut* (David 1972; Dowd 1992) was used to estimate, for each 30 m × 15 m × 10 m block, the proportion of 5 m × 5 m × 5 m blocks above the mining cut-off grade together with the mean grade above this cut-off grade; the data used for these estimates were the grade values selected from the 30 m × 15 m drilling grid. This approach can be extended to cover a range of cut-off grades so that a grade-tonnage curve is estimated for each 30 m × 15 m × 10 m block. Other estimation methods such as indicator kriging, probability kriging, multigaussian kriging and disjunctive kriging could also be used.

The five block grade models ('true' 30 m × 15 m × 10 m block grades, estimated 30 m × 15 m × 10 m block grades, 'true' 5 m × 5 m × 5 m block grades, estimated 5 m × 5 m × 5 m block grades, estimated recoverable 5 m × 5 m × 5 m block grades) were then converted to revenue block models using the data in Table 2.

Optimal open pits for each of the block models summarized in Table 3 were designed using the Lerchs–Gossmann algorithm.

Table 2. *Parameters used in optimal open pit designs*

Mining cost (ore)	£3.00 per tonne
Mining cost (waste)	£2.00 per tonne
Processing cost	£3.50 per tonne of feed
Gold price	£6.85 per g
Recovery	95%
Cut-off grade	0.75 g/t
Pit wall slopes	55°

Table 3. Block models used for pit design

Block model 1	'true' grades of 30 m × 15 m × 10 m blocks
Block model 2	grades of 30 m × 15 m × 10 m blocks estimated from sample grades on a 30 m × 15 m drilling grid
Block model 3	'true' grades of 5 m × 5 m × 5 m blocks
Block model 4	grades of 5 × 5 m × 5 m blocks estimated from sample grades on a 30 m × 15 m drilling grid
Block model 5	recoverable reserves for each 30 m × 15 m × 10 m block estimated from sample grades on a 30 m × 15 m drilling grid and based on a recoverable unit of 5 m × 5 m × 5 m

Table 4. Comparison of pit designs based on the block models in Table 3

	Block model 1	Block model 2	Block model 3	Block model 4	Block model 5
Total ore (tonnes × 10^6)	10.738	11.949	8.980	12.111	9.621
Mean grade (g/t)	5.78	4.53	6.43	4.32	6.08
Total waste (tonnes × 10^6)	15.133	19.224	11.611	20.054	13.256
Waste : ore ratio	1.41 : 1	1.61 : 1	1.29 : 1	1.66 : 1	1.38 : 1
Net pit value (£ × 10^6)	308.93	240.95	298.32	224.66	296.47

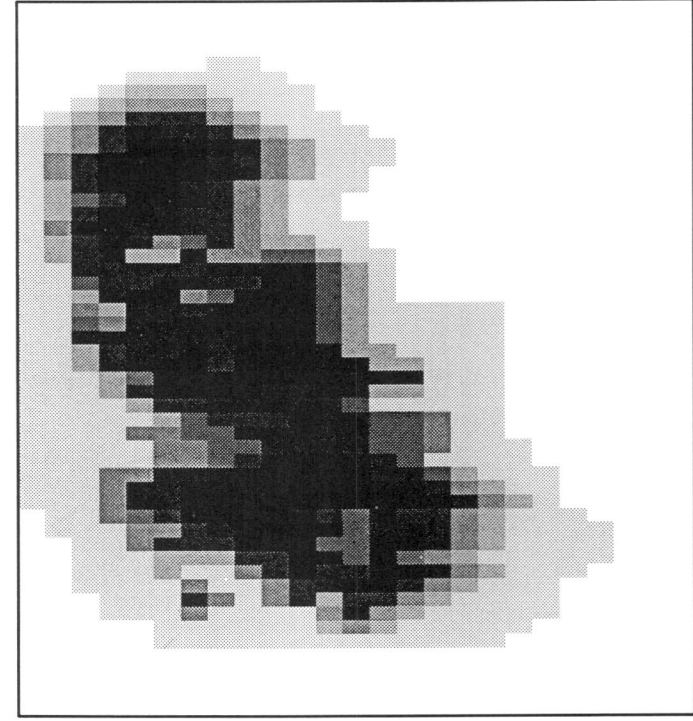
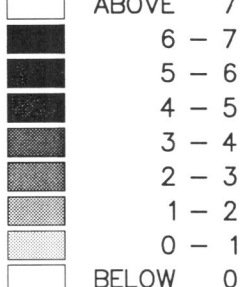

Fig. 4. Optimal open pit design for block model 1.

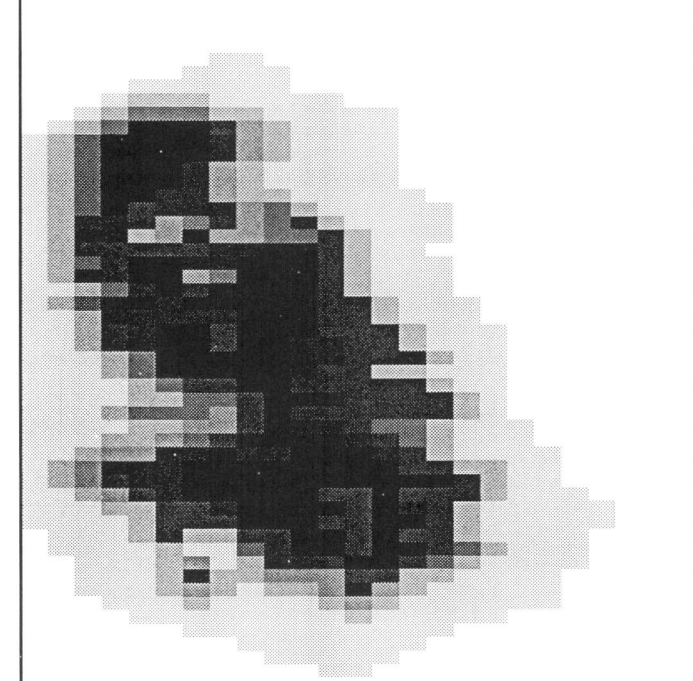

Fig. 5. Optimal open pit design for block model 2.

Comparison of results

The numerical results of the optimal pit designs for each of the revenue block models are summarized in Table 4.

Note the considerable discrepancies between the pits for the two estimated block grade models (2 and 4) and the corresponding pits for the true grade models (1 and 3) and that the discrepancies are considerably greater for the smaller block size. The recoverable reserves block model (no. 5) gives a pit design which is significantly closer to the true optimum (model 3) than the $5\,\text{m} \times 5\,\text{m} \times 5\,\text{m}$ estimated block grade model.

As an example of the effects of block size, and of accuracy of estimates, on pit design the designs for block models 1 and 2 are shown in Figs 4 and 5. Note that no attempt has been made to smooth the pit contours or to impose minimum access constraints (other than the implied minimum of an individual block).

At this stage it would be possible to calculate the actual ore within the pits for block models 2, 4 and 5. However, these results would not be particularly helpful as they would not take into account the additional grade control sampling which would take place during mining and which would lead to more accurate discrimination between ore and waste on the basis of the selective mining unit. The important figures in the evaluation stage are those reported in Table 4. The figures for block models 2, 4 or 5 would be used in feasibility studies and cash flow calculations. Net present values of these three pits would almost certainly show significantly larger differences than the net pit values in Table 4 because of the different locations of the estimated high grade blocks in the pit and hence the different times at which they would be mined.

Conclusions

Whilst the specific details of these results cannot be taken as general they do illustrate the general discrepancies which can arise between optimal pit designs based on estimated block grades and the true (but unattainable) optimal pit design based on true block grades. In addition, they illustrate the effects of using different block sizes to design pits. Allowing for edge blocks, the 5 m × 5 m × 5 m block grade model contains more than 160 000 blocks whereas the 30 m × 15 m × 10 m model contains approximately 5000. As computing time for the Lerchs–Grossmann algorithm is approximately proportional to the square of the number of blocks the 5 m × 5 m × 5 m block grade model will incur a significantly higher computing cost and yield results which have doubtful validity. The impact of the information and support effects on pit design is clearly demonstrated by the examples presented here. In particular, it is futile to subdivide the orebody into blocks which are significantly smaller than the drilling grid in an attempt to model the perceived variability of grades or orebody shape.

The best way of incorporating the effects of selective mining units (smu) on pit design is to use recoverable block grade models based on one of the many geostatistical techniques available for estimating the smu-recoverable reserves of a planning block. In the example presented here each block in the model is 30 m × 15 m × 10 m and is accompanied by the estimated grade-tonnage curve based on a 5 m × 5 m × 5 m smu. The smu effect is included at no additional pit design computing cost; there is, however, an additional computing cost for the estimation of the block grade model.

Simulation is a powerful means of assessing the impact of information and support effects on pit design. It can also be used for other forms of sensitivity and risk analyses by varying geostatistical and geotechnical parameters, costs, prices, recoveries, selective mining units and blasting patterns.

References

CAI, W. L. 1992. Sensitivity analysis of 3-D model block dimensions in the economic open pit limit design. *23rd Symposium on the application of computers and operations research in the mineral industries (APCOM)*. AIME, Littleton, Colorado, 475–486.

DAVID, M. 1972. Grade tonnage curve: use and misuse in ore reserve estimation. *Transactions of the Institution of Mining and Metallurgy* **81**, 129–132

——, DOWD, P. A. & KOROBOV, S. 1974. Forecasting departure from planning in open pit design and grade control. *12th Symposium on the application of computers and operations research in the mineral industries (APCOM)*. Colorado School of Mines, **2**, F131–F142.

DEVROYE, L. 1986 *Non-uniform random variate generation*. Springer-Verlag, Berlin.

DOWD, P. A. 1992. A review of recent developments in geostatistics. *Computers and Geosciences* **17**, 1481–1500

—— 1992. Geostatistical ore reserve estimation: a case study in a disseminated nickel deposit. *In:* ANNELS, A. E. (ed.) *Case histories and methods in mineral resource evaluation*. Geological Society, London, Special Publications, **63**, 243–255.

—— & DAVID, M. 1976. Planning from estimates: sensitivity of mine production schedules to estimation methods *In:* GUARASCIO, M., DAVID, M. & HUIJBREGTS, C. D. (eds) *Advanced geostatistics in the mining industry*. NATO ASI SERIES C: Mathematical and Physical Sciences, **24**, Reidel Pub. Co. Dordrecht, Netherlands, 163–183.

—— & ONUR, A. H. 1992. Optimizing open pit design and sequencing. *23rd Symposium on the applicaiton of computers and operations research in the mineral industries (APCOM)*. AIME, Littleton, Co., 411–422.

FRANCOIS-BONGARÉON, D. & GUIBAL, D. 1982. Algorithms for parameterising reserves under different geometrical constraints. *17th Symposium on the applicaiton of computers and operations research in the mineral industries (APCOM)*. AIME, New York, 297–310.

—— & MARECHAL, A. 1976. A new method for optimum pit design: parameterisation of the final pit contour. *14th Symposium on the application of computers and operations research in the mineral industries (APCOM)*. AIME New York.

JOURNEL, A. G. & ALABERT, F. 1989. Non-gaussian data expansion in the earth sciences. *Terra Nova*, **1**, 123–134.

—— & —— 1990. New method for reservoir mapping. *JPT February 1990*, 212–218.

—— & HUIJBREGTS, C. 1978. *Mining Geostatistics*. Academic Press, New York.

LERCHS, H. & GROSSMANN, I. F. 1965. Optimum design of open pit mines. *CIM Bulletin* **58**, 47–54.

MATHÉRON, G. 1975a. *Paramétrage des contours optimaux*. Note Géostatistique No 128, Centre de Géostatistique et de Morphologie Mathématique, Internal Report N-403, Fontainebleau, France.

—— 1975b. *Compléments sur le paramétrage des contours optimaux*. Note Géostatistique No. 129 Centre de Géostatistique et de Morphologie mathématique, Internal Report N-401, Fontainebleau, France.

—— 1975c. *Le paramétrage technique des reserves*. Note Géostatistique No. 134, Centre de Géostatistique et de Morphologie mathématique, Internal report N-453, Fontainebleau, France.

STUART, N. J. 1992. Pit optimization using solid modelling and the Lerchs–Grossmann algorithm. *International Journal of Surface Mining and Reclamation* **6**, 19–30.

WHITTLE, J. 1989. *The facts and fallacies of open pit optimization*. Whittle Programming Pty Ltd, North Balwyn, Victoria, Australia.

Dilution in underground bulk mining: implications for production management

M. J. SCOBLE[1] & A. MOSS[2]

[1] *McGill University, Montreal, Quebec, Canada H3A 2A7*
[2] *Golder Associates Ltd, Vancouver, British Columbia, Canada V5C 6C6*

Abstract: The minimization of dilution represents a major opportunity for quality improvement in Canadian underground metal mining. The characteristics of the principal forms of dilution, planned and unplanned, are reviewed. These are controlled by quality factors relating to exploration, mine design and stoping practice. The scope for quality improvement and the need for technology development is analysed in the context of underground bulk mining. The design principles for a production management system which tightly monitors and controls dilution are proposed. This revolves around the reconciliation of accurate and timely data on ore reserves, stope fragmentation, cavity morphology, rockmass integrity, and production statistics. Particular priority will require to be placed on continued sensor development for measurement of cavity geometry, borehole deviation, grade and tonnage. This is intended to exploit a rapidly evolving underground communications technology and forms part of a rationale for future computer integrated mining.

The recession of a decade ago caused Canadian underground mines to adopt workforce rationalization, mechanization and new, predominantly bulk, mining methods in efforts to improve productivity and reduce costs. The advent of the current recession saw a response which has turned more to new technology and total quality improvement (Loring *et al.* 1992). Key measures of quality are *recovery* and *dilution*. Recovery relates to the effectiveness of mining design, measured by the proportion of the known orebody to have been recovered. Dilution relates to mining efficiency, measured by the proportion of waste introduced as part of the mining process. The two measures are interdependent: e.g. a particular recovery target may only be attainable at the expense of a certain level of dilution. Mine design is a process of compromise involving dilution, recovery, productivity and cost criteria. Recent metal price reductions have underscored further the need to understand and thence minimize dilution.

Though excessive dilution and lack of recovery are two of the commonly quoted reasons for mine failure, there has been limited effort until recently by industry to understand better the factors that control dilution and recovery. This is due to the complexity of the problem, poor accessibility to data and the level of judgement required for ore estimation and dilution prediction. Until technology provides the tools to study and understand the basic processes more effectively then design and control will tend still to be based on empirical techniques and intuition.

Production control traditionally has been based on weight of rock hoisted rather than weight of metal hoisted. This tended to encourage inefficient mining practice. Despite dilution's significance, most mines are unable to collect adequate data to calculate its value with any precision, whether to monitor and control production or to optimize stope design. Recent developments in underground communications now offer the potential to improve dramatically data collection, transmission and analysis of mine production monitoring and control systems. Such communications capability needs to be exploited for quality improvement through the monitoring and control of dilution. The task remains, however, to identify the data, sensors and analysis required for the design of such systems. A more scientific approach to mine production management is now being sought.

This paper attempts to clarify the nature of dilution in bulk mining; its definition, causes and impact. It examines the principal controlling factors associated with the exploration, development and stoping phases. The integration of dilution monitoring and control into future mine production management systems is then considered, particularly in the context of recent developments in underground communications technology. It concludes by considering the sensor development required for monitoring cavity morphology and integrity, blast-hole deviation, grade and tonnage.

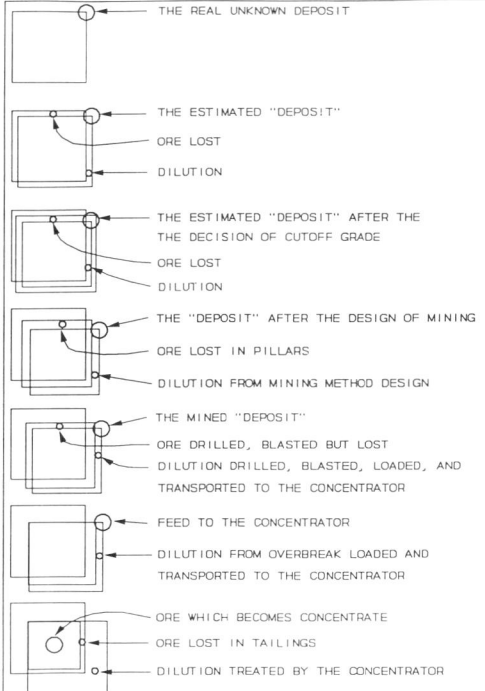

Fig. 1. The sequence of ore loss and dilution affecting reserves (after Elbrond 1986).

Ore reserves and dilution

Dilution, the reduction in grade of a mine reserve, arises in several forms through the sequence of exploration, mining and processing operations. Figure 1 illustrates conceptually how ore loss and dilution are affected by a sequence of design operations (initial ore estimation, application of cutoff grade, and mining design); followed by a sequence of production operations (mining and processing). Various definitions of dilution exist; it is prudent to be cautious in comparing statistics from various sources. The two most common are based on tonnage as follows:

Dilution, D_1 = waste/ore; or D_2 = waste/(ore + waste).

However, as noted by Wright (1983), it is very difficult to determine the tonnage of waste in the mined material, therefore grade is used as an indirect measure of dilution. Assuming that the waste material carries zero grade, then dilution can be estimated from:

D_1 = (stope grade − head grade)/stope grade

or

D_2 = (stope grade − head grade)/head grade

where stope grade is the estimated grade for the stope block, and the head grade is that of the material actually drawn from the stope. The *geological reserve* is the estimate of the ore contained within geological limits as specified by any particular cutoff grade. It can be considered to account for the dilution arising from internal waste, errors in ore estimation and from the application of the cutoff grade (Fig. 2). This reserve tends to be used solely as an inventory of the ore. It is the basis for calculating the *mining* or *mineable reserve*, which is the ore contained within the designed mining limits. This is established during the design process and accounts for the limitations of the mining method, i.e. the recovery and dilution which are unavoidably associated with the particular method designed. This dilution may be termed the *planned dilution*. It arises from the material below the cutoff grade that lies within the designed stope boundaries (mining lines) as

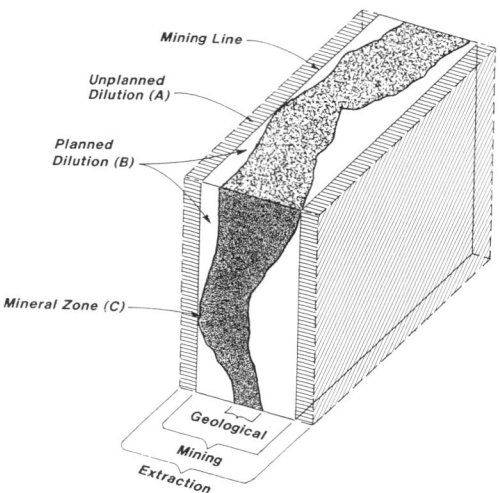

Fig. 2. Planned and unplanned dilution.

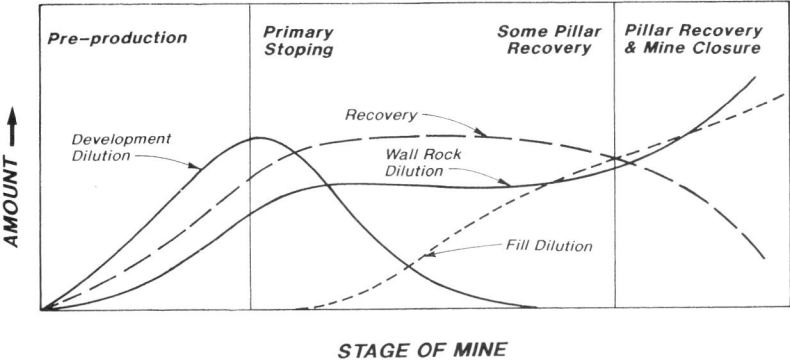

Fig. 3. Concept of dilution as a function of mine age (after HBM & S 1990).

determined by the selectivity inherent to the particular mining method and the complexity of the orebody morphology. The variability in planned dilution and recovery according to stoping method is evident in Table 1. These are broad estimates only; precise prediction of planned dilution is difficult and related to site-specific factors.

Table 1. *Quality ranges of stoping methods*

Stoping method	Recovery	Planned dilution %
Blasthole	90–95	5–20
Room-and-pillar	75–85	15–20
Shrinkage	90–95	10–15
Cut-and-fill	93–97	10–30
Sublevel caving	80–90	15–20
Block caving	80–90	15–20

Alternatively, the following relationship has been suggested by O'Hara (1980) for preliminary estimation of planned dilution as a percentage, D, according to orebody width (W) and dip (X):

$$D = 100/(K (W^{0.5} \sin (X)))$$

where K varies according to stoping method (i.e. blast hole = 100; shrinkage = 60; cut-and-fill = 45; room-and-pillar = 70). If stope walls are regular and competent then dilution was estimated to be only 0.7 of the above, whereas if the walls are weak or irregular then it could be as high as 1.5 times these values.

The final modification to the underground resource creates what may be termed the *extraction reserve*. This is the tonnage and grade of the ore expected to be actually extracted from the stopes for delivery to the mill. It is based on the mining reserves adjusted according to the ore loss and dilution expected in the mining process and tends to be used for short term production forecasting. Ineffective drilling and blasting may result in fragmentation within or beyond the mining lines, i.e. underbreak or overbreak. *Unplanned dilution* is that additional material below cutoff grade that is derived from three possible sources: blast overbreak, which may arise directly from the inadvertent fragmentation of rock outside the mining lines; or secondly, the sloughing of unstable wallrock, related to the following factors: rockmass geomechanical quality and stresses, blast damage and time. Thirdly, backfill may inadvertently be excavated from the stope (in cut-and-fill stoping), or it may occur as blast overbreak or subsequent slough in any adjacent backfilled stope walls (in delayed fill blasthole stoping). Unplanned dilution is a measure of the exploration, design and practice quality.

Mine design often requires compromise between costs, productivity, recovery and planned dilution. The economic significance of dilution is such that it is imperative that it be monitored accurately and that mining tactics be established to ensure the means exist to adapt and minimize unplanned dilution. Whilst some planned dilution may be claimed to be unavoidable, mining should aim to avoid any unplanned dilution. *Total dilution* is the sum of planned and unplanned ilution, i.e. all waste beyond the defined orebody (the mining reserve). Such waste may have a grade, albeit sub-cutoff. *Overpull* is the ratio of the actual tonnage of material (ore and waste) extracted from a stope to the tonnage in the mining reserve.

As a mine ages, from pre-production development through primary stoping to pillar recovery and eventual mine closure, then the tendency will be for variation in dilution and recovery (Fig. 3). During early mining, dilution can arise from processes associated with development, lack of knowledge of the orebody, and the general learning curve associated with operating a mine in a new environment. As primary stoping evolves then dilution should stabilize. In the final production phase, as pillar recovery intensifies and ground conditions deteriorate, then dilution can be expected to increase at the expense of recovery. This influence of age on the ability to control dilution should be taken into account when planning the mining sequence.

Quality factors

In addition to the complexity of the deposit, dilution will be controlled by the quality of exploration, mine design and stoping practice. It may therefore be considered in terms of the level of quality achieved within these three categories.

Exploration quality

The exploration quality factors relate to the quality of ground characterization, through mapping, drilling and sampling intended to establish:

- the grade distribution, ore limits and ore body morphology;
- the rock density distribution;
- the petrological, structural and geomechanical characteristics of the rockmass;
- the selection of cutoff grade and ore classification.

Ore delineation is generally accomplished by mapping, sampling and analysis of development excavations and core from diamond drilling. The analysis of drill chippings or sludge, typically at 50% of the cost of core drilling, is a good source of local supplementary data from exploration of blast-hole drilling. These holes also provide the opportunity to undertake borehole geophysical logging to characterize boundaries and weakness zones, ore types, grade distribution and waste (if mineralogy provides sufficient contrast between ore types and waste) and rockmass quality ahead of excavations. For example, Inco Ltd and Fisher Ltd have jointly developed a borehole tool to map ore intersections in blast holes (Inco pers. comm., 1992). It can be used in wet or break-through holes up to 50 m long, at dips from -45 to $-90°$. It transmits high energy magnetic pulses from a coil at 100 Hz. After each pulse is transmitted, the unit then listens using the coil as a receiving antenna. Almost all metals and some minerals have been found to be detectable by such pulse indications. A 'multiprobe' logging system was successfully used to locate ore boundaries in open stope blast holes at the Zincgruvan and Malmberget mines (Luleå University 1991). The Geological Survey of Canada is similarly developing multiparameter borehole logging technology (Killeen 1991). It has also developed a new spectral gamma-gamma borehole logging tool intended to make *in situ* assays of mineralization. Further potential relates to ground-penetrating radar in boreholes for defining structure and cavities. Seismic and radiowave tomography is also being researched for three-dimensional mapping in underground mines. Another supplement to exploration data is from the monitoring and interpretation of blast-hole drilling performance parameters; trials have attempted to locate ore-waste contacts and define ore grades (Schunnesson 1990).

According to Pentilla (1989) geostatistics can prove that there is an exploration drilling layout, unique to each deposit, beyond which more drilling will not improve the precision of the ore estimation. It is also the view that usually much more exploration is required to determine the ore boundaries accurately. Each deposit justifies early resolution of a specific plan to delineate adequately the ore-body morphology, in addition to the conventional tonnage and grade estimation.

Studies at the Viscaria copper mine showed how dilution and ore loss were influenced by geological complexity in horizontal section as well as the availability of exploration information (Puhakka 1990). Planned dilution was seen to decrease from 15% to 5% when drilling was intensified from 25 to 7.5 m spacing between sections. It was less for ore bodies with complex irregular wall geometries than for those with straight regular ore boundaries. In evaluating the economics of different spacings, assuming unplanned dilution was 10%, it was found that there was an optimum spacing of 12.5 m.

In order to evaluate the accuracy of ore location and its relation to dilution in vertical section, Inco's Thompson Mine studied geological sections mapped in detail from old cut-and-fill stopes, representative of its five major orebodies (Braun 1991). These sections enabled construction of artificial drill intersections over a range of 34, 17, and 8.5 m vertical spacing. These intersections were given to mine geologists to interpret ore outlines, which then were given to mine engineers to establish mining lines and to

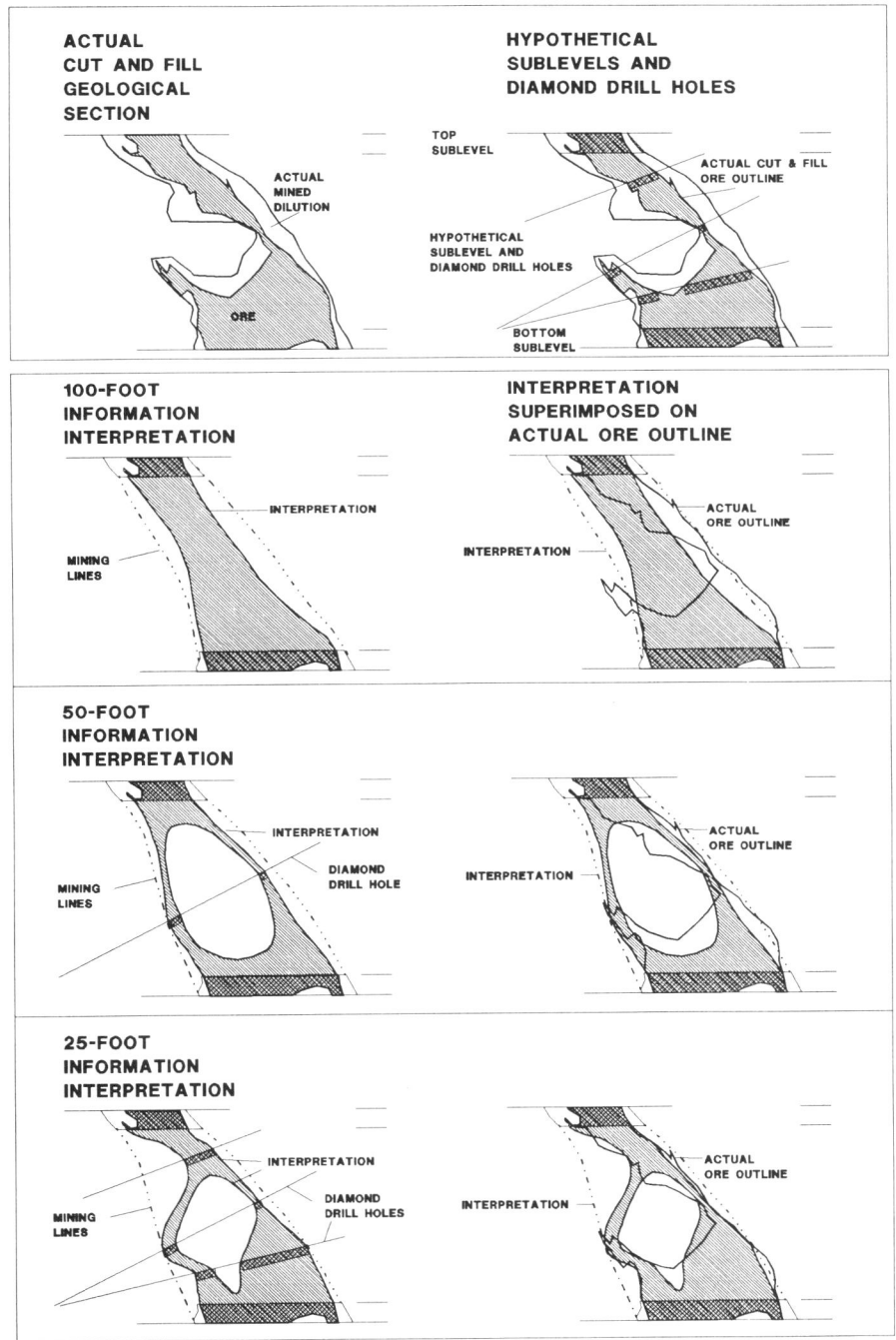

Fig. 4. Ore delineation and information density (after Braun 1991).

estimate unplanned dilution. Figure 4 shows this procedure on a typical cross-section. It shows how ore loss (not drastically affected here) and planned dilution (significantly affected) may vary according to information density. Calculated average ore contact accuracies varied from 2.3 m on either side of the orebody (34 m spacing) down to 0.9 m (8.5 m spacing). The average planned dilution varied from 52% (34 m spacing) to 43.5% (8.5 m spacing); the prior, actual cut-and-fill dilution had averaged 33%. The ore loss ranged from 9.4% (34 m spacing) to 3.0% (8.5 m spacing). Using these results it was seen that intensifying the spacing from 34 to 8.5 m resulted in an average reduction of 8.5% in dilution and 6.4% in ore loss. This translated, for a typical 33 000 t stope block, to a revenue benefit of $98 000 (the value of the ore lost plus the savings from avoiding the costs of mining and processing the waste). This was equivalent to 5 to 10 times the cost of normal 8.5 m spaced drilling). An important point made in conclusion was also that such an analysis, if undertaken early in the planning of an orebody, could assist in the design of the development layout, so as to also meet the requirements of the intended exploration strategy.

Mine design quality

The mine design quality factors relate to the effectiveness in designing the following:

- stoping method, sub-level interval, stope-pillar layout, equipment, backfill;
- stope design (geometry, dimensions, sequence, support-type, density, location);
- drilling-blasting design (blasthole pattern, diameter, sequence, explosive type and distribution).

The selected mining method and equipment will govern the extent of both planned and unplanned dilution. The design of the sub-level interval is a trade-off between the cost and time required for pre-production development and the cost benefit of mining more selectively. Both reducing the sub-level interval and increasing the span of the drill drifts can improve quality in: the knowledge of orebody morphology; the control over drilling and blasting; and the responsiveness of mining lines to ore-body complexity. These quality improvements are more pronounced as the ore-body morphology becomes more complex in vertical section.

Incentives to increase the size of stopes, equipment and blasts relate to the economies of scale which affect improved costs and productivity. For example, in changing from conventional 50 mm diameter blast-hole stoping to 165 mm diameter in-the-hole (ITH) blast-hole stoping, Heath Steel Mines Ltd in New Brunswick reported that: stope development costs decreased 5%; drilling yielded 6.2 t per m (ITH) versus 0.7 t per m (conventional); total drilling and blasting costs per tonne improved 60% (Ladner 1979). If, however, unplanned dilution is increased because of blast damage, drilling inaccuracy or wall instability then the economic impact of the dilution that results must be matched against the potential productivity and cost benefits to determine the optimum design. The ore-body at the Whalesback Mine, for example, was considered to justify high dilution, in order to achieve 100% recovery (Graham 1968). Experience indicates that 1 m of dilution due to blasting overbreak in blast-hole stopes is typical, with more in narrow ore bodies (due to the effects of confinement). Blast damage at the slot tends to be greater due to the higher powder factors required. Any damage initiated at the slot tends to unravel from the hanging wall as the stope is mined out. As an additional precaution, slot raises should be driven away from the hanging wall. There is much interest in reducing blasting overbreak and potential sloughing by using low density explosives. The relative merits of vertical retreat stoping with large diameter ITH drilling versus slot-and-dash stoping with tophammer and tube drilling are currently a design issue. In narrow stoping areas where stress-related hole closure dictates the use of larger diameter blast holes, and thus a high powder factor, column charges of low density explosives, e.g. polystyrene/Anfo blends or low density nitroglycerin based slurries, have been replacing current products.

Sloughing from exposed stope surfaces will contribute to dilution and can limit recovery. The benefits of mining and backfilling at higher and consistent rates, minimizing wall exposure life, are well recognized for dilution and ground control. The size, geometry and sequence of stoping, together with support requirements, are design variables that are the subject of rock mechanics study. Three-dimensional, numerical modelling techniques (based on boundary element, finite element and displacement discontinuity methods) are being increasingly used to evaluate the influence of such stope design variables on the redistribution of rockmass stresses (Grant *et al.* 1993; Quesnel & Chau 1993; Wiles & Nicholls 1993). Such tools are valuable when effectively calibrated against data monitored from the mine by observation and instrumentation. The extent and location of

Table 2. *Parameter ranges for unplanned dilution estimation*

Parameter	Value	Rating	Source
Wall dip (factor C)	Vertical	8	Based on graphs developed by Mathews
	70–90	7	
	55–70	5	
	30–55	3	
	0–30	1	
Rock Quality (Q')	Massive	>50	Based on the Q rockmass rating system, Barton (1976)
	Mod. jointed	25–50	
	Jointed	10–25	
	Foliated	1–10	
	Very foliated	<1	
Orientation Based on graphs (factor B)	Planar contact: structures		developed by Mathews
	parallels wall	0.5	
	dips from wall	0.3	
	dips into wall	1	
	Complex contact: structures		
	parallels wall	0.3	
	dips from wall	0.3	
	dips into wall	0.8	
Sub-level interval (factor E)	20–30 m	1	Underground observation
	15–20 m	1.5	
	10–15 m	2	
	5–10 m	3	
Ore width (factor D)	>20 m	1	Underground observation
	10–20 m	0.9	
	5–10 m	0.8	
	<5 m	0.5	

zones of reduced confinement in the vicinity of stopes are of particular concern as locations for sloughing.

Current analytical methods for estimating the amount of slough are limited. An alternative approach is to use the stability graph method, based upon an empirical relationship linking stope cavity geometry, rock quality and structural stability (Golder Assocs. 1981; Bawden et al. 1989; Potvin et al. 1989) The actual magnitude of sloughing is a function of several factors, principally time, rock-mass geomechanical and structural integrity, ore-body dip and variability, size and shape of the stope, stress distribution in the rock-mass, blast vibration levels and the type and quantity of support installed. Undercutting of geological structure in the hanging-wall presents a particular problem as the stope is mined out, often due to changes in structure orientation, lack of knowledge of where these occur and constrained selectivity due to the sub-level interval. An empirical relationship between the most important of those factors that influence unplanned dilution and the quantity of dilution that occurs is proposed here. Initially, a very simple approach has been adopted, which with time is being modified as experience is gained. The Mathew's method (Golder Assocs. 1981) has been taken as the basis for this approach since it is based on a proposed relationship between the size of an exposed wall, a measure of rock quality and wall performance (i.e. stability). It is suggested, on the basis of experience, that:

unplanned dilution = f_n (stope size, wall dip, sub-level interval, drilling and blasting, rock quality and ore-body complexity)

Mathew's method is based upon:

wall stability = f_n (stope size, wall dip, rock quality, stress and structure orientation)

Relating the common parameters in these two functions gives:

unplanned dilution = f_n (S, Q', A, B, C, D, E,)

where:

S = hydraulic radius (wall area/wall perimeter);
Q' = a measure of rock quality, based on Barton (1976);

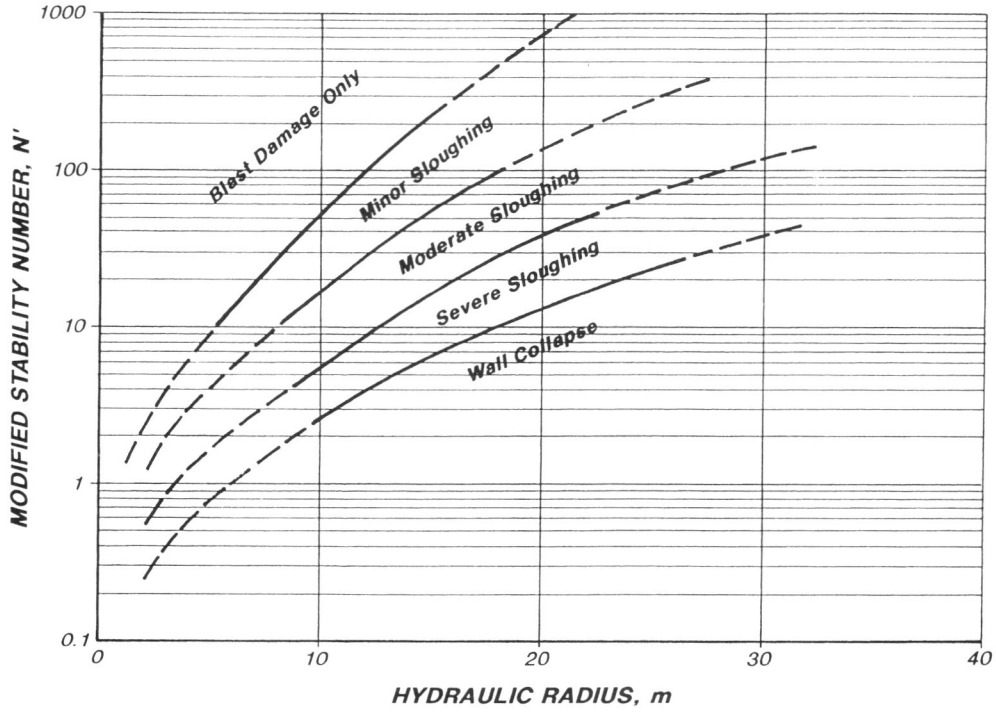

Fig. 5. Empirical estimation of unplanned dilution.

A = a stress factor (= 1 for relaxed hangingwalls);
B = a structural orientation adjustment factor, incorporating the effects of contact complexity;
C = a factor used to rate wall dip;
D = a factor to rate the effects of blasting practice, it is based on ore width to account for the amount of confinement and blasting difficulty;
E = a factor used to rate sub-level interval, within this factor are assumptions with regard to blasthole deviation and ore geometry knowledge.

Table 2 provides the description and ratings currently being used, only to illustrate the direction of this ongoing research. Rock quality should be based upon actual Q' values. Factors D and E are currently the subject of particular study. The above factors are combined to give a modified stability number, N', as follows:

$$N' = Q' \times A \times B \times C \times D \times E$$

N' can then be plotted against the shape factor, i.e. hydraulic radius S, on a dilution graph (Fig. 5) to derive a qualitative indication of the amount of wall dilution.

Some studies indicate that dilution from backfill sloughing can reach 20% in large mechanized cut-and-fill stopes, where backfill is inadvertently loaded during the mucking process. In bulk mining with delayed backfilling, however, dilution occurs as slough from backfill exposures due to blast damage, undercutting, or inadequate mechanical integrity. Experience indicates that, with good quality cemented rockfill, backfill dilution may be of the order of 3–5%. Dilution with cemented hydraulic backfill tends to be greater for an equivalent stope width, typically about 10%. The design of backfill should resolve the following: the physical and mechanical target properties required to maintain stability; and placement techniques to avoid segregation.

Stoping practice quality

The stopping practice quality factors relate to the quality achieved in:

- blast-hole drilling precision (set-up, devia-

tion, surveying);
- explosives loading and performance monitoring;
- support installation and performance;
- backfill preparation, transport and placement;
- production monitoring;
- workforce motivation.

Stoping practice should aim to avoid overbreak. Particular recognition of the significance of drilling accuracy, particularly in promoting overbreak, has recently been evident in Canada (Hendricks et al. 1991). In general, lack of quality in drilling accuracy can result in:

- reduced drilling productivity (arising from the need for more redrilled holes, and reduced performance in deviated holes);
- higher drilling and blasting costs;
- poorer fragmentation, lower mucking productivity from excessive fines or oversize;
- reduced ore recovery where deviation results in inadequate explosive distribution inside planned stope limits;
- dilution arising from direct overbreak where blast holes deviate into waste walls and the increased blast damage which may promote further, delayed wall failure.

Three sources of drilling inaccuracy exist: set-up collar location; set-up alignment; and drilling trajectory deviation. The amount of deviation is a function of hole size, drilling equipment, hole length, operating procedures, maintenance quality, and geological structure. Experience at a number of mines has shown deviations of 3–6% during drilling and up to 3% from errors in drill setup. Holes of concern are often those tending to cross-cut the dominant structure, i.e. at 40–50° dip. Deviation results in their leaving the stope and damaging the wallrock. These also do not break through to any sill excavation to permit conventional survey monitoring of accuracy. It is not uncommon on some mines to see as much as 25% of blast holes requiring redrilling. The lack of prior efforts towards quality improvement in drilling accuracy relates both to a lack of awareness as well as to the inadequacies of available survey instrumentation.

Platford et al. (1991) reported trials which clearly demonstrated the benefits to dilution control of maintaining high quality stoping practice. Three blocks were mined under controlled conditions at Inco's Thompson mine, where typically the ore zone is 3–6 m wide, dipping at an average of 65°. The blocks, 20 m high on dip and 17 m long on strike, were extracted by longitudinal retreat. The geology and ground condition were similar in all the blocks. Drilling and mucking equipment were also constant. Table 3 shows the details of the stoping practice and performance for each block. Block 35 was mined by conventional practice, as a crater-based block. The remaining two blocks (34 and 33) were mined under different designs but again with close control to ensure accurate drilling, controlled loading and blasting. The swell was only mucked from these two blocks so as to provide additional wall support and reduce overbreak. Data were acquired on the quality of the fragmentation, including the amount of oversize and secondary blasting. Monitoring was also undertaken of blast vibration and the overall stope stability. The attention to quality in both design and practice was evident in the substantially reduced dilution, improved fragmentation, and reduced energy factor and drilling costs.

Another factor considered to affect sloughing is exposure time. Efficient practice is where possible to muck consistently and quickly once a block is fully blasted. If backfill is to be placed then this should be undertaken without delay. The stoping schedule should be based upon the backfilling capability in order to avoid excessive exposure times for cavities. During the blasting phase it is most effective to muck only the swell volume of the fragmentation (see Table 3). In addition, further dilution (and the potential for reduced recovery) may occur if the period over which a stope is to be mined is extended. At the Thompson Mine of Inco Ltd a data base of geotechnical and mine operational data was established to record in detail the mining of over 100 vertical crater retreat blocks (CANMET 1990). Analysis of the data base indicated that the longer a stope wall is exposed, then the more likely it is to fail and generate overpull.

Production management

Production management is one of several interactive systems within mine management (Fig. 6). In the not too distant future, mines will be serviced by effective underground communications systems which will be the backbone for mine-wide voice, video and data transmission. Such systems will support the integration of widespread data sources for computer-based monitoring and control, with the potential to emulate computer-integrated manufacturing (CIM). The concept of CIM is the automated flow of information among engineering, produc-

Table 3. *Mining trials, blocks 33, 34, 35 (after Platford et al. 1991)*

Block number	35	34	33
Design reserve (tonnes)	4160	10955	7548
Drilling			
Spacing (m)	1.5	1.5	1.5
Burden (m)	2	2	1.8–3*
Number of holes	36	72	52
Blasting			
Crater blasts per lift	6	8	7
Vertical m per blast	2.4	2	2.8
kg explosive	12.3	12.3	12.3
Total kg			
Aquamex	1370	1262	266
Low density aquamex	0	0	184
Amex	2455	3492	1735
Lomex	0	0	1880
Mucking			
Type	†	‡	‡
Days to complete	12	20	17
Stand up time (days)	19	44	56
Tonnes mucked			
ore	3060	7997	5954
rock	1905	168	1551
backfill §	1105	0	0
Dilution, %‖			
Over ore zone : planned	45.7	2.6	24
Over design: unplanned	19.3	0	6.2
Oversize (chunks per 1000t)	21	1	11
Energy factor (kg explosive per tonne)	0.75	0.52	0.32
Drilling cost ($ per tonne blasted)	1.75	1.73	1.69

* Distance at the toe; † mucked every blast; ‡ mucked swell only; § rock was stored as backfill; ‖ calculated over designed tonnes.

tion and various support groups in a factory. Goldhar (1988) defined it as a combination of computer-aided design and computer-aided manufacturing and flexible manufacturing systems including robotics. The use of computers and communications links aims to create the efficiency of high degrees of integration without the rigidities associated with mechanical integration.

Mining production monitoring and control aims to meet the following objectives: to ensure that mine (and mill) targets are met according to measures of quality and quantity, as defined in the mine plan; to update the mine model (geological excavation) and to permit re-evaluation of the mine plan, according to actual production data; and to interact with the other component mine management systems.

Production management takes responsibility for underground exploration, development and stoping. It has traditionally been intuitive and non-scientific, with little priority placed on quality. This has arisen mainly because of the inability to generate reliably accurate and timely information. Communications and sensor technology development now offer the capability for the first time to manage the mining production process scientifically. Management is dependent upon the following quality control information: grade and tonnage of ore (*in situ*, fragmented and in transit), productivity, dilution, recovery, equipment availability–utilization, costs (labour, materials, maintenance, power etc.), safety, development and stoping excavation advances, environmental conditions and rockmass behaviour. Management exerts control by reference to mine plan targets relating to quality (in product, safety and environmental protection), productivity and economics.

Dilution control is accomplished through a

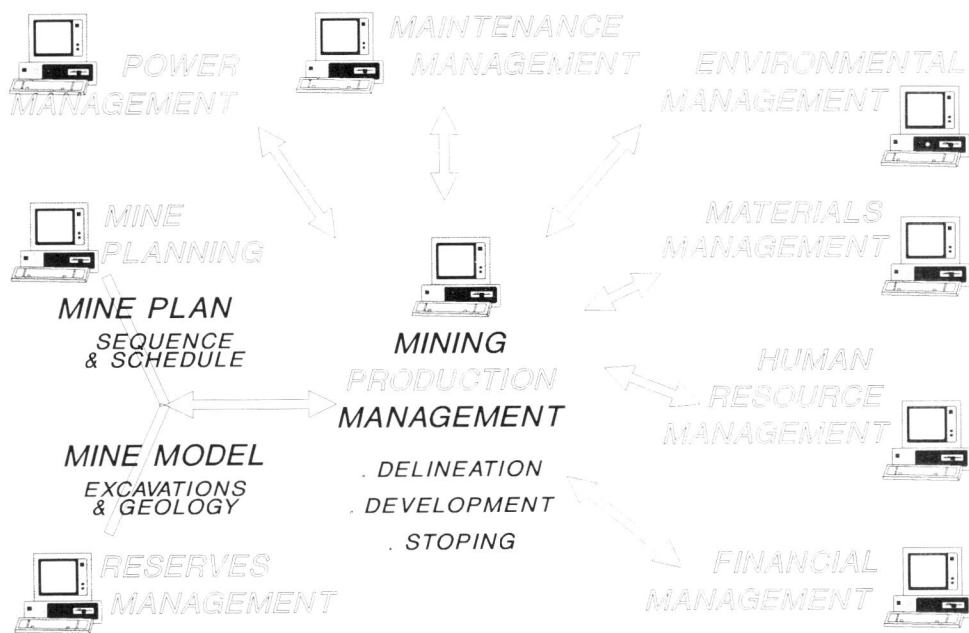

Fig. 6. Mining production management systems.

process of reconciliation of information from the five sources below. This permits an evaluation of mining performance which can be referred to the mine plan objectives. The reconciliation allows a scientific analysis to explain the events of the production period under review, before determining the control response. The reconciliation process matches data for the period from:

- *the mine model*—the predicted tonnes-grade for the planned excavation volume;
- *the drilling and blasting activity*—the location and volume of rock actually drilled and loaded, plus indications of its grade distribution;
- *the stope cavity*—geometry (the actual location and volume of rock excavated); integrity (geomechanical quality and stability); evidence of remnant ore in the wallrock;
- *the muckpile*—geometry and grade (the volume and grade distribution of the fragmented rock in the stope cavity and drawpoints);
- *the loading equipment*—the tonnage and grade removed from the stope.

The objective should be to collect reliable and precise data during stope exploration, development and production, in order to exert control on each stope. The rationale for control should be to optimize the quality of exploration, design, drilling and blasting. Once, however, the rock fragmentation process is underway then there is little flexibility in the mining method itself to accommodate control measures. Developing tactical flexibility is an important area for future stoping method research. Once into production, then control relates mainly to blending and interaction with other stopes and development in governing the feed to the mill. Some flexibility is provided by the capability to install support, or modify any remaining drilling and blasting, but the key is to ensure that the original design and practice are carried through to the highest of quality. Dilution control is also critically dependent upon the attitude not only of management but of the complete workforce. The concept that the quality of tonnes of ore hoisted outweighs the importance of the number of tonnes hoisted is paramount.

Sensor development

Real time monitoring of stope production parameters should be an ambition so as to enable interaction with mine planning, probably based on a GIS and 3-dimensional CAD system, as well as with production control. A comprehensive underground communications system was installed at the Copper Cliff North Mine of Inco Ltd, capable of voice, video and data transmission from stationary or mobile sources from all parts of the mine (Baiden & Scoble 1992). Real time production control, however, awaits further sensor development to enable capture of the critical input data and to exploit the full potential of such communications.

Drilling

Surveying sensors are required to monitor the accuracy of drill set-up coordinates, collar alignment, and trajectory deviation. Some initial efforts have been made to integrate trajectory sensors into the drill string (Pathak & Dias 1987; Jenkins & Ball 1991). Drill set-up and collaring in practice are controlled primitively. Borehole tools exist to measure trajectory after drilling, as a separate exercise, but these are still limited in accuracy, reliability, cost or productivity (Hendricks et al. 1991).

Monitoring while drilling

Monitoring while drilling has received some success in surface mines for characterizing mineralization, rockmass quality and structure (Pollitt et al. 1991). It is based on the interpretation of the variation with depth of various drilling performance parameters, e.g. thrust, rotary speed, torque, penetration rate, vibration.

Explosives monitoring

This should confirm that the explosives were loaded and distributed as per plan, and that their initiation and performance was of acceptable quality.

Cavity surveying

The geometry of the blasted cavity should be surveyed after blasts and subsequently over time whilst mucking still proceeds. This enables the quality of the design and practice to be evaluated. The control of structure and blast damage may be evident in the cavity morphology. The Noranda Technology Centre have developed a laser-based cavity survey system (Miller et al. 1992). Such sensing needs to be feasible by access through development excavations or boreholes. The integration of vision, using a video camera with the survey tool, would also enhance the data interpretation by allowing the discrimination of any mineralization or symptoms of instability that were evident in the cavity surface.

Fragmentation sensing

Assessment of the size and shape distribution of the fragmented rock in the stope cavity and drawpoints serves to evaluate the quality of the blasting operation and its influence on downstream productivity and economics. Image analysis techniques have been developed to evaluate fragmentation (Hunter et al. 1990). A potential, as yet unrealized, is to link the fragmentation block shape, size distribution and petrology to its source location, possibly from blast overbreak or sloughing. The identification, measurement and removal of sloughed backfill within the fragmentation is difficult and justifies further attention.

It is important to develop effective grade sensing for fragmentation, based on XRF or other analytical techniques. Current practice of grab sampling and/or estimation of grade from visual estimation of sulphides content in muckpiles or vehicles is inadequate.

Loading performance

It is important to emulate the advances made in equipment performance monitoring technology in surface mines (Scoble et al. 1991). Accurate load sensing is needed for load-haul-dump (LHD) and truck haulage at the stope. This should be combined with productivity and health monitoring also to service production, financial and maintenance control. Grade sensing of LHD or truck loads would also contribute to dilution control.

The stope environment has dictated that reliable and precise measurement of the following is rarely possible: mineral width, grade, tonnage, design grade and tonnes, actual tonnes and grade. This relates also to any other variables on which they are based, such as % overpull. In particular, any variable that is based on grade is likely to be imprecise. These include mineral tonnes, design tonnes and actual tonnes, which are calculated from volumetric measurements and grade measurements (ore estimates) through the use of tonnage factor formulae. Actual tonnage is likely to be less precise than the others because the volumetric measurements in this case are traditionally based on bucket, truck or skip counts. According to Pentilla (1989), geostatistics do not offer the means to predict dilution, but account for statistical inaccuracies in ore reserve estimation. Grade reduction occurs because of dilution and ore losses. Ore estimation traditionally overestimates grade and underestimates tonnage. A unique historical reduction factor per mine can work if accurate ore reserve estimation and high exploration quality is achievable. It should be an aim, however, to develop technology for production control which renders such needs redundant.

Conclusions

Dilution has a significant effect on mine economics and is now widely recognized as an opportunity for quality improvement. Planned dilution is a consequence of the complexity of the orebody and lack of sensitivity of the stoping method. Unplanned dilution arises from a combination of exploration, design and stoping practice quality factors. The underground exploration plan and investment should be evaluated in terms of the ore-body complexity, mining method and cost-benefit analysis. New exploration technology related to borehole geophysics and tomography offers the potential to improve the quality of ore-body delineation. Design methods by empirical stability assessment and numerical stress modelling techniques are advancing, but calibration against observed excavation behaviour is important. Cablebolt technology should contribute further to stope stability control. It has been clearly demonstrated that tight control over stoping practice can dramatically improve dilution control standards.

Sensor development will be key to understanding dilution processes as well as their control. This will enable improved monitoring of drilling/blasting, fragmentation, cavity morphology and integrity, rockmass stress and deformation, and materials handling. The control rationale involves reconciliation of exploration, design and stoping data relating to tonnage and grade. The data will also be available for integration into a real-time mine production management system, based upon a mine-wide underground communications system. This is the potential backbone for a comprehensive computer integrated mining system, within which production management is formalized.

The standard of control will also be governed by the flexibility designed into the stoping system itself, i.e. its ability to adapt to unanticipated events, relating for example to ground instability or grade variation. There is a need for more concerted efforts to develop improved stoping methods. Dilution control is also dependent upon attitudes and motivation, in order to apply effectively tactics and technology.

The authors wish to acknowledge several geologists and engineers whose assistance has helped to shape this paper. Particular acknowledgement is given to personnel of Inco Ltd, Ontario and Manitoba Divisions, and the Hudson Bay Mining and Smelting Co. Ltd, Manitoba. The views expressed are entirely those of the authors.

References

ASHCROFT, J. W. 1991. A total quality improvement opportunity. *93rd Annual General Meeting*, Canadian Institute of Mining and Metallurgy, Vancouver.

BAIDEN, G. & SCOBLE, M. 1992. Mine-wide information system development. *Bulletin of the Canadian Institute of Mining and Metallurgy*, **85**, 65–70.

BARTON, N. 1976. Recent experiences with the Q-system of tunnel support design. *Proceedings of the Symposium on Rock Exploration*, Johannesburg.

BAWDEN, J. W., NANTEL, J. & SPROTT, D. 1989. Practical Rock engineering in Optimization of Stope Dimensions—Application and Cost Effectiveness. *Bulletin of the Canadian Institute of Mining and Metallurgy*, **82**, 63–70.

BRAUN, D. 1991. Ore interpretation and its relationship to dilution. *93rd Annual General Meeting*. Canadian Institute of Mining and Metallurgy, Vancouver.

CANMET, 1990 *VBM design guidelines, summary*. Canada–Manitoba Mineral Development Agreement. Final report, DSS file no. 14SQ. 23440-4-9147-3. CANMET, Ottawa, 62.

ELBROND, J. 1986. Ore losses, rock dilution and recovery. *In: Proceedings of the Symposium on Estimation Design and Operation; Ore Reserve Estimation Method, Models and Reality*. Canadian Institute of Mining and Metallurgy, 130–134.

GOLDER ASSOCIATES 1981. *Prediction of stable excavation spans of mining depths below 1000 metres in hard rock*. Report to CANMET, DSS Serial No. 05480-0081.

GOLDHAR, J. 1988. In the factory of the future—innovation in progress. *In: Proceedings of the 5th Conference on Computer Integrated Manufacturing, Toronto*, 22–29.

GRAHAM, E. P. 1968. Whalesback, an example of low-cost, high-dilution mining. *Bulletin of the Canadian Institute of Mining and Metallurgy*, **61**, 847–853.

GRANT, D. R., POTVIN, Y. & ROCQUE, P. 1993. Three-dimensional stress analysis techniques applied to mine design at the Golden Giant Mine. *In: Proceedings of the 1st Canadian Symposium Numerical Modelling Applications in Mining Geomechanics, Montreal.* McGill University.

HBM & S LTD 1990. *Dilution workshop.* Inco–HBM & S Homestake–CCARM, Thompson, Manitoba.

HENDRICKS, C., SCOBLE, M., BOUDREAULT, F. & SZYMANSKI, J. 1991. Blasthole stoping: drilling accuracy and measurement. *In: International Conference on Research in the Mining Industry, Univ. Nottingham,* Institute of Mining and Metallurgy, London, 15.

HUNTER, G. C., MCDERMOTT, C., SINGH, A. & SCOBLE, M. J. 1990. Image Analysis Techniques for Measuring Blast Fragmentation. *International Journal of Mining Science and Technology*, **11**, 19–36.

JENKINS, P. P. & BALL, T. G. 1991. Horizontal drilling Techniques in British Coal Mines. *Transactions of the Institute of Mining and Metallurgy*, **100**, A11–21.

KILLEEN, P. 1991. Borehole Geophysics: taking geophysics into the third dimension. *GEOS*, Energy, Mines and Resources, Canada, **20**, 2.

LADNER, E. 1979. In-the-hole Drilling at Heath Steele Mines Ltd. *Bulletin of the Canadian Institute of Mining and Metallurgy*, 59–65.

LORING, J., HAYWARD, G. & MACLEAN 1992. The use of TOI tools in the implementation of new technology. *In: 94th Annual General Meeting, Canadian Institute of Mining and Metallurgy, Montreal.*

LULEÅ UNIVERSITY 1991. *Teaching and research activities.* Department of Mining and Underground Construction, Luleå University, 20–21.

MILLER, F., JACOB, D. & POTVIN, Y. 1992. Cavity Monitoring System: Update and Applications. *In: 94th Annual General Meeting, Canadian Institute of Mining and Metallurgy, Montreal.*

O'HARA, T. A. 1980. Quick Guide to the Evaluation of Orebodies. *Bulletin of the Canadian Institute of Mining and Metallurgy*, **73**, 87–99.

PATHAK, J. & DIAS, M. F. 1987. Microprocessor-Controlled Down-The-Hole Drill for Enhancing Productivity and Accuracy in Underground Bulk Mining Methods. *IEEE Transactions on Industry Applications*, **IA-23**, 6.

PENTILLA, V-J., 1989. *Waste Rock Dilution and its Economic Importance.* Meeting of Mine Geologists of NMD 28.–29.9.1989. Unpubl. Report, Outokumpu Ltd.

PLATFORD, E. R., MUNDY, D. & BONYAI, C. 1991. Dilution control at Inco's Thompson mine, Manitoba Division. *In: Canadian Institute of Mining and Metallurgy, Underground Operator's Conference., Val d'Or.*

POLLITT, D., PECK, J. & SCOBLE, M. 1991. A Pattern Recognition Technique for Lithology Characterization by Drill Performance Monitoring. *Bulletin of the Canadian Institute of Mining and Metallurgy*, **84**, 951, 25–30.

POTVIN, Y., HUDYMA, M. & MILLER, H. D. S. 1989. Rib Pillar Design in Open Stoping. *Bulletin of the Canadian Institute of Mining and Metallurgy*, **82**, 31–36.

PUHAKKA, R. 1990. *Geological Waste Rock Dilution*, Finnish Association of Mining and Metallurgical Engineers, Research Report No. A94.

QUESNEL, W. & CHAU, P. 1993. Mine design validation utilizing a 3-D finite element model at la Mine Doyon. *In: Proceedings of the 1st Canadian Symposium on Numerical Modelling Applications in Mining Geomechanics, Montreal.* McGill University, 29–39.

SCOBLE, M., PECK, J. & HENDRICKS, C. 1991. A Study of Surface Mine Equipment Monitoring. *International Journal of Surface Mining*, **5**, 111–116.

SCHUNNESSON, H. 1990. *Drill process monitoring in percussive drilling.* Licenciate Thesis, Luleå University of Technology, Sweden.

WILES, T. & NICHOLLS, D. 1993. Modelling discontinuous rockmasses in three dimensions using MAP3D. *In: Proceedings of the 1st Canadian Symposium on Numerical Modelling Applications in Mining Geomechanics, Montreal.*

WRIGHT, E. A. 1983. Dilution and mining recovery—review of the fundamentals. *Erzmetall*, **36**, 23–29.

Quantifying differences between computer models of orebody shapes

E. J. SIDES

International Institute for Aerospace Survey and Earth Sciences, Kanaalweg 3, 2628 EB Delft, The Netherlands

Abstract: Several different methods have been used for the representation of orebody shapes in computer systems including serial-slice, gridded seam, regular block, mathematical functions, solid geometry, and boundary representation models. The accuracy and precison of such models are affected by factors such as errors associated with the original data used, natural variations, errors in data capture and processing, etc.

In order to asses the relative importance of such factors it is necessary to quantify differences in accuracy and precision between different model structures. To meet this objective a new technique which relies on statistical study of a series of sample lines, intersecting two or more different orebody models, is proposed.

Application of this technique on orebody models at the Neves–Corvo copper–tin deposits in southern Portugal suggested that in certain circumstances regular block models can give very poor estimates of the total volume of materials of different geological types. In addition it was shown quantitatively that shape interpretations close to the plane of drill-sections are better, in terms of accuracy and precision, than interpretations for zones falling between the drill-section planes.

Over the past three decades, computerized modelling techniques have gained widespread acceptance in the field of orebody modelling for reserve estimation purposes. Nowadays most large mines use sophisticated software packages for creating, storing, editing and analysing the orebody models used as the basis for mine design and production scheduling. Initially such packages were often developed in-house for individual mines, or companies; however, nowadays several integrated geological modelling and mine planning packages are sold commercially (see Gibbs 1991).

Such packages offer a wide range of techniques for modelling orebody shapes and grade variations, as well as a great variety of input and output options. Selection of the most appropriate package, or method, to use on a particular deposit is consequently a difficult task. In addition to the technical capabilities offered, the main criteria used when selecting a package of this type are often the degree of user friendliness, the types of graphic output produced, and interfaces to existing databases or other applications programs.

Most such packages contain options for geostatistical estimation of orebody grades, whereby the uncertainties associated with interpolation and extrapolation of spatially correlated values, can be quantified. Users of such packages often forget that both the input data and the types of structures used for creating, storing and manipulating the computer models may also contain inherent limitations to their accuracy and precision. This aspect is often neglected because of the lack of agreed procedures for estimating the errors associated with predicting the location of orebody contacts, and volumes of different geological or mining units, within a deposit.

This paper suggests a simple approach for quantifying such errors, and illustrates its application with examples from the Neves–Corvo deposits in southern Portugal. The work presented is described in more detail by Sides (1992b).

Before describing the studies that were carried out, a review of the principal types of structures used for computer modelling of orebody shapes is presented. This is followed by a discussion of possible types and sources of error inherent in such models.

Computer modelling of orebody shapes

Sides (1992b) identified the following two fundamentally different types of approach which can be used for creating and handling three-dimensional representations of orebodies.

(1) *Geometric modelling:* for dealing with properties which are best represented by sharp boundaries. In such cases modelling concentrates on the identification and interpretation of the boundaries. These are then used to define the volumes occupied by material with different

Table 1. *Summary of the main types of model structure used for geometric modelling (modified after Sides 1992b)*

Model type	Description
Serial-slice	Polygonal shapes are interpreted on, and digitized from, a series of regularly spaced parallel sections or plans. Each polygon is normally assigned a volume of influence halfway to the next section.
Gridded-seam	Multiple sub-parallel surfaces are represented by grids of elevations and/or thicknesses.
Block	The volume being modelled is divided into a series of regular rectangular blocks in all three dimensions.
Mathematical	Discontinuities are represented as global, or piecewise, mathematical functions such as polynomial trend surfaces, cubic B-splines, etc.
Solid	Complex objects are represented by groups of geometric primitives (e.g. planes, cubes, spheres, etc.) thus defining volumes with common properties.
Boundary representations	Volume boundaries are represented by a collection of points (vertices), and planar surfaces (facets).

geological characteristics. The divisions represented are essentially thematic, and are usually identified by name or code (nominal or ordinal measurement scales) rather than by numeric values on a continuous measurement scale (interval or ratio measurement scales).

(2) *Volumetric modelling:* for dealing with attributes which vary more or less continuously within the volumes occupied by individual geological units (as defined by geometric modelling). Such properties can normally be quantified from sample measurements made on a continuous numeric scale. These are usually analysed and estimated using geostatistical or other geomathematical techniques.

Volumetric modelling methods usually require definition of the volumes of the main geological units present as a preliminary step. Such methods have received a lot of attention over recent years, particularly in association with the application of geostatistical estimation techniques (e.g. David 1977; Isaaks & Srivastava 1989). In such cases there are clearly established methods for deriving estimation variances (associated with ore reserve block grade estimates), which take into account the effect of aspects such as block size, position and orientation, as well as the spatial distribution of the sample values used.

This paper concentrates on the use of geometric modelling methods to represent orebody shapes, and discusses the identification and quantification of errors associated with such models. Such methods are important not only in the representation of the volumes of different geological units, but also in the representation of designed and actual mine openings. The main types of computer model structure used for geometric modelling are summarized in Table 1 and Fig. 1.

These techniques are discussed in more detail in Sides (1992b).

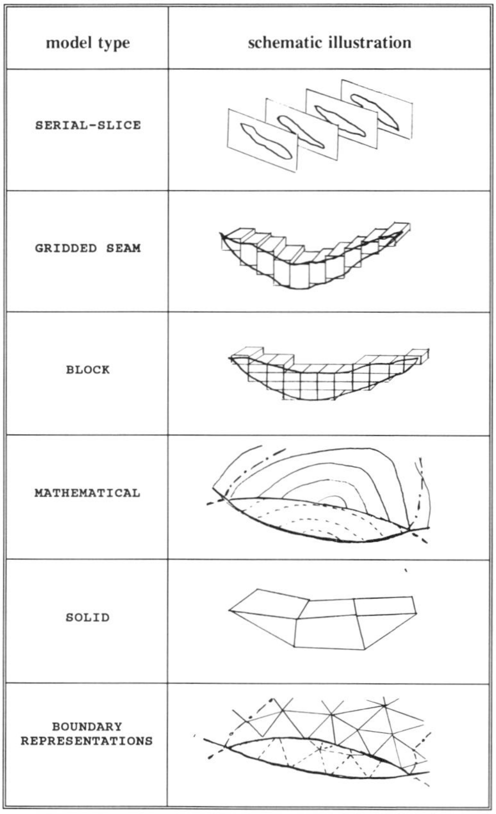

Fig. 1. Types of model structure used for computer representation of orebody shapes.

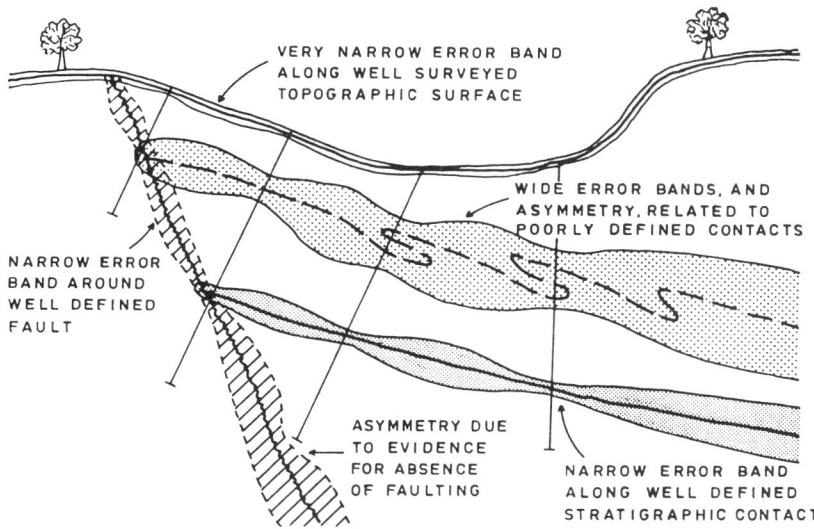

Fig. 2. Schematic cross section showing locational error bands associated with different discontinuities in a hypothetical deposit (from Sides 1992b).

Error types and sources

In general terms an error can be defined as a difference between a measurement, or prediction, and the most reliable estimate of the (unknown) true value (Merks 1985). In this context it is important to distinguish between inaccuracies (caused by incorrect assumptions, or the use of inappropriate methods), and the statistical uncertainties (related to the errors inherent in sampling, analysis and modelling). The identification and control of errors associated with ore reserve estimation are discussed in general terms in Sides (1992a).

By analogy with similar work done in relation to geographical information systems (Burrough 1986), five main sources of error associated with the three-dimensional modelling of orebody shapes can be recognized (Sides 1992b).

(a) *Inaccuracies associated with original data:* e.g. gross errors such as incorrect drillhole survey or coordinate measurements; invalid geological interpretations; etc.

(b) *Sampling and analytical errors:* e.g. uncertainties associated with the identification of boundaries of different geological units; limitations to the precision of survey instruments; etc.

(c) *Errors due to natural variation:* e.g. the roughness (irregularity) of boundaries. Such variations can often be analysed and predicted using geostatistical techniques (e.g. David 1977), as well as using fractal geometry (Goodchild 1980).

(d) *Errors in data capture:* e.g. mistakes made during input of information to databases by digitizing or keyboard entry.

(e) *Computer processing errors:* e.g. uncertainties related to software and/or hardware limitations; inherent limitations of model structures; etc.

Errors, such as those outlined above, result in inaccuracies and uncertainties associated with predictions based on the resultant computer models. From the point of view of modelling shapes (in order to estimate volumes of different materials present), three important aspects can be identified, namely:

- boundary recognition;
- prediction of boundary location;
- estimates of total volume (or area).

The schematic cross section presented in Fig. 2 shows an error band associated with each of the interpreted boundaries. This illustrates the impact of such errors in terms of locational uncertainty. Clearly identifiable and measurable features such as the surface topography normally have a very narrow error band. More irregular and/or less clearly identifiable features have much wider error bands. The width of such error bands can be expected to increase with distance from sample points, as illustrated in Fig. 2.

Quantifying locational and volumetric errors

In order to identify and control such errors, it is essential to establish empirical methods of quantifying differences between different shapes. Although there are many published reconciliation studies which present comparisons of predicted and actual ore tonnage, relatively little work has been done on comparison of predicted and actual locations of ore mineralization.

Houlding (1991) suggested two ways in which the differences (errors) between a polygon represented using two different modelling techniques (a set of linear vectors, and a regular block scheme) can be quantified, namely:

(a) Area error (difference in total areas obtained using the two methods, as a percentage of the true polygon area);
(b) Boundary error (half the total misclassified area as a percentage of the true polygon area).

Neither of these measures allows the quantification of the locational accuracy and precision with which the boundaries of the polygon are represented.

An approach which can provide such measures is illustrated in Fig. 3. Here a set of pairs of points is generated which can then be analysed statistically. The selection of test line spacing and orientation is obviously a critical part of this process. Locational errors might be expected to show different sets of lines. Consequently, the orientation and spacing of the grid lines selected for study should be similar to the size of planned mine openings and the direction of mining attack. The application of this approach is illustrated by the studies described below.

Effects of using different model structures

A series of tests was carried out to compare the results obtained using two different types of model structure to represent a stratabound massive sulphide deposit. The deposit studied was the Lombador deposit, at the Neves–Corvo mine in southern Portugal. A review of the geology of these deposits, and a description of the reserve estimation methods used at the mine, is given by Richards & Sides (1991).

The Lombador deposit is a massive sulphide lens, up to 50 m thick, which was discovered in 1988 by surface drilling. Following the methodology used on the other deposits at Neves–Corvo, initial reserve estimates were based on a gridded seam model. These gridded models were derived from triangulated boundary representations of the footwall and hanging-wall contacts of each ore lens (based on drillhole intersections and geological interpretations digitized from transverse cross sections and isobath plans). This approach was adopted since it was considered that regular block models would introduce significant uncertainty in the estimates of ore volumes. The studies described below were carried out to try to verify this hypothesis in a quantifiable manner.

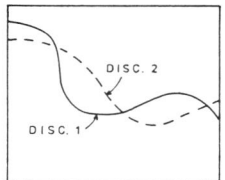

A. SELECT THE DISCONTINUITIES TO BE COMPARED

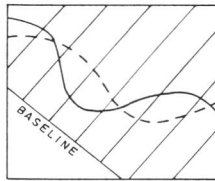

B. SELECT TEST LINE ORIENTATION AND SPACING, AND ESTABLISH A SET OF TEST LINES WHICH INTERSECT BOTH DISCONTINUITIES

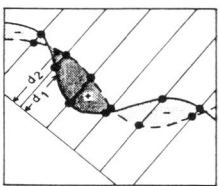

C. CALCULATE CO-ORDINATES OF THE PAIRS OF INTERSECTION POINTS WHICH THE TEST LINES MAKE WITH THE DISCONTINUITIES, AND OUTPUT TO FILE

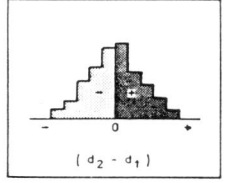

D. PERFORM STATISTICAL ANALYSIS OF DIFFERENCES BETWEEN THE DISCONTINUITIES, AS MEASURED ALONG THE TEST LINES

Fig. 3. Method used for analysing the locational differences between two different discontinuities (from Sides 1992b).

The gridded seam model used for reserve estimation was adopted as a reference against which regular block models were compared.

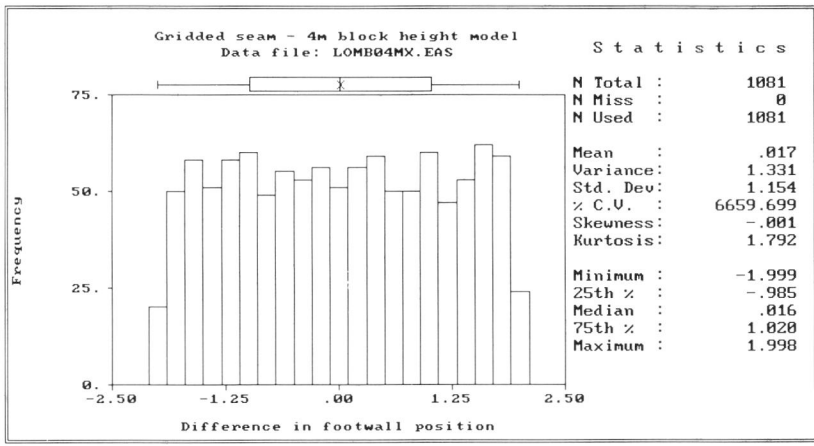
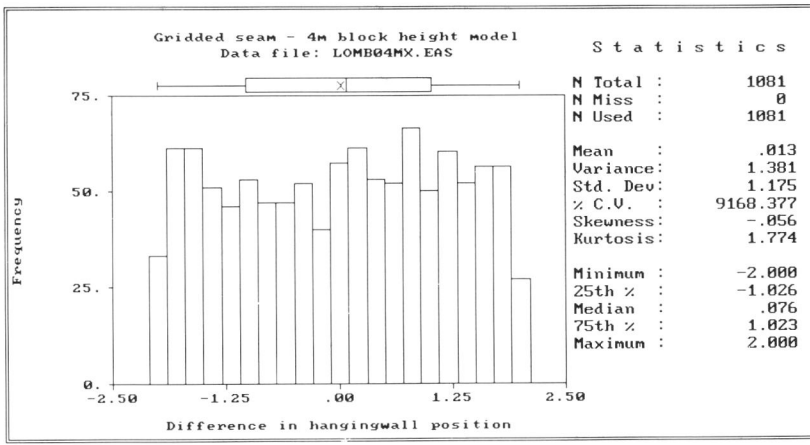
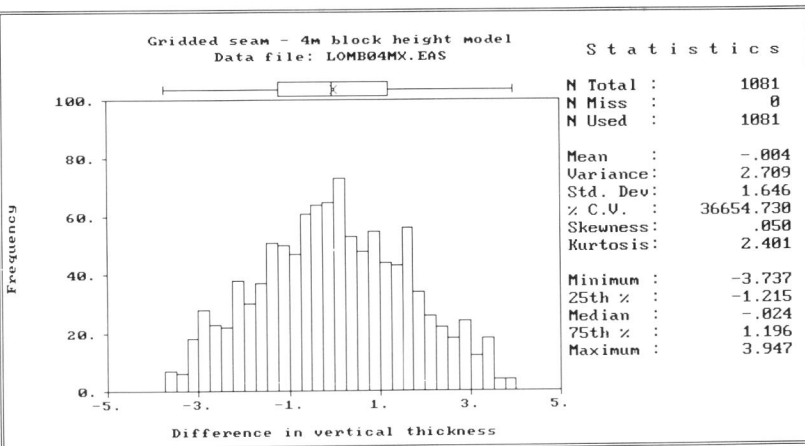

Fig. 4. Comparison of gridded seam and 4 m block height models for the Lombador deposit. Differences in footwall (upper) and hanging-wall (middle) positions, and vertical thickness (lower) are shown (compiled from Sides 1992b).

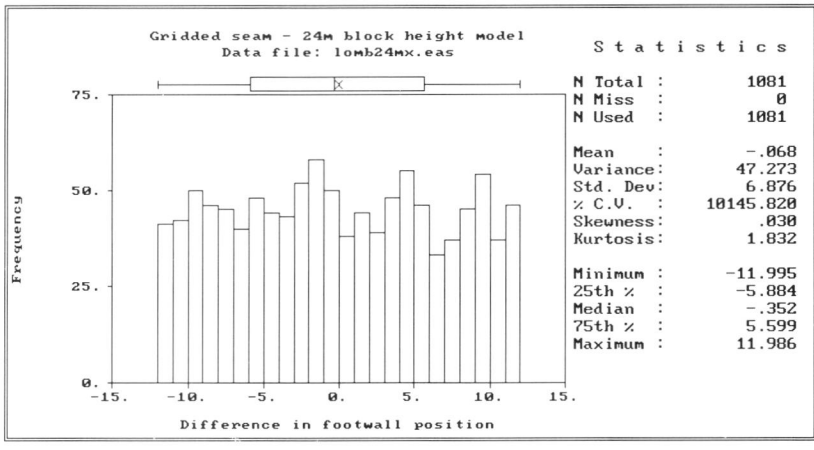

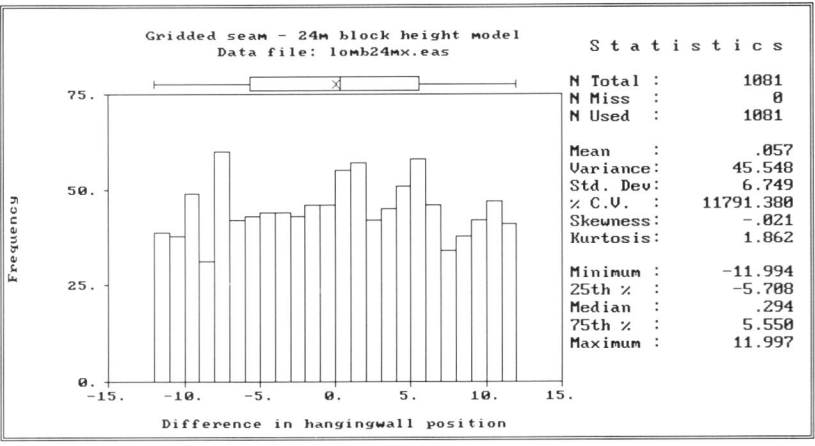

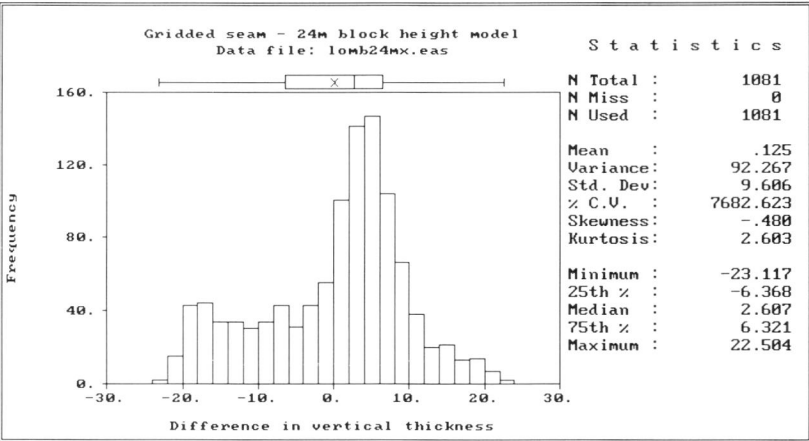

Fig. 5. Comparison of gridded seam and 24 m block height models for the Lombador deposit. Differences in footwall (upper) and hanging-wall (middle) positions, and vertical thickness (lower) are shown (compiled from Sides 1992b).

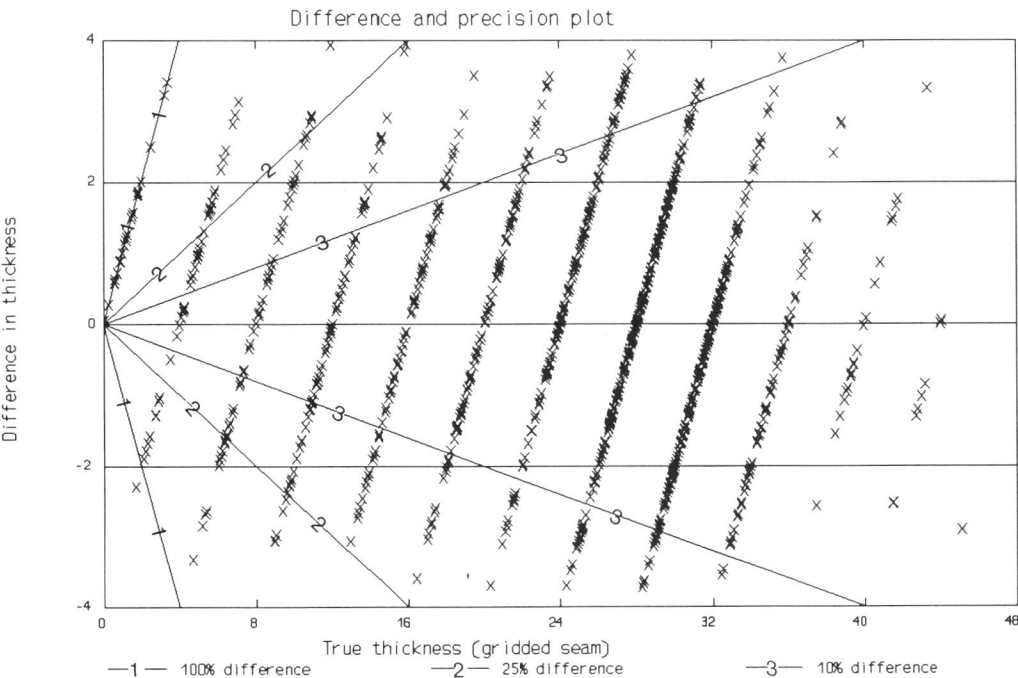

Fig. 6. Plot of difference (block model gridded seam) against true thickness (gridded seam) for 4 m block height model of the Lombador deposit.

Block models with different height blocks, but the same lateral dimensions as the grid used for the gridded seam model, were generated using the information stored in the gridded seam model. This allowed comparable values, for the gridded seam model and a corresponding regular block model, to be extracted at each grid point. Output files containing pairs of estimates for the elevation of footwall and hanging-wall of the main ore lens, and also its vertical thickness, were generated for different block height models.

The results obtained for 4 m and 24 m block height models are summarized graphically in Figs 4 and 5, respectively.

The differences between the footwall and hanging-wall elevations show similar patterns in both cases with distributions which can be characterized as rectangular in form. This reflects the fact that the difference between predicted thicknesses shows a very different pattern, with that for the 4 m block height model having a near-Gaussian distribution, and that for the 24 m block height model a more complicated bimodal distribution.

These patterns can be explained by reference to the distribution of values obtained when the throws of two dice are combined. In such tests the values for each die will show a rectangular distribution, since each value from 1 to 6 has an equal chance of being selected. However, when the values for two dice are combined a symmmetrical distribution with a modal value of 7 (1 in 6 probability), and tails at 2 and 12 (each with a 1 in 36 probability) is obtained.

In terms of the graphs presented in Fig. 4, this means that the actual elevation values are independent of the block height and start elevations. However, when the two elevation values are combined to give an estimate of the thickness a near-Gaussian distribution is obtained. The breakdown of this pattern in Fig. 5 suggests that the values for top and base elevation are no longer independent, reflecting the fact that the block height is now comparable to the thickness of the unit being modelled.

The variance of the thickness distributions shown in Figs 4 and 5 give an indication of the precision of the estimates of orebody thickness. These are obviously closely related to the

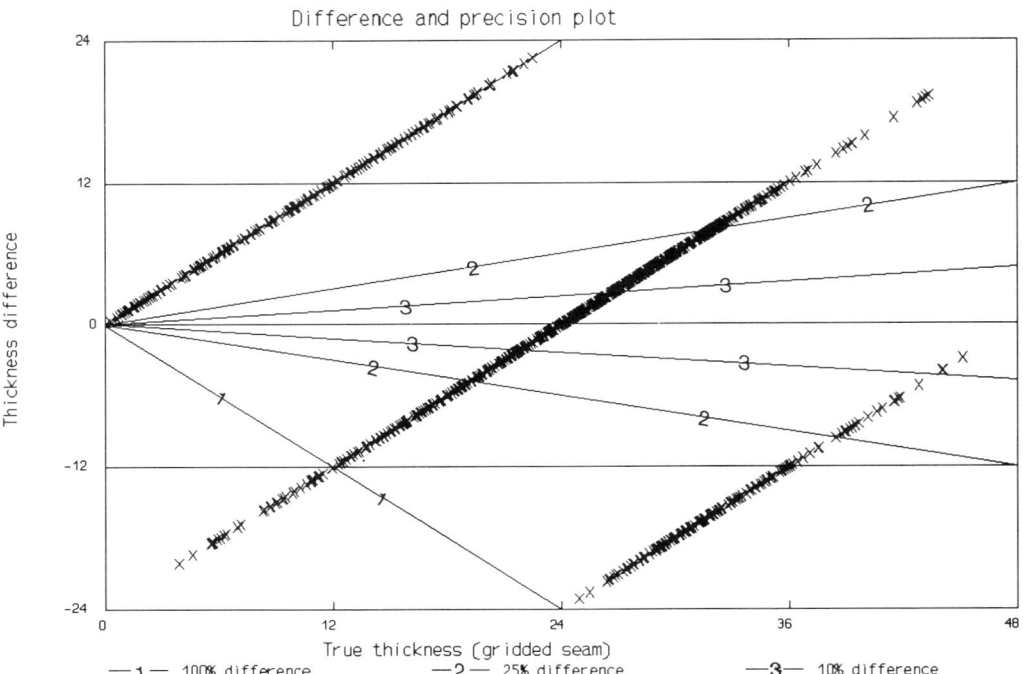

Fig. 7. Plot of difference (block model gridded seam) against true thickness (gridded seam) for 24 m block height model of the Lombador deposit.

resolution of the two different models (i.e. 4 m and 24 m respectively), and can be used to quantify precision in terms such as 'thicknesses estimated using this model are expected to be within X metres of the true value at a 95% confidence level (where the value of X can be determined from the standard deviation of the distribution of thickness values)'.

In order to highlight differences in the precision of the estimates over the full range of thicknesses, the data were summarized in a series of plots of difference versus true thickness. Following trials using different types of graphical presentation (see Sides 1992b), the results presented in Figs 6 and 7 were obtained. These graphs plot the error (block model estimate minus the gridded seam estimate of thickness) against the gridded seam thickness (taken to be the best estimate of the true thickness, since it was derived directly from triangulated boundary representations having a much better resolution than the grids used). Isolines of percentage difference (i.e. error divided by true thickness expressed as a percentage) are also shown.

In both cases the values for the 1081 grid points tested fall along a set of parallel diagonal lines. This pattern reflects the fact that the error at any grid point can be one of two values:

(a) error_1 = remainder (true thickness / block height);
(b) error_2 = block height error 1.

The maximum absolute error is equal to plus, or minus, the height of a single block. Extreme errors are less common than those close to zero, reflecting the conclusions arrived at during consideration of Figs 4 and 5 (i.e. the errors should be normally distributed with a mean of zero). Consequently the points plotted tend to be more frequent close to the centre of each diagonal line. The relationship of the parallel diagonal lines of data points to the isolines of percentage difference illustrates how asymmetrical error distributions could be predicted in the case of the 24 m block height model.

The exact distribution of points along each diagonal line will depend on both the statistical distribution of the true thickness values and on how wide a range of values the diagonal line

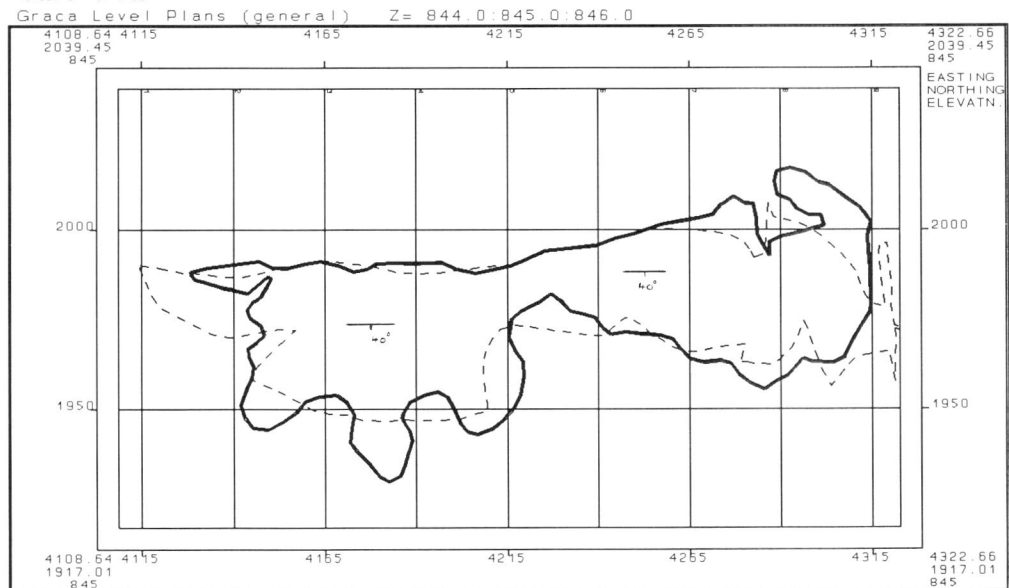

Fig. 8. Comparison of orebody outlines for different reserve interpretations of the 845 level in the Graça orebody. (Dashed line = 1988 indicated reserve interpretation; solid line = 1992 measured reserve interpretation.) (After Sides 1992b.)

crosses. By combining the known distribution of errors, the effects of changing block height can be predicted in advance. This aspect is the subject of continuing studies.

Influence of drillhole sections on shape prediction

Statistical theory recommends that random sampling should be applied in the selection of samples for use in predicting the characteristics of a given population (Sprent 1981). Nevertheless it has long been accepted that the taking of geological samples for evaluation purposes is best done either on regular grids or along parallel sets of sections (Popoff 1966). Amongst other reasons, this practice is based on the realization that the additional uncertainties associated with trying to interpret the geometry of geological discontinuities from randomly sited, or oriented, drillholes would outweigh any statistical advantages. Additionally, drillholes which intersect important discontinuities at a shallow angle give much less representative samples than those which intersect the main discontinuities orthogonally.

Consequently, inclined tabular orebodies are often sampled by fans of drillholes, sited on transverse vertical sections, at regular intervals along strike. In such cases it is reasonable to assume that interpretation of the orebody shape in the planes of the drillhole sections should be better (i.e. more accurate and more precise) than the predictions of the shape in zones falling between section planes. A study was carried out to try to test this hypothesis using the method described earlier.

Data for this study were obtained from two different geological interpretations of the Graça orebody at the Neves–Corvo mine (Richards & Sides 1991; Richards & Ferreira 1992). The first interpretation, used in the estimation of indicated reserves in 1988, was based on underground drilling on transverse sections at 25 m spacing. The second interpretation, used in the estimation of measured reserves in 1992, was based on re-interpretation of the geology using stope mapping data, on level plans at 2 m vertical intervals. At this stage most of the ore within the volume studied had been extracted. The values obtained from these two sets of interpretations are termed the indicated and measured values, respectively, in the discussion given below.

The orebody outlines compared were interpreted on, and digitized from, level plans at 2 m vertical intervals between the 840 and 860 levels. An example of one of these level plans is shown in Fig. 8.

Table 2. *Differences in accuracy and precision of indicated reserve interpretations of the Graça orebody shape, as obtained by comparison against measured orebody shape (recalculated after Sides 1992b)*

Feature	Position	Accuracy		Precision	
		Mean	Median	Standard deviation	3rd–1st quartile
Footwall location	Along sections (<6.25 m)	2.97	1.59	4.91	3.56
	Between sections (⩾6.25 m)	2.91	1.61	4.26	3.32
Hanging-wall location	Along sections (<6.25 m)	−0.94	0.33	7.69	6.43
	Between sections (<6.25 m)	−2.70	−2.32	9.15	9.34
Horizontal thickness		4.07	1.95	8.55	9.05
	Between sections (⩾6.2 m)	6.12	5.51	8.98	12.15

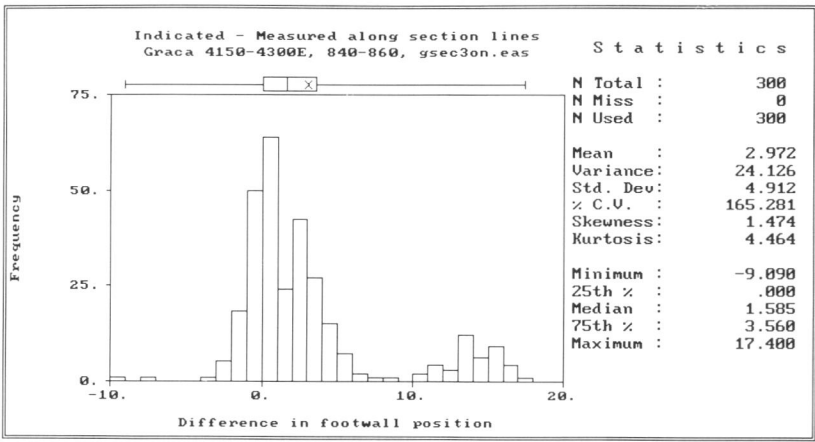

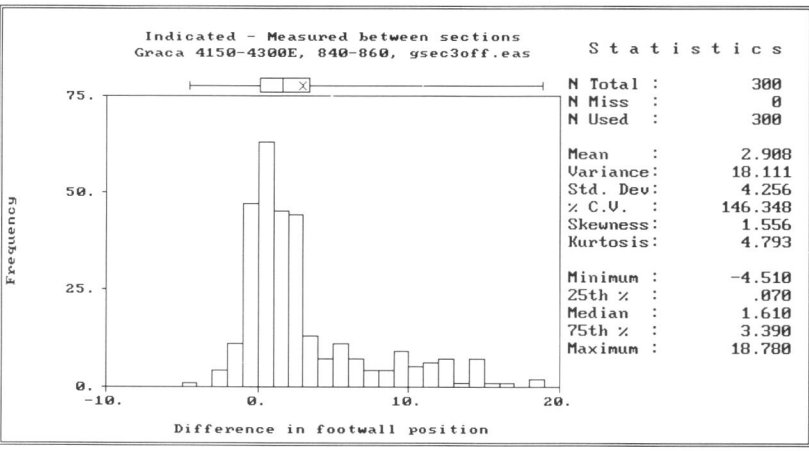

Fig. 9. Differences in prediction of the location of the footwall of the Graça orebody along section lines (upper), and between section lines (lower) (compiled from Sides 1992*b*).

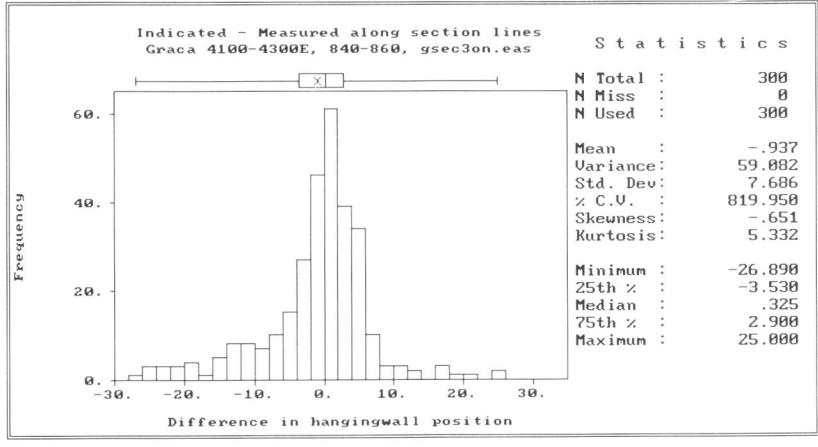

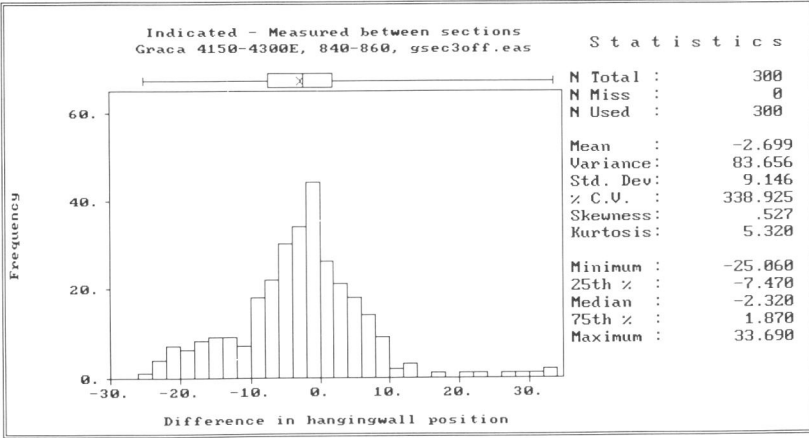

Fig. 10. Differences in prediction of the location of the hanging-wall of the Graça orebody along section lines (upper), and between section lines (lower) (compiled from Sides 1992b).

Differences between the two sets of orebody shapes were obtained by determining the positions at which they intersected a set of north–south lines at 2.5 m intervals, between 4150E and 4300E. This resulted in a set of 600 comparisons which were split into two groups corresponding to zones close to drillhole sections (projection distances ≤ 6.25 m) and zones falling between section lines (projection distances > 6.25 m). Graphs showing comparisons of the differences in the prediction of the orebody footwall and hanging-wall positions, and the total horizontal width of the mineralized unit, are shown in Figs 9, 10 and 11. A summary of the overall differences in terms of accuracy (as reflected by the mean and median differences), and precision (as reflected by the standard deviation, and the difference between 1st and 3rd quartiles) is given in Table 2. The differences in footwall and hanging-wall positions are based on measuring the distance of the contact positions from the centre of the measured orebody shape (along each test line). The differences shown are derived by subtracting the measured value from the indicated value. In general the results presented support the hypothesis that predictions of orebody shape close to the planes of sampling and interpretation are more accurate and more precise than predictions for zones falling between section planes. This is particularly true for the hanging-wall position and also for the horizontal thickness of the orebody.

Nevertheless, the opposite is true for predictions of the footwall position, although the differences are not great. This is due to the fact

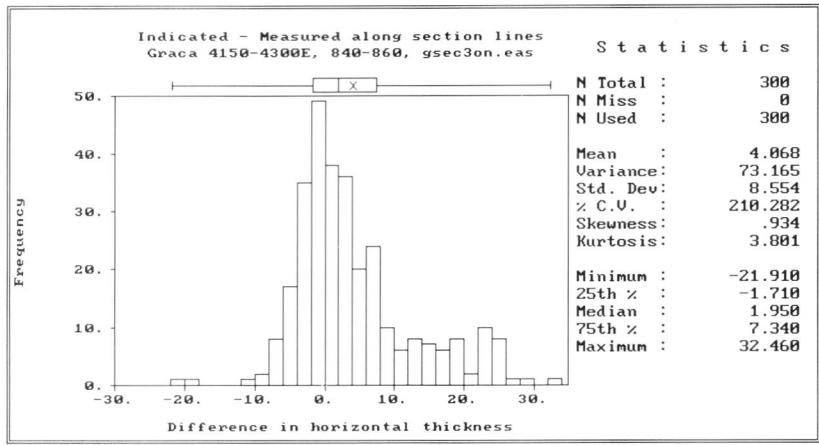

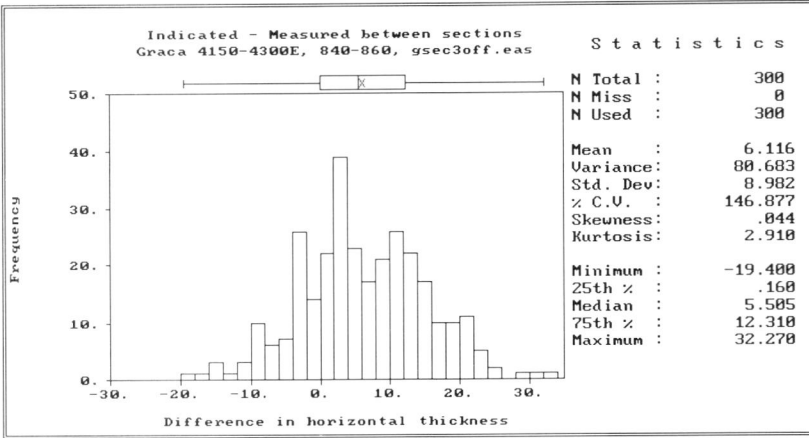

Fig. 11. Differences in prediction of the horizontal thickness (north–south direction) of the Graça orebody along section lines (upper), and between section lines (lower) (compiled from Sides 1992b).

that in the eastern part of the zone studied considerable structural complexity, which had not been predicted by the underground drilling, was encountered during mining (see Fig. 8). This resulted in a bimodal pattern of errors for the prediction of footwall position (Fig. 9). Consequently, conclusions based on the overall statistics (presented in Table 2) may be unreliable.

Table 2 also highlights the fact that there is, on average, a bias in predicting the position of the two orebody contacts. Interpretation of these differences is complicated by the fact that there is also an overall bias in the prediction of the horizontal thickness (since the reference line used is based on the centre of the measured orebody shape). Overall it would appear that the predicted location of the footwall is slightly south of its true location, and that of the hanging-wall slightly north of its true location. These biases work in opposite directions so that the overall effect is to give an even larger bias in the estimates of the horizontal thickness of the orebody.

Conclusions

The importance of geometric modelling techniques in the representation of orebody shapes has been illustrated. The need for empirical measurements of the differences between different predictions of orebody shapes, in order to identify and control errors inherent in the modelling process, has been stressed. A simple method for determining locational accuracy and precision has been proposed and used in two applications, namely:

- the comparison of computer models of the same orebody shape using different model structures;
- the comparison of predicted orebody shapes determined at different stages in the evaluation of a single deposit.

The following conclusions can be made as a result of the studies presented:

- under certain circumstances the use of regular block models may cause significant uncertainties in the prediction of orebody volumes;
- it can be shown quantitatively that interpretations of orebody shape close to the plane of drillhole sections are better, in terms of accuracy and precision, than interpretations for zones falling between sections;
- the information revealed by the studies justifies the attempts made to quantify errors in orebody shape predictions;
- further studies are required in order to identify other methods of quantifying locational accuracy and precision, and to apply such techniques in different situations.

The method used to quantify errors is entirely empirical in nature, and may appear somewhat simplistic. Further work is therefore required to provide a more theoretical basis for the determination of such errors. It is hoped that the results presented here will stimulate others to carry out similar studies on other deposits.

The work presented here is based largely on a PhD research project completed recently (Sides 1992b). The support of RTZ Consultants Limited (UK), Somincor SARL (Portugal) and Rio Tinto Minera SA (Spain) during the course of this research work is acknowledged. Permission of Somincor to present the results of studies carried out on the deposits at Neves–Corvo is gratefully acknowledged. Comments made by an anonymous referee are also appreciated.

References

BURROUGH, P. A. 1986. *Principles of geographical information systems for land resources assessment.* Monographs on Soil and Resources Survey, **12**, Oxford University Press, Oxford.

DAVID, M. 1977. *Geostatistical ore reserve estimation.* Elsevier Scientific Publishing Company, Amsterdam.

GIBBS, B. L. 1991. Mining Industry Software. *In: Mining Annual Review—1991.* Mining Industry Publications, 197–205.

GOODCHILD, M. F. 1980. Fractals and the accuracy of geographical measures. *Mathematical Geology*, **12**, 85–98.

HOULDING, S. W. 1991. Computer modelling limitations and new directions—Part 1. *CIM Bulletin*, **84**, 75–78

ISAAKS, E. H. & SRIVASTAVA, R. M. 1989. *An introduction to applied geostatistics.* Oxford University Press, New York.

MERKS, J. W. 1985. *Sampling and weighing of bulk solids. Series on bulk materials handling*, **4**. Trans. Tech. Publications, Clausthal-Zellerfeld, Germany.

POPOFF, C. C. 1966. *Computing reserves of mineral deposits: principles and conventional methods.* Bureau of Mines Information Circular **8283**, United States Department of the Interior, Washington, USA.

RICHARDS, D. G. & FERREIRA, A. V. M. M. 1992. Metal zoning in Graça orebody, Neves–Corvo. Minerals Industry International, **1005**. (Paper presented at the IMM 16th Annual Commodity Meeting—Copper, London, 2nd December 1991.)

—— & SIDES, E. J. 1991. The evolution of reserve estimation strategy and methodology at Neves–Corvo. *Transactions of the Institute of Mining and Metallurgy*, **100**, B192–B208.

SIDES, E. J. 1992a. Reconciliation studies and reserve estimation. *In:* ANNELS, A. E. (ed.) *Case Histories and Methods in Mineral Resource Evaluation.* Geological Society, London, Special Publications, **63**, 197–218.

—— 1992b. *Modelling three-dimensional geological discontinuities for mineral evaluation.* PhD thesis, Imperial College, University of London.

SPRENT, P. 1981. *Quick Statistics: an introduction to non-parametric methods.* Penguin Books, England.

A review and evaluation of the costs of exploration, acquisition and development of copper and gold projects in Chile

PATRICK GORMAN

MRDL, 90 Colney Lane, Cringleford, Norwich, Norfolk NR4 7RG, UK

Abstract: In recent years Chile has become the focus of activity for many multinational mineral resource companies. Undeniably they are attracted by the geological potential, business/investment rules and lifestyle, all of which are excellent by most standards.

In view of the continued interest in Chile a detailed profile of 36 gold and 27 copper projects and advanced prospects was prepared from a number of public sources of data. From this list a representative selection of about 10 projects was made and the costs associated with those projects were reviewed and evaluated. Based on the data a modest reserve containing 31 tonne (1 M oz) of gold might be discovered in Chile at a cost of US$ 225 kg^{-1} Au (US$7 oz^{-1}) or acquired for US$ 772 kg^{-1} Au (US$24 oz^{-1}) and then a mine constructed for US$ 1522 kg^{-1} Au (US$ 47 oz^{-1}). The total average cost taken from initial reconnaissance to start-up of the mine is equivalent to US$ 25 723 kg^{-1} gold per year (800 oz^{-1} Au per year). Historically a medium-sized copper deposit in Chile containing 1 M t of fine copper might have been discovered for US$ 6.5 t^{-1} Cu (0.3 US cents lb^{-1}) or acquired for US$ 10.3 t^{-1} Cu (0.5 US cents lb^{-1}) and then constructed for US$ 64 t^{-1} Cu (2.9 US cents lb^{-1} Cu). The total cost to mine start-up is equivalent to US$ 3969 t^{-1} Cu per year (US$ 1.80 lb^{-1} Cu per year). The recent purchases by Emablos/Minorco of one-third of Collahuasi and Placer of 50% of Zaldivar have increased the prices paid to acquire copper resources to US$ 60 and US$ 93 t^{-1} Cu *in situ* (2.7 to 4.2 US cents lb^{-1}) respectively. On the basis of the above figures, the cost of acquiring a gold project in Chile is typically equivalent to US$ 24 oz^{-1} or 7.3% of the price of gold at US$ 330 per ounce. Copper project acquisitions previously averaged the equivalent of 0.5% of the price of copper at 95 US cents per lb, but this has reached 4.5% with the price paid for Zaldivar. These guideline values have been prepared to give explorationists, acquirers and developers useful empirical guidelines to help them to decide whether they have the budget to consider entry to Chile or whether to look elsewhere.

Many multinational mining companies are either investing or are considering investing in Chile. Why do mining companies think that Chile is *the place* to be? There are world class mines in Chile and, better still, they crowd around the low, break-even cost sector of the business. For example, BHP Minerals and their partners at Escondida have constructed a lucrative project and they deserve all the credit for a number of reasons:

- they committed to Chile in the late 1970s and early 1980s;
- they went out to explore for copper in an aggressive manner;
- they raised the finance and brought it on stream at the right cost and at the right time.

How many parties are willing to raise their hands now and admit 'we took a look at Escondida and said thanks, but no thanks!'? There are also world-class projects waiting in the wings, such as Collahuasi, which attracted an immense amount of interest, judging by the estimated 17 parties which originally swarmed over the Chevron share. There is no doubt that developing anything close to home is difficult, if not impossible, these days. You have only got to go to a bar in Denver and ask about the mining business to find out that the Clinton administration has few friends in the mile high city and those that remain are seriously thinking of packing-up and going to Chile, Mexico and elsewhere. Once you have visited Chile, you have no difficulty in finding other good reasons to stay, such as: the mining tradition; other companies' successes; political and economic stability; a life style and climate that is attractive to the companies' executives and their families; the good infrastructure and availability of services and a well-trained workforce. Let us not forget that the potential to discover economically attractive copper and gold deposits is good, since Chile has been blessed with the right geological setting.

Table 1. *Gold project cost data*

Project	Reserve M oz	Explore US$M	Acq US$M	Develop US$M	Total US$M
La Coipa[1]	5.4†	30.0	148	253	431
El Indio	1.8†	8.0	0	200	208
Marte[2]	1.6	30.0	0	97	127
El Hueso[3]	0.9	9.0	56.7	28.6	94.3
San Cristobal	0.6	5.5	9.5	27	42.0
Choquelimpie	0.7†	4.5	8.0	32.7	45.2
Can Can	0.4†	5.0	8.5	12*	25.5
Fachinal	0.9†	7.2	5	30*	42.2
Lobo	4.0	8.0	0	130*	138.0
Refugio	3.3	16.5	0	130*	146.5
Andacollo[4]	1.0	24.0	4.8	35*	63.8
Total	20.6	147.7	240.5	975.3	1363.5
Group average	1.9	13.4	34.4	88.7	124.0

[1] La Coipa Phase I and II.
[2] Marte includes losses of US$ 60 M in start-up year 1990.
[3] Homestake paid US$ 56.7 M for a 10 year lease.
[4] Chevron expenditure accounts for US$ 20 M of exploration.
* Estimate.
† Gold equivalent.

Table 2. *Copper project cost data*

Project	Reserve M oz	Explore US$M	Acq US$M	Develop US$M	Total US$M
Escondida[1]	28.6	115	0	845	960
Zaldivar[2]	2.1	35	18	500*	553*
Quebrada Blanca[3]	2.3	40	26	360	426
Leonor	0.6	4*	9*	70*	83*
Pelambres[4]	0.6	0	10	60	70
Lince[5]	0.26	16	9	39	64
Ivan/Zar	0.13	5*	0	30*	35*
Las Luces	0.11	6*	0*	23*	29*
Candelaria[6]	5.3†	28*	40*	470*	538
Co Colorado	1.3	15*	30*	245	290
Total	41.10	264	142	2642	3048
Project average	4.11	26.4	14.2	264.2	304.8

[1] Excludes the US$ 200 M expansion hydro-metallurgical plant.
[2] Potential reported reserve is currently 4.8 M t Cu.
[3] Cominco acquisition for US$ 26 M in 1988.
[4] Excluding Anaconda expenditure of US$ 81 M (1979–1983).
[5] Excludes the Luksic group buy back from Chemical Bank US$ 18.9 M and Outokumpu US$ 36 M.
[6] Sumitomo purchase 20% for US$ 40 M (1991).
* Estimate.
† Copper equivalent.

Analysis

Reviewing the history and relating costs per tonne of copper or troy ounces of gold *in situ* enables a mining executive to get a feeling for budget purposes of what it now takes to be successful in Chile at various entry levels. This could involve discovering a new deposit through exploration or alternatively involve a joint venture or acquisition of an existing project. If one does either it is also useful to know how much capital might be needed to develop it and

Table 3. *Gold project development and production data*

Project	Reserve M oz	Discovery to start-up period (years)	Planned mine life (years)	Production k t oz/yr
La Coipa	5.4	9	12	300†
El Indio	1.8	5	10	300†
Marte[1]	1.6	2	10	100
El Hueso	0.9	5	12	70
San Cristobal	0.6	4	7	60
Choquelimpie	0.7	2	4	100†
Can Can	0.4	5	6	30†
Fachinal	0.9	10+	10	80*†
Lobo	4.0	5+	16*	250*
Refugio	3.3	4+	10*	233
Andacollo	1.0	7+	8*	110
Total	20.6			1703
Group average	1.9	5+	10	155

[1] Designed at 100 000 t.oz/yr, actual production in 1990 was 28,000 t.oz/yr.
* Estimate.
† Gold equivalent.

Table 4. *Copper project development and production data*

Project	Reserve M t Cu	Discover start-up period (years)	Planned mine life (years)	Production k t Cu/yr
Escondida[1]	28.6	10	52	320
Zaldivar	2.1	10+	20	110
Quebrada Blanca	2.3	6	14	75
Leonor	0.4	2+	10	28*
Pelambres	0.6	7	20	25*
Lince	0.26	5	10	20
Ivan/Zar	0.13	7+	10	10*
Las Luces	0.11	4+	7	10*
Candelaria	5.3 †	10+	34+	130 †
Co Colorado[2]	1.3	12+	25	40
Total	41.10			768*
Group average	4.11	7+	20.2	76.8*

[1] Excludes the expansion by 80 000 tCu/yr.
[2] Production design of 60 000 tCu/yr.
* Estimate.
† Equivalent copper.

how much time it might take. In an effort to provide this information, a detailed historical, technical and commercial profile of 36 gold and 27 copper projects or advanced prospects was compiled from a number of public sources of data. Attempts were made to cross reference and corroborate, where possible, the data collected. From these data sheets the information on reserves, expenditures, development and operational parameters was extracted and is presented in Tables 1 to 4 inclusive. The objective was to develop some useful empirical guidelines to keep in mind when venturing into Chile. Since this exercise is not mathematically rigorous the data have been analysed in project terms and not the pluses and minuses which make the picture more or less favourable for interim owners. If this is done and the incremental upside

Table 5. Analysis of gold project costs

Project	Discovery cost US$/oz[1]	Acquisition cost US$/oz[2]	Development cost[1] US$/oz	Total cost[1] US$/oz/yr
La Coipa	6	27	47	1437†
El Indio	5		111	693†
Marte	19		61	1270
El Hueso	10	63	32	1347
San Cristobal	9	16	45	700
Choquelimpie	6	11	47	452†
Can Can	13	21	30*	850*†
Fachinal	8	6	33*	528*†
Lobo	2		33*	552*
Refugio	5		39*	629*
Andacollo	24	5	35*	580*
Weighted average	7	24	47*	800*

[1] 20.6 M oz.
[2] 9.9 M oz.
* Estimate.
† Gold equivalent.

Table 6. Analysis of copper project costs

Project	Discovery cost[1] US c/lb Cu	Acquisition cost[2] US c/lb Cu	Development cost[3] US c/lb Cu	Total cost US$/tpaCu
La Escondida	0.2		1.3	3000
Zaldivar	0.8	0.4	10.8*	5027
Quebrada Blanca	0.8	0.5	7.1	5680
Leonor	0.4	1.0	7.9*	2964*
Pelambres		0.8	4.5*	2800*
Lince	2.8	1.6	6.8	3200*
Ivan/Zar	1.8		10.5*	3500*
Las Luces	2.5		9.5*	2900*
Candelaria	0.2	0.3	4.0	4138
Co Colorado	0.5	1.0	8.6*	7250*
Weighted average	0.3	0.5	2.9*	3969*

[1] 40.50 Mt. reserve.
[2] 12.26 Mt reserve.
[3] 41.10 mt reserve.
* Estimate.

potential of the acquisition (which can be substantial) and the effect of inflating currencies (whose influence can also be substantial) are ignored we get the results detailed in Tables 5 and 6. The eleven gold and ten copper projects encompass the large to the small and from this one can extract some interesting data in US$ per tonne of copper metal or troy ounce of stated reserve. In addition one could repeat the analysis by reviewing the data base to look at factors such as location, scale, process flowsheet, mining system and other parameters of interest.

Gold projects

Based on the group surveyed, if a mining company wished to control a gold project containing 1 M oz, it would need to either

Mineral Resouces Evaluation II: Methods and Case Histories

Special Publication No. 79

ERRATA

P. 214, para 3, lines 10-12: the units should read l s^{-1} m^{-1} per metre.

p.217, last line of main text shoud read: risk to the stability of the high wall.

The Publishers apologize for any inconvenience caused.

spend an average of US$ 7 M in exploration activities or more than triple it to US$ 24 M and buy it outright. Could the company have done better elsewhere or not? I seem to recall that the consensus amongst the exploration managers in the late 1980s was that you were doing fine if you found your own project for US$ 5 oz^{-1} or bought someone else's reserves for US$ 10 oz^{-1} and were working in Australasia or in Central/West Africa. Only if you wanted USA or Canadian reserves might you have considered paying US$ 25 oz^{-1}. I understand that current expectations are closer to US$ 65 oz^{-1} in Canada provided you US$ 35 oz^{-1} in Australia and the USA. Acquiring Chilean reserves therefore appears to be attractive at US$ 24 oz^{-1}. The data also show the value of spending exploration dollars not acquisition dollars. One can afford three US$ 7 oz^{-1} exploration failures for each success. If two are found, one of them could be sold to recoup expenses. Or perhaps it is preferable to buy now for an acquisition price of US$ 24 oz^{-1} before stocks run out. The main deposits that are left to be developed and which rival the gold content of La Coipa and El Indio, are Lobo and Refugio, both good exploration value for money. The Anglo–Cominco group appear to have decided that Lobo is not for sale. Perhaps their philosophy is that if you can close up the larder of major gold deposits until the gold price improves again, it would be attractive to have a Lobo in the larder. Amax Gold and Bema have begun raising finance for Refugio. Excluding small-scale mining activities by prospectors, but including change of ownership and false starts, a typical Chilean gold project takes an average of five years to get from discovery to start-up. Would you prefer to find an El Indio district for US$ 5 oz^{-1} and offering upside potential or buy an El Hueso at US$ 63 oz^{-1} which has a ten year lease over a well defined, limited area but comes with a starter plant? The value of the upside at El Indio was part of the reason why Bond's Dallhold group bought the project from St Joe in 1987 for an estimated US$300 M and then Lac bought the equity and debts in 1989 and now controls up to 83% of the project at a price of US$374 M. Based on the original 1.8 M oz reserves to first start-up in 1980, the extra cost paid exceeded the Homestake price for El Hueso by several fold but among the projects included were Nevada, Mahoma, Sancarron, Tambo and others. The Wendy Norte resource alone appears to have potential for a 1 M oz reserve and results from Nevada are encouraging.

The overall US 24 oz^{-1} cost to acquire ounces is equivalent to 7.3% of a gold price of US$ 330 and is a substantial price to pay. At present, this price appears to be reasonable especially when one considers that the Clinton administration has tabled a royalty of 12.5% on production from unpatented claims on public lands (10% of major USA production) although this may be reduced to 8% and is talking of a gold surtax for production from private lands.

Copper projects

Historically, if a mining company wished to a control a copper project containing 1 M tonne copper metal *in situ,* it would need to spend 0.3 US cents lb^{-1} of copper or an average US$ 6.5 M in exploration funds or alternatively spend a total of US$ 10.3 M to purchase the discovered resource. The cost of success or failure can be considerable if a company is trying to find a large project. The reward for finding an Escondida reserve containing almost 30 M tonne of copper metal is worth the US$ 115 M spent, but there are no guarantees as we all know. Although the pre-development component represented 12% of the final capital cost at Escondida, in terms of revenue stream, it is equivalent to only a 0.2% Gross Royalty at a copper price of US$ 0.95 per pound. These costs have dramatically changed in recent years, especially for the big projects. In 1985, Shell and Chevron agreed to fund a US$ 45 M pre-development programme over an 8 year period in return for a one-third share of Collahuasi each. The recent Minorco and Emablos purchase of Chevron's one-third share of Collahuasi for US$185 M is equivalent to 0.9 US cents lb^{-1} Cu based on a 9 M tonne Cu reserve for the project. Of course to the incoming party, the cost is actually 2.7 US cents lb^{-1} Cu for the reserve that they own. In industry circles there has been criticism of the price paid – but it seems that the Minorco/Emablos groups wanted access to one of the largest undeveloped projects on the market and presumably started from the assumption of finding out what the seller wants and not what the project's NPV was on a purely technical basis alone. However, if recent reports are accurate and the reserve at Collahuasi reaches 2000 M tonne of ore at 1% (20 M tonne Cu) then the Minorco/Emablos price reduces to 1.2 US cents lb^{-1} Cu. Placer's US$100 M purchase of 50% of Zaldivar from Outokumpu (plus arranging financing) is equivalent to 2.1 US cents lb^{-1} Cu, although the cost to Placer for its share of copper is actually 4.2 US cents lb^{-1} Cu. However, geological reserves in the Zaldivar claim areas are reported to be 4.8 M tonne Cu and not 2.1 M tonne Cu and this would reduce

Placer's purchase price to 1.9 US cents lb^{-1} Cu. Therefore, whether it's big or small copper projects you are after, the cost in Chile, is now firmly in the region of 3 to 4 US cents lb^{-1} of copper with the advantage that if one selects a larger project with upside potential then significant reductions in costs per lb of copper are achievable.

Discussion

The value of 4 US cents pound now being paid for copper reserves is only equivalent to 4.5% of the revenue stream at a copper price of US$ 0.95 lb^{-1}. In the case of gold, an acquisition cost of US$24 oz^{-1} is equivalent to 7.3% of the revenue stream at a gold price of US$330 per ounce. One aspect that is currently slowing down copper and gold exploration in Chile and is contributing to driving up prices, is the fact that the companies which hold mineral claims in Chile are not required to undertake a minimum annual work commitment on their claims or lodge the results of this work in the public domain. There is currently no obligation other than to pay the holding fee in March of each year. This has allowed certain companies to pursue a policy of 'passive exploration', where ground is coveted for extended periods of time without much taking place. If the claims are dropped the information that has been developed is not required to be made available to the incoming party and involves extra time, effort and expense to repeat data collection. This issue is being discussed throughout the country and a solution will eventually be found. In my opinion, Chile still offers good potential within its borders and is an ideal stepping stone to investigate adjacent countries, such as Argentina and Peru. In Chile the prices are already high and will probably get higher. I suggest spending what money is available in applying good technical skills and aggressively seeking out resources. Throughout Chile there are companies who have the personnel, techniques and backing to increase the odds of success. What Chile needs now is more of these types of companies to explore for the next wave of projects suitable for development.

Mining project finance and the assessment of ore reserves

J. O'LEARY

Montagu Mining Finance Limited, 10 Lower Thames Street, London EC3R 6AE, UK

Abstract: The estimation of ore reserves is one of the most critical activities in the mining investments decision process. All other activities will depend eventually on how well grades and the associated tonnages are computed. Unreliable estimates will affect the financial viability of a project and while errors may not be 'life threatening' to high profit margin projects errors on the scale commonly encountered can do irreparable damage to a low margin business. In response to this, numerous classification systems have been devised, the purpose of which is to impart from the technical decision makers to the financial decision makers (in industry or the financial community) the confidence that may be had from a reserve estimate.

Project finance is a difficult and misunderstood mechanism for providing finance to major infrastructure projects. This is particularly true for the mining industry where project finance is common. By definition, project finance is that form of finance that the sponsor (typically a mining company) has segregated from the general assets and corporate obligations of that company. The project borrowings will be securitized, typically by the project assets and repayments will be derived from the project cash flows. That is to say there is little or no recourse to the borrower should the project fail. This form of finance can be attractive for the banks in that interest margins can be three to four times higher than for corporate borrowing, but attendant with their reward is the assumption by the banks of considerable risk.

As repayment will be derived from project cash flow the major risks being assumed are that project revenues will not be up to that forecast or that project cost will be in excess of that forecast or both and that the cash flow is insufficient to meet the interest and principal repayments. Whereas there is considerable experience in forecasting costs, revenue is more problematical in that it is dependent mainly on the commodity price and the estimate of quantity of the metal produced. The commodity price has its own special problems that are beyond the scope of this paper. The quantity of metal is directly related to the forecast of mineable tonnes and grade i.e. the statement in any one period of the mineable reserve.

The confidence of a reserve estimate is thus of direct interest to a bank involved in a mining project finance. Traditionally the confidence was supplied through a classification system and the terms 'proved', 'probable' and 'possible' and their equivalents are a means of expressing the uncertainty (geological and other) associated with a reserve estimate. There has been, however, much criticism of these traditional statements of confidence and many of the criticisms of the traditional classification schemes centre around the subjective nature of the definitions of assurance and their misuse and abuse. Recently the need for quantitative definitions has arisen in view of the trend to mining low-grade deposits where a high degree of confidence in a reserve estimate is vital. Many argue that because a quantitative approach can be taken then it should. With the development of geostatistical theory starting in 1966, estimation techniques have evolved where the expected error of the estimate can be calculated as an integral part of the estimation procedure. There is therefore the possibility of not so much modifying the current classifications but developing a new scheme aimed at the financial community where the levels of confidence are related both to the estimation error and to the time period for which the estimation is related.

The estimation of ore reserves is one of the most critical activities in the mining investment decision making process. All other activities will eventually depend on how well the grades and tonnages of mineral deposits are computed.

Poor and unreliable estimates will affect the financial viability of a project. True mineral grades are often 10% to 20% different from the estimated grades. Errors on this scale may not be damaging on deposits where there is a high profit margin but on low profit margin deposits, such as some of the large, long life, low grade deposits, such errors could be critical.

Reserves are often quoted in such a way as to

imply an accuracy in prediction of one in a hundred to one in a thousand and it may be forgotten that the calculation is by projecting values from a small volume to a much larger volume. There is a need, therefore, to place some sort of ranking on quotations of reserves to reflect these uncertainties inherent in reserve estimation. In the past this has been satisfied by classifications such as 'proven', 'probable' and 'possible', which have been recommended for use in the mining industry by various professional organizations. In recent years these definitions have been criticized in the light of the availability of some of the newer statistically based techniques of estimation, in particular geostatistics. In order to appreciate the problems involved in classification some of the objectives are considered.

The main purpose of any classification is to enable information to be transmitted easily and clearly from geologists and engineers to decision makers in industry or government. At this stage two main objectives can be differentiated. On one side the classification must provide information at a local mine level, and on the other a more general objective associated with regional and national resources.

Governments require such information to assess the present and potential resources of a country or region. This is needed for long term planning, provision of infrastructure and land use planning. It was to aid these objectives that the general resource classifications, such as USBM/USGS guide-lines, were developed. The terminology used for resources, however, is commonly used in the mining industry.

For use in industry the objectives are more detailed. There are three objectives:

(1) to give a measure of the amount of ore that can be reasonably relied upon to meet the production requirements of a project;
(2) to provide as far as possible a standard procedure for estimating this amount;
(3) to provide a means of expressing the uncertainties (geological and other) associated with an estimate.

Current classification systems use the qualitative terms 'proven', 'probable' and 'possible', and 'measured', 'indicated' and 'inferred' in an attempt to meet these objectives.

These definitions are used in effect as a planning tool in devloping a project. As the investigation of a deposit progresses from discovery to production, reserves are continuously assessed to discover if enough material of the required quality is available for the project to proceed. The accuracy required as the development progresses will have to increase. Similar progression is required for accuracy in cost estimates for project evaluation and ongoing re-evaluation of the reserves of an operating mine in the light of actual production statistics for ongoing planning and investment.

The specifications outlined above are especially relevant to organizations investing in mining operations. Investment companies may require that a quantity of the reserves should fall into the 'proven' and 'probable' classes and this quantity should be able to support the project until the loans are paid back. Equity investors will have similar requirements.

Many of the criticisms of the traditional classification schemes centre around the subjective nature of the definitions of assurance and their misuse and abuse. Recently the need for quantitative definitions has arisen in view of the trend to mining low grade deposits where a high degree of confidence in a reserve estimate is vital. Many argue that because a quantitative approach can be taken it should. With the development of geostatistical theory starting in 1966, estimation techniques have evolved where the error of the estimate can be calculated as an integral part of the estimation procedure. There is therefore the possibility of modifying the current classifications or developing a new scheme where the levels of confidence are related to the estimation error.

Classification systems

Broadly speaking classification schemes fall into two categories; those used in the mining industry and those originally designed for classifying resources. Many consider that resource definitions are not suitable for use in the mining industry. Some of the official guide-lines on resource and reserve classification are included in Appendix 1. These are not exhaustive but give an idea of the variation in terminology and scope of the definitions used.

Resource classification

The first official classification scheme was that of the US Bureau of Mines in 1943. This used the terms measured/indicated/inferred to define geologic assurance. Since then it has been modified twice, in 1974 and 1980. The system is probably the best known of the resource classifications. Recovery is introduced by defining 'reserve' as including only recoverable material. The term 'reserve base' was intro-

Fig. 1. United Nations classification of resources (from Schanz 1980).

duced for *in-situ* (demonstrated) resources from which reserves are estimated. Under the USBM/USGS classification qualifiers such as '*in-situ*' and 'recoverable' are not recognized.

The West and East German recommendations and those of the Soviet Union differ in denoting the degree of assurance by letter categories rather than in descriptive terms. The terminology for economic feasibility is also altered. In the West German classification the currently usable proportion of total resources is denoted by the term 'mining resources' and 'potential resource' as those which may be used in the future but do not meet current requirements for utilization. These are equivalent to the economic and sub-economic divisions respectively of the North American definitions.

In an attempt to formulate an international resource classification the United Nations formed a special committee to investigate the subject. Three basic resource categories were defined based on the degree of geological assurance (Fig. 1). These classes represent all the *in-situ* quantities of possible economic significance over the next 20 to 30 years. Each of the categories is subdivided on the basis of economic feasibility. The term 'reserve' was excluded from the classification to avoid confusion as some languages (e.g. Russian) do not differentiate between resource and reserve. To denote *in-situ* resources the committee decided to use capital letter ('R'). The lower case ('r') would indicate the recoverable equivalent of R-1-E.

Of these different systems the USBM/USGS definitions are probably the most widely used.

Classification of geological assurance

These largely developed from the original IMM classes of 'in-sight', 'probable' and 'possible' which were developed in 1902. Original mining terms concentrated on geometric requirements, so 'proved' ore had to be *exposed* on three sides. The USBM/USGS introduced their definition in 1943 because they found it impossible to classify national resources using the old, more restrictive terms. Despite the differences in objectives of resource and reserve classifications, the definitions of assurance appear to be very similar. Both types of classification use very general terms to indicate the increasing accuracy that is expected as the categories progress from possible (inferred) to proven (measured). No specific requirements for sample spacing are given (with some exceptions) that being left to the engineer or geologist to define.

Definitions recommended specifically for use by the mining industry have been issued by the Australian IMM, the American Society of Mining Engineers and the Association of Professional Engineers of Ontario (1986). These use the terms 'proven', 'probable' and 'possible', which have been accepted for reporting purposes by the US Securities and Exchange Commission and the Melbourne Stock Exchange. To the knowledge of this author, the APEO (1986) definitions are the only ones which recommend that the projections from sample points defining each class be stated when quoting reserves, though most state that the data used for analysis be described.

Some attempts have been made to assign quantitative limits to various classes. For example the 1976 definitions of assurance issued by the USBM/USGS included a specification in the 'measured' class that

> tonnage and grade are judged to be accurate within limits which are stated, and no such limit is judged to be different from the computed tonnage or grade by more than 20%.

However, no guidance is given on that confidence level at which this classification is to be made.

Project finance

Various sources of finance are available to sponsors seeking to develop or expand mining projects. Some fall within the definition of project finance, whilst others rely upon innovative and sometimes unusual financial structures being put in place.

Project finance is a specialized form of finance that covers a wide range of financing structures, which all share one common feature: the finance is of a limited recourse nature. That is, financing depends primarily on the successful operation of the project, and not on the credit support of the borrower, or the value of any security taken over the project assets.

An economic viability test should demonstrate that the future project cashflows are capable of sustaining the repayment of capital and interest, whilst an independently conducted feasibility study should demonstrate the technical viability of the project.

Project finance lenders seek higher margins to reflect their risk exposure, and higher fees due to the invariably complex nature of the loan and security arrangement, and documentation. Margins and fees will also reflect the type of financial structure that is put in place.

The major risks that are being assumed by banks are as indicated in Table 1.

Table 1. *Categories of mine project finance risk*

Within the company's control	Outside company & bank's control	Within the bank's control
Operating	Reserve	Syndication
Technical	Commodity	Funding
Cost/economic	Transportation	Legal
Management	Environmental	
Sponsor	Political	
Design	Force majeurel	
Completion		

For this the typical return to the banks is about 1.5% over their cost of funds (Table 2).

Table 2. *Average interest rates for mining project transactions*

Financial requirement	Interest rate LIBOR +
Acquisition	2.7%
Refinance	2.6%
Project finance	1.5%
Working capital	1.0%
Trade finance	0.9%
Revolving credit	0.5%
General corporate	0.4%

Project finance is a difficult and misunderstood mechanism for providing finance to major

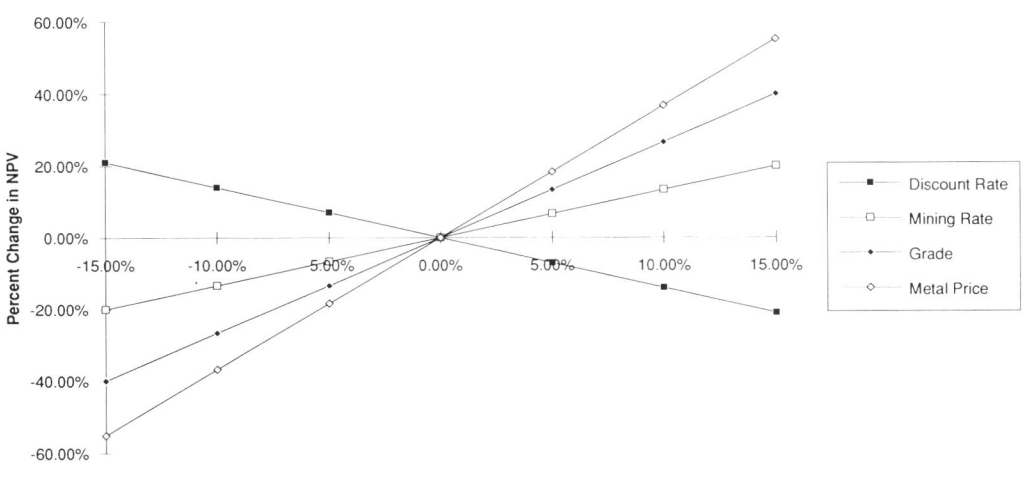

Fig. 2. Sensitivity of a typical project return to its basic assumptions.

infrastructure projects. This is particularly true for the mining industry where project finance is common. By definition project finance is that form of finance that the sponsor (typically a mining company) has segregated from the general assets and corporate obligations of that company. The project borrowings will be securitized, typically by the project assets and repayments will be derived from the project cash flows. That is to say there is little or no recourse to the borrower should the project fail. This form of finance can be attractive for the banks in that interest margins can be three or four times higher than for corporate borrowing, but attendant with their reward is the assumption by the banks of considerable risk.

As repayment will be derived from project cash flow the major risks being assumed are that project revenues will not be up to that forecast or that project costs will be in excess of that forecast or both and that the resultant cash flow is insufficient to meet the interest and principal repayments of any project borowing. Whereas there is considerable experience in forecasting costs, revenue is more problematical in that it is dependent mainly on the commodity price and the estimate of the quantity of the metal produced, which in turn is directly related to tonnage and grade. Figure 2 shows the sensitivity of a typical project return to its basic assumptions. A 10% variation in the realized grade from that forecast results in a 30% change in the project's 'value'. The situation is considerably worse for the forecast metal price. The commodity price has its own special problems that are beyond the scope of this paper. The forecast of mineable tonnes and grade, i.e. the statement in any one period of the mineable reserve should be of considerable interest to the geologist and engineer.

The problem for the financial community is that the classification systems are not set up in a manner which allows them to assess the confidence in a future cash flow stream and furthermore the classification rules are interpreted differently by different consulting engineers and sponsor companies. An example is a major North American gold producer that defines 'proved' ore as 'a block estimated on the basis of at least one sample no more than 75 m from the block centre', whereas a South American copper producer defines a block confidence ±25% at the 95% confidence level on an individual block basis to qualify in the same category.

Project risk

Given the foregoing it is not difficult to see that the risks being assumed by the bank in a project financing are fundamentally related to the quality of the estimate of the reserves. In the majority of projects the cash flows are more sensitive to the grade than to any other factor excluding the commodity price and in many cases this sensitivity can be two to three times as large as, for example, capital cost. It is common practice nowadays for a consulting engineering group or a sponsor mining company to quote uncertainties alongside their estimates of capital and operating costs for a project. Recent experience has indicated that they are prepared to commit themselves to plus or minus 15% at the 90% confidence level when presenting an S1 study. At the same time most mining feasibility studies have at their base a reserve which uses geostatistical or similar techniques such as inverse distance weighting. Inherent in these techniques are the concepts of kriging variance and estimation variances which in most cases are calculated and never used again. It is most unusual for a consulting engineer or a sponsor mining company to quote confidence in the tonnage or grade profiles presented in feasibility studies except in the more general terms even though poor assessment of grades will have a large impact on the project should they be in error.

Typically, cash flows are estimated annually with large negative cash flows up front followed by positive cash flows in later years when a mine reaches full production. An example of a typical project cash flow is given in Appendix 2. In almost all cases such cash flows are presented in annual increments and ignore the basic requirements of both the mining company itself and the financial institutions which is for more information in the early years of the project and less information in later years. This is especially true of projects with large working capital requirements where it is very difficult to assess the working capital requirement unless a project is examined in monthly or perhaps quarterly increments in the early years. Most projects are not sensitive to changes in their assumptions say 20 years in the future. An example of an alternative cash flow summary is also presented in Appendix 2. In this alternative the time slice, for which the forecast is made, becomes larger as the forecast period becomes more distant and a confidence prediction made for each period. In this example the time slice has been adjusted to make the prediction confidence a constant as far as practical.

Kriging is one tool amongst many which allows the user to establish the uncertainties associated with any volume; more commonly it is associated to regular rectangular blocks but this need not be the case and it can be applied to any volume, irregular or regular. The better solution for financial institutions, particularly lending banks, is as described in the alternative cash flows, that is to have a series of production increments presented as part of the cash flows which cover short periods for the initial years and followed by increasingly larger periods as we move further into the future. The first year of a mine's life may be forecast monthly, in the first quarter, followed by quarterly in the remaining three quarters. Years two and three may be forecast half yearly with annual plans for years four and five. Eventually the forecast may be in five year increments when beyond year 15 or 20. During the planning process it is possible to assess the uncertainty of any such increment by using relative variance plans created as part of a standard geostatistical estimation. Additionally it is possible using standard techniques to approximate the standard error of such increments. This can be followed through with a full calculation of the estimated variance of a particular increment once these have been finalized. The resulting cash flow would not only present schedules of tonnes, grades, prices, costs and cash flows but also a schedule of uncertainties associated with such a prediction. Such a scheme would obviate the need for any traditional classification systems and present the banks with a clear picture of the uncertainty associated with cash flow predictions.

Appendix 1: selected classification systems

A: US Bureau of Mines and US Geological Survey

(Geological Survey Circular 831, 1980)

Measured. Quantity is computed from dimensions revealed in outcrops, trenches, workings or drill holes; grade and/or quantity are computed from results of detailed sampling. The sites for inspection, sampling and measurement are spaced so closely and the geological character is so well defined that size, shape, depth and mineral content of the resource are well established.

Indicated. Quantity and grade and/or quality are computed from information similar to that used for measured resources but the sites for inspection are farther apart or otherwise less adequately spaced. The degree of assurance, although lower than that for measured resources, is high enough to assume continuity between points of observation.

Demonstrated. A collective term for the sum of measured and indicated reserves or resources.

Inferred. Estimates are based on an assumed continuity beyond measured and/or indicated resources for which there is geological evidence. Inferred resources may or may not be supported by samples or measurements.

B: Australian Institute of Mining and Metallurgy

(*Reporting of ore reserves*, Joint committee of the Australian IMM and Australian Mining Industry Council 1985)

Proved ore reserves are those in which the ore has been blocked out in three dimensions by excavation or drilling, but include in addition minor extensions beyond actual opening and drill holes where the geological factors that limit the ore body are definitely known and where the chance of failure of the ore to reach those limits is so remote as not to be a factor in the practical planning of mine operations.

Probable ore reserves cover extensions near at hand to proved ore where the conditions are such that ore will probably be found but where the extent and limiting conditions cannot be so precisely defined as for proved ore. Probable ore reserves may also include ore that has been cut by drill holes too widely spaced to assure continuity.

Possible ore (not reserves) is that for which quantitative estimates are based largely on broad knowledge of the geologic character of the deposit and for which there are few samples or measurements. The estimates are based on an assumed continuity or repetition of which there is geologic evidence; this evidence may include comparison with deposits of similar type.

C: terminology adopted by Joint Committee comprising:

American Institute of Mining, Metallurgical and Petroleum Engineers (AIME)
Society of Economic Geologists (SEG) and
American Institute of Professional Geologists (AIPG)
(from Banfield & Havard 1975).

Proven ore an ore reserve so extensively sampled that the tonnage, grade, geometry and recoverability of the ore within the block or blocks of ground under consideration can be computed with sufficient accuracy so that the uncertainties involved would not be a factor in determining the positive feasibility of a mining operation.

Probable ore an ore reserve for which sufficient continuity of dimensions and grade can be assumed for preliminary financial planning, but for which the risk of failure in continuity is greater than for proven ore.

Possible reserves mineralized material of which the dimensions and grade are based on geological correlation between samples so widely spaced or so erratic that additional exploration is required to establish whether ore reserves are present.

D: classification of the deposit resources of solid mineral raw materials of the German Democratic Republic, January 1962

(Fettweis 1979, p.365)

Class A. If they have been investigated in such a way that the setting, form and structure of the bodies of raw material, the qualities of the raw material, the different kinds of raw material and their special distribution have been determined, barren areas and areas departing from the specified conditions have been delimited within the bodies of raw material, and the technological properties of the raw material and the mining technological factors which determine the conditions for mining work are known. The outline of the resources has to be determined by successful exploration of work.

Class B. If they have been investigated in such a way that the most important peculiarities of the setting, form and structure, qualities of different kinds of raw material have been determined without details as to their distribution, the conditions and character of barren areas departing from the specified conditions within the bodies of raw material have been established, and the most important technological properties of the raw material and the main mining technological factors determining the conditions for mining work are known. The limits of the resources have to be determined by successful drilling or other exploration work; a limited extrapolated zone can be included in the calculation block where there is constant thickness and constant quality of the body of raw material.

Class C_1. If they have been investigated in such a way that the basic features of the setting, form and structure, qualities and different kinds of raw material have been determined and those of the technological properties of the raw material and the mining technological factors determining the conditions for mining work are known. The outline of the resources has to be determined by exploration work and/or with the aid of interpolation and extrapolation of geological, geophysical and other data.

Class C_2. If they have been investigated in such a way that the setting, form and structure of the bodies have been approximately determined according to the results of individual drilling and prospecting operations or outcrops and other geological or geophysical data. In deposits being investigated for the first time the quality and technological properties of the raw material must be determined by studying individual samples. By reference to known deposits the properties of the raw material can be established theoretically or on the basis of analogies. The outline of the resources must be determined by isolated natural or artificial openings and/or with the aid of interpolation and extrapolation of geological, geophysical and other data.

E: classification of the deposit resources recommended by the committee of the German Mining Engineers and Metallurgists Society 1959

(Fettweis 1979, p.362)

Proved (Class A). The contours of the resource are entirely known or their continuity is proved by exploratory workings accordingly close together. Error tolerance ±20%; degree of assurance >90%.

Probable (Class B). The contours of the resource are incompletely known or their connection with proved resources has been found by exploratory workings sufficiently close together. Error tolerance ±20%; degree of assurance 70–90%.

Indicated (Class C_1). The presence of the resource has been shown by widely spaced exploratory workings or proved geophysical indications. Error tolerance ±30%; degree of assurance 50–90%.

Inferred (Class C_2). The presence of the resource has been shown by individual exploratory workings or can be assumed from the geological position and geophysical or chemical indications. Error tolerance ±30%; degree of assurance 30–50%.

Indicated and inferred classes are grouped under the term possible.

These general rules are supplemented by guide-lines which determine for different deposits how small the spacing of the exploratory workings or boreholes has to be. They also suggest what shall be required of the indications for the terms 'entirely known', 'incompletely known', 'proved indications', etc. to be accurate.

F: Uranium Resource Appraisal Group and energy, mines and resources, Canada

(Sabourin 1984)

Measured ore refers to ore for which tonnage is computed from dimensions revealed in outcrops, trenches, workings or drillholes and for which the grade is computed from adequate sampling. The sites for inspection, sampling and measurement are so closely spaced and the geological character so well defined that the size, shape and mineral content are well established. The tonnage and grade should refer to ore recoverable by mining with due regard for dilution.

Indicated ore refers to ore for which tonnage and grade are computed partly from specific measurements, samples or production data and partly from projection for a reasonable distance on geological evidence. The openings or exposures available for inspection, measurement and sampling are too widely or inappropriately spaced to outline the ore completely or to establish its grade throughout.

Inferred ore refers to ore for which quantitative estimates are based largely on a broad knowledge of the geological character of the deposit and for which there are few, if any, samples or measurements. Estimates are based on assumed continuity or repetition for which there is geological evidence; this evidence may include comparison with deposits of similar types. Bodies that are completely concealed but for which there is some geological evidence may be included. Estimates of inferred ore should include a statement of the specific limits within which the inferred material may lie. These limits vary depending upon the characteristics and knowledge of the orebodies.

G: proposed United Nations resource classification

(from Schanz 1980)

Category R-1 encompasses the *in-situ* resources that have been examined in sufficient detail to establish their mode of occurrence, size and essential qualities within individual ore bodies. The major characteristics relevant to mining and processing, such as the distribution of ore grade, the physical properties that affect mining, the mineralogy and deleterious constituents, are known mainly by direct physical penetration and measurement of the ore body combined with limited extrapolation of geological, geophysical and geochemical data. Quantities should have been estimated at a relatively high level of assurance, although in some deposits the estimation error may be as high as 50%. The primary relevance of such estimates is in the planning of mining activities.

Category R-2 provides for estimates of *in-situ* resources that are directly associated with discovered mineral deposits but, unlike the resources included in category R-1, the estimates are preliminary and based largely upon broad geological knowledge supported by measurements at some point. The mode of occurrence, size and shape are inferred by analogy with nearby deposits included in R-1, by general geological and structural considerations and by analysis of direct or indirect indications of mineral deposition. Less evidence can be placed on estimates of quantities in this category than those in R-1; estimation errors may be greater than 50%. The estimates in R-2 are relevant mostly for planning further exploration with an expectation of eventual reclassification to category R-1.

Category R-3 resources are undiscovered but are thought to exist in discoverable deposits of generally recognized types. Estimates of *in-situ* quantities are made mostly on the basis of geological extrapolation, geophysical or geochemical indications or statistical analogy. The existence and size of any deposits in this category are necessarily speculative. They may or may not be discovered within the next few decades. Estimates for R-3 suggest the extent of exploration opportunities and the somewhat longer-range prospects for raw material supply. Their low degree of reliability should be reflected by reporting in ranges.

H: the Institution of Mining and Metallurgy definitions of reserves and resources

1 Mineral reserve is that portion of a mineral resource on which technical and economic studies have been carried out to demonstrate that it can justify extraction at the time of determination and under specified economic conditions.

1A Proved mineral reserve is that portion of a measured mineral resource as defined on which detailed technical and economic studies have been carried out to demonstrate that it can justify extraction at the time of the determination and under specified conditions.

1B Probable mineral reserve is that portion of a measured and/or indicated resource as defined on which sufficient technical and economic studies have been carried out to demonstrate that it can justify extraction at the time of the determination and under specified economic conditions.

2 Mineral resource is a tonnage or volume of rock or mineralization or other material of intrinsic economic interest the grades, limits and other appropriate characteristics of which are known with a specified degree of knowledge.

2A Measured mineral resource is that portion of a mineral resource for which tonnage or volume is calculated from dimensions revealed in outcrops, pits, trenches, drill-holes or mine workings, supported where appropriate by other exploration techniques. The sites used for inspection, sampling and measurement are so spaced that geological character, continuity, grades and nature of the material are so well defined that the physical character, size, shape, quality and mineral content are established with a high degree of certainty.

2B Indicated mineral resource is that portion of a mineral resource for which quantity and quality are estimated with a lower degree of certainty than for a measured mineral resource. The sites used for inspection, sampling and measurement are too widely or inappropriately spaced to enable the material or its continuity to be defined or its grade throughout to be established.

3 Mineral potential describes a body of rock or mineralization or other material or an area for which evidence exists to suggest that it is worthy of investigation but to which neither volume, tonnage nor grade shall be assigned.

Appendix 2: Cash flow models

Extract from a typical cash flow model

Year	1994	1995	1996	1997	1998	1999
Period	1	2	3	4	5	6
Production						
East						
Tonnes	1839	1839	1839	1800	900	0
Grade	1.37	1.32	1.32	1.31	1.31	0
West/central						
Tonnes	0	591	1221	3600	4500	5400
Grade	0.00	1.32	1.27	1.08	1.10	1.06
Total						
Tonnes	1839	2430	3060	5400	5400	5400
Grade	1.37	1.32	1.30	1.15	1.14	1.06
Copper recovery	0.92	0.92	0.92	0.92	0.92	0.92
Copper concentrate grade	0.38	0.38	0.385	0.385	0.4	0.4
Payable metals produced						
Copper	22 246	28 339	35 177	55 029	54 305	50 738
Gold	61	77	95	148	140	131
Silver	2728	3475	4255	6657	6315	5900
Metals prices						
Copper	0.950	0.950	0.950	0.950	0.950	0.950
Gold	330.000	330.000	330.000	330.000	330.000	330.000
Silver	3.500	3.500	3.500	3.500	3.500	3.500
Gross value of payable metals						
Copper	$46 591	$59 353	$73 675	$115 253	$113 737	$106 265
Gold	$643	$819	$1003	$1570	$1489	$1391
Silver	$307	$391	$479	$749	$711	$664
Total	$47 541	$60 563	$75 157	$117 572	$115 936	$108 320
Smelting and refining charges	$13 037	$17 047	$21 061	$33 222	$32 200	$30 268
Net smelter return	$34 504	$44 533	$55 495	$87 176	$86 785	$81 325
Operating costs						
Mining	$8457	$11 903	$18 006	$21 312	$17 145	$12 744
Plant	$6830	$8720	$10 412	$16 701	$17 047	$17 047
Eng & services	$1250	$1652	$2080	$3672	$3672	$3672
G & A	$2825	$2825	$2825	$2825	$2825	$2825
Concentrate transport	$1061	$1359	$1666	$2611	$2479	$2320
Total	$20 423	$26 459	$34 989	$47 121	$43 169	$38 607
Operating profit	$14 694	$19 577	$25 592	$44 252	$43 616	$42 717

Suggested cash flow model

Year	1994	1994	1995	1996	1997 to 1998	1999 to 2001
Period	1	2	3	4	5	6
Production						
East						
Tonnes	919	919	1839	1800	1800	0
Grade	1.37	1.32	1.32	1.31	1.31	0
West/central						
Tonnes	0	296	1221	3600	9000	16 200
Grade	0.00	1.32	1.27	1.08	1.10	1.06
Total						
Tonnes	919	1215	3060	5400	10 800	16 200
Grade	1.37	1.32	1.30	1.15	1.14	1.06
Copper recovery	0.92	0.92	0.92	0.92	0.92	0.92
Copper concentrate grade	0.38	0.38	0.385	0.385	0.4	0.4
Payable metals produced						
Copper	11 123	14 170	35 177	55 029	108 611	152 214
Gold	30	39	95	148	281	393

Year	1994	1995	1996	1997	1998	1999
Period	1	2	3	4	5	6
Silver	1364	1737	4255	6657	12 629	17 700
Metals prices						
Copper	0.950	0.950	0.950	0.950	0.950	0.950
Gold	330.000	330.000	330.000	330.000	330.000	330.000
Silver	3.500	3.500	3.500	3.500	3.500	3.500
Gross value of payable metals						
Copper	$23 295	$29 677	$73 675	$115 253	$227 473	$318 796
Gold	$322	$410	$1003	$1570	$2978	$4173
Silver	$153	$196	$479	$749	$1421	$1992
Total	$23 770	$30 282	$75 157	$117 572	$231 872	$324 961
Smelting and refining charges	$6518	$8523	$21 061	$33 222	$64 401	$90 804
Net smelter return	$17 252	$22 267	$55 495	$87 176	$173 570	$243 974
Operating costs						
Mining	$4229	$5951	$18 006	$21 312	$34 290	$38 232
Plant	$4511	$5456	$10 412	$16 701	$31 557	$46 066
Eng & services	$625	$826	$2080	$3672	$7344	$11 016
G & A	$2825	$2825	$2825	$2825	$2825	$2825
Concentrate transport	$531	$680	$1666	$2611	$4959	$6959
Total	$12 720	$15 738	$34 989	$47 121	$80 975	$105 098
Operating profit	$5145	$8031	$25 592	$44 252	$92 595	$138 876
Confidence						
Grade	4%	3%	3%	5%	5%	6%
Tonnage	5%	4%	3%	3%	4%	4%

References

ASSOCIATION OF PROFESSIONAL ENGINEERS OF ONTARIO 1986. *Guidelines for Professional Engineers Reporting on Mineral Properties*. Toronto, Canada.

AUSTRALIAN INSTITUTION OF MINING AND METALLURGY AND AUSTRALIAN MINING COUNCIL 1985. *Reporting of Ore Reserves*.

BANFIELD, A. F. & HAVARD, J. F. 1975. Let's Define Our Terms in Mineral Valuation. *Mining Engineering*, July, 74–78.

FETTWEISS, G. B. 1979. *World Coal Resources: Methods of Assessment and Results*. Developments in Economic Geology, **10**, Elsevier.

SABOURIN, R. L. 1984. Application of a geostatistical method to quantitatively define various categories of resources. *In:* VERLY, G. *et al.* (eds) *Geostatistics for Natural Resource Characterisation Part 1*, Reidel, 201–215.

SCHANZ, J. J. 1980. The United Nations endeavour to standardise mineral resource classification. *Natural Resources Forum*, **4**, 307–313.

US BUREAU OF MINES AND US GEOLOGICAL SURVEY 1976. *Mineral Resource Classification Systems of the USBM and USGS*. USGS Bulletin **1450-A**.

—— 1980. *Principles of a Reserve/Resource Classification for Minerals*. USGS circular **831**.

The optimal design of quarries

P. A. DOWD

Department of Mining and Mineral Engineering, University of Leeds, Leeds LS2 9JT, UK

> This paper describes the various general approaches to the optimal design of pits and quarries. Three types of deposit and corresponding numerical models can be distinguished and in each case a particular method of optimization can be used. In the most general case computer methods must be used to construct optimal pit/quarry outlines on the basis of a regular, three-dimensional revenue/grade block model. The paper includes a Fortran algorithm for a simple implementation of the Lerchs–Grossmann method for optimal pit/quarry design.

Over the past 30 years a great deal of effort has been expended in the metalliferous mining industry on the optimal design and scheduling of open pits, i.e. on the shape of the pit (the total amount of ore and waste to be mined) and the sequence in which individual parcels of ore and waste are to be mined. Optimization requires a criterion and this is usually defined as maximum net profit or maximum net present value.

The optimal design of pits and quarries should not be confused with computer aided design (CAD) techniques although computers are usually required to implement optimal designs. CAD design of pits and quarries is now supplied as an option in most quarry and pit evaluation packages (e.g. SURPAC). These packages are essentially computerized versions of hand drawn methods of pit design and there is no intention of providing an optimal design in the sense that any specified criterion is minimized or maximized.

Most software suppliers now offer the ability to interface with a proprietary open pit design package from a third-party supplier. The most well-known of these are the Whittle 3-D and 4-D packages (Whittle 1989). However, for low cost industrial minerals operators the costs of such packages are often prohibitive.

The purpose of this paper is to introduce the concepts of optimal quarry design and to stimulate the use of optimizing techniques by supplying a basic computer program for the implementation of general optimization.

Optimal pit/quarry design

Optimal design begins by building a model of the mineralization. For this purpose three types of deposit and corresponding models can be distinguished.

(1) Stratigraphically confined deposits of uniform quality or grade consisting usually of a single stratigraphic unit. In this case the model consists of the estimated ore boundaries.

(2) Stratigraphically confined deposits in which the ore quality or grade varies throughout the deposit. For a single stratigraphic unit the model will consist of the estimated average grade between the footwall and hangingwall for specified vertical increments (perhaps corresponding to bench heights).

(3) Disseminated deposits or stratigraphic deposits which are erratic and/or consist of multiple units. This is the most general case. The model consists of the estimated grades of blocks into which the deposit has been subdivided

Each deposit model is then converted to a revenue model by applying costs and prices to the ore and waste tonnages within the different ore outlines.

The most common optimization criterion is to design the quarry in such a way that the extracted ore and associated waste will yield the maximum net profit. Ideally, the optimization criterion should be the maximum net present value. However, it is not possible to use this criterion as a direct part of an optimizing procedure because it involves an insoluble circular argument. In order to define the net present value of a parcel of ore or mineral, the time at which the ore is mined must be known; however, the time at which a parcel of ore is mined will not be known until the quarry is designed but the quarry cannot be designed until the present values of all parcels of ore are defined. A common approach to this problem is to design the quarry on the basis of maximum net profit and then schedule the quarrying operation within this shape so as to achieve maximum net present value.

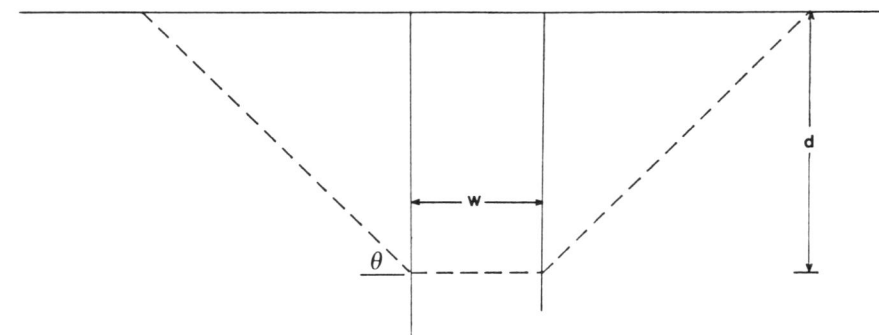

Fig. 1. Simple, stratigraphically defined deposit.

Table 1. *Derivation of optimum mining depth for case illustrated in Fig. 1*

	S	selling price per tonne of ore
	C_o	cost of mining one tonne of ore
	C_w	cost of mining one tonne of waste
	C_t	cost of processing one tonne of ore
	g_o	specific gravity of ore
	g_w	specific gravity of waste
	d	depth of mining
	l	strike length of orebody
	w	width of orebody
	θ	wall slope of pit
	r	processing recovery
Tonnage of ore mined	$T_o =$	$d \times w \times l \times g_o$
Tonnage of waste	$T_w =$	$2 \times \tfrac{1}{2} \times d \times d/\tan\theta \times l \times g_w$
Profit	$=$	$S \times r \times T_o - C_o \times T_o - C_t \times T_o - C_w \times T_w$
	$=$	$d \times w \times l \times g_o \times (rS - C_o - C_t) - d^2 \times C_w \times g_w \times l/\tan\theta$

Differentiating profit with respect to depth and setting to zero gives optimum mining depth:

$$d_{opt} = \frac{w \times \tan\theta \times g_o \times (rS - C_o - C_t)}{2 \times g_w \times C_w}$$

The approach to optimization for each of the three categories of deposit is summarized in the following sections.

Stratigraphically confined deposits of uniform quality or grade consisting usually of a single stratigraphic unit

The simplest case is in the mining of dipping, stratigraphically defined structures of uniform quality or grade as shown in Fig. 1. As the pit is deepened more and more waste must be removed. Here the pit shape can be defined as a function of the net value of mining ore and waste down to a given depth. Once the pit/quarry slopes are defined the object is to determine the depth which gives the maximum profit. Simple calculus can be used to determine the optimal depth and thus optimal pit shape. To illustrate this consider the simple case shown in Fig. 1.

Assume that the ore has constant width w and a strike length of l. Table 1 shows the derivation of the optimal mining depth. Similar, if more complex, formulas can be derived for morerealistically shaped and oriented stratigraphic deposits or sequences.

Stratigraphically confined deposits in which the ore quality or grade varies throughout the deposit

When the ore is not of constant grade the determination of the optimum quarry is not as simple.

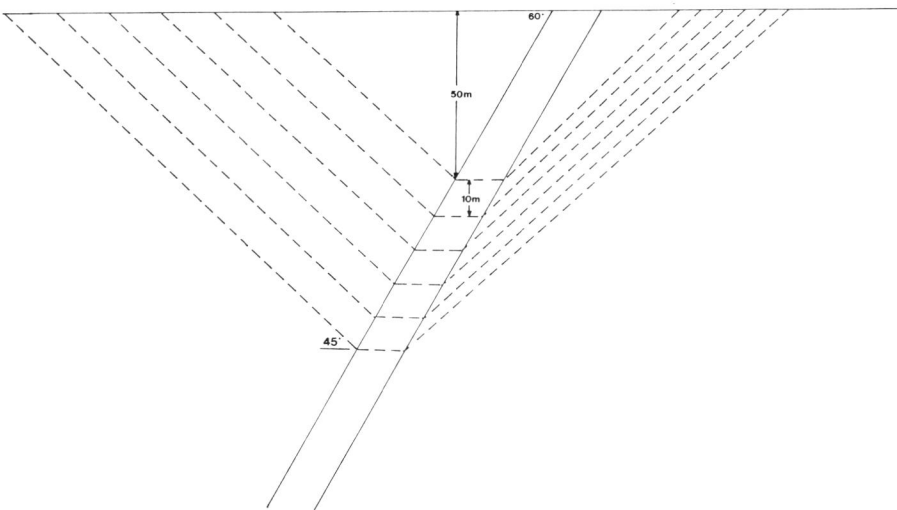

Fig. 2. Cross-sectional view of fluorspar vein.

Table 2. *Calculation of revenue for pit optimization example.*

Depth (m)	Ore volume (m³)	Waste volume (m³)	stripping ratio	Grade (%CaF$_2$)	recoverable tonnes	Value of ore (t)
50	693	2501	3.6:1	35.0	668	93 520
60	139	1099	7.9:1	40.0	153	21 420
70	139	1301	9.4:1	45.0	172	24 080
80	139	1503	10.8:1	50.0	191	26 740
90	139	1705	12.3:1	42.0	161	22 540
100	139	1907	13.7:1	38.0	146	20 440
110	139	2109	15.2:1	30.0	115	16 100
120	139	2311	16.6:1	32.0	123	17 220

Table 3. *Net profit calculations for pit optimization example*

Depth (m)	Ore mining costs			Waste mining costs			Proc. transp. cost	Total costs inc o/h	Ore value Table 2	Net profit
	Cost (£ t⁻¹)	tonnes	Est. cost	Cost (£ t⁻¹)	tonnes	Est. cost				
50	4.50	2010	9045	3.50	5252	18 382	20 049	52 225	93 520	41 295
60	4.60	403	1854	3.60	2308	8309	4596	16 235	21 448	5213
70	4.70	403	1895	3.80	2732	10 382	5169	19 191	24 126	4935
80	4.85	403	1955	4.00	3156	12 625	5745	22 358	26 810	4452
90	5.00	403	2016	4.15	3581	14 862	4824	23 869	22 517	−1352
100	5.15	403	2076	4.30	4005	17 222	4503	26 179	21 016	−5316
110	5.30	403	2136	4.45	4429	19 709	3447	27 821	16 086	−11 735
120	5.50	403	2217	4.60	4853	22 324	3675	31 038	17 156	−13 882

As an example consider the cross-sectional view of a fluorspar vein shown in Fig. 2. The vein is dipping at 60° and the quarry wall slopes are 45°. An initial quarry is taken to a depth of 50 m and additional vertical increments of 10 m are considered. A selling price of £140.00 per tonne is assumed with total processing costs of £25.00 per tonne of finished product, transport charges of £5.00 per tonne of finished product and overheads of 10% of total costs. The average fluorspar grade of the initial quarry and the average grades of the successive 10 m vertical increments are shown in Table 2 together with the corresponding volumes (per metre of strike length) of waste and ore. Mill recovery is taken as 95% and the specific gravity of ore and waste are 2.9 t m^{-3} and 2.1 t m^{-3} respectively.

The total *in situ* values of the ore in the initial quarry and each of the 10 m increments are shown in Table 2. Mining costs per tonne of ore and waste are shown in Table 3. The net profit calculations for the initial quarry are as follows:

Ore tonnage   = 693×2.9 = 2010 tonnes
Waste tonnage   = 2501×2.1 = 5252 tonnes
Recovered tonnage   = 2010×0.35×0.95 = 668.3 tonnes
Revenue from recovered
  fluorspar   = 140.00× 668   = £93 520
Ore mining cost   = 2010×4.50   = £9045
Waste mining cost   = 5252×3.50   = £18 382
Processing cost   = 668.3×25.00   = £16 708
Transport cost   = 668.3×5.00   = £3 342
  Total costs   = £47 477
  plus 10% overheads   = £52 225
  Net profit   = £41 295

Similar calculations can be done for each 10 m increment; the results are shown in Table 3.

From Table 3 it is seen that the optimal depth of the quarry is 80 m (or somewhere between 80 m and 90 m). The total net profit could, however, begin to increase at lower depths if the grade of the deposit increases sufficiently. The mining increments, or pushbacks could be made to coincide with a year's production, in which case the net profit increments could be discounted to give a NPV.

Disseminated deposits or stratigraphic deposits which are erratic and/or consist of multiple units

In the general case the solution is not as simple as those described under the previous two headings. In these cases the ore may not be confined to observable geological boundaries and/or it may be disseminated through the host rocks; alternatively the occurrence of the ore may be stratigraphically controlled but only on a small or erratic scale, e.g. fractures and erratic, small veins. The orebody is usually represented in the form of a block model: the orebody is subdivided into a regular, rectangular array of blocks each of which has an estimated grade value. As the estimated block grades are obtained from samples (usually a drilling grid) this will effectively limit the lower size of the blocks.

Each block is assigned an estimated revenue value calculated from the estimated block grade, costs and prices. The object now is to identify the combination of blocks which satisfy mining constraints and which, when extracted, will yield the maximum profit. This is illustrated by the simple two-dimensional example in Fig. 3.

The problem now is to determine, from the vast number of possible pit or quarry shapes, the one which maximizes the total profit.

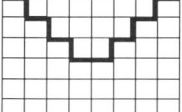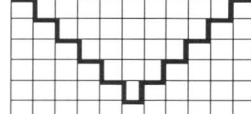

Fig. 3. Two possible quarry shapes from the same rectangular array of blocks; each quarry will have a different profit value.

Moving cones

This is the simplest method of determining the optimal quarry shape and is the most widely used of the heuristic algorithms. Each block is assigned a cone which is defined by the quarry wall slopes in all directions around the block; if the wall slopes change significantly with direction the cone will be a pseudo cone. This cone is called a *removal cone* as it defines all blocks which must be mined in order to mine the block on which the cone is positioned. A pit or quarry is now seen as a combination of cones as shown by the simple two-dimensional example in Fig. 4 in which the combined removal cones for blocks X, Y and Z form a quarry/pit outline.

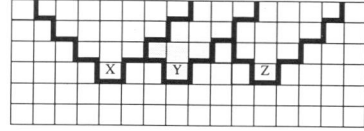

Fig. 4. Quarry outline formed by the combined removal cones of three blocks.

The value of a removal cone is the total value of the blocks within it. The optimal quarry shape is then the combination of removal cones with the highest total value.

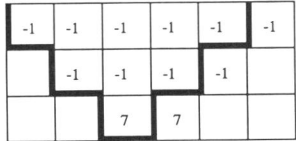

Fig. 5. Removal cone for first positive block.

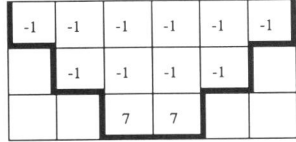

Fig. 6. Removal cone for combined positive blocks.

There is a serious problem with moving cone methods as illustrated by the simple example in Fig. 5 in which the numbers in the blocks are the net profits obtained by mining them. The removal cone for the first positive block, as shown in Fig. 5, has a value of $7-8=-1$ and the block is rejected. Similarly, the removal cone of the second positive block has a value of -1 and it too is rejected. However, if the two blocks are considered together as shown in Fig. 6 then the removal cone for the pair of blocks has a value of $7+7-10=+4$ and the two blocks can be profitably mined.

It is thus not sufficient to consider the removal cone of each block independently of all other removal cones which intersect it. Various techniques have been proposed to overcome this problem but the numerous forms of the moving cone method remain heuristic algorithms for which rigorous proofs of optimization are not possible and for which a counter example of non-optimization can usually be found.

The Lerchs–Grossmann algorithm

The Lerchs–Grossmann algorithm (Lerchs & Grossmann 1965) converts the three-dimensional grid of blocks in the orebody model into a directed graph. Each block in the grid is represented by a vertex which is assigned a mass equal to the net revenue value of the corresponding block. The vertices are connected by arcs in such a way that the connections leading from a particular vertex to the surface define the set of vertices (blocks) which must be removed if that vertex (block) is to be mined. A simple, two-dimensional example is shown in Fig. 7.

Vertices connected by an arc pointing away from a vertex are termed successors of that vertex, i.e. the vertex y is a successor of the vertex x if there exists an arc directed from x to y. The set of all successors of x is denoted Γx. For example, in Fig. 7, $\Gamma x_9 = \{x_2, x_3, x_4\}$. A closure of a directed graph, which consists of a set of vertices X, is a set of vertices $Y \subset X$ such that if $x \subset Y$ then $\Gamma\ x \in Y$. For example, in Fig. 7, $Y = \{x_1, x_2, x_3, x_4, x_5, x_8, x_9, x_{10}\}$ is a closure of the directed graph. The value of a closure is the sum of the mases (revenue values) of the vertices in the closure. Each closure defines a possible pit; the closure with the maximum value defines the optimal pit.

This method is the only method which can be proved rigorously, mathematically always to lead to the correct optimal solution. However, a number of recently published new methods (Dowd & Onur 1992, 1993) have also made similar claims but they remain to be independently verified.

Mining sequence planning

Ideally, optimal quarry/pit limits should be determined on the basis of optimizing net present value. However, the problem, as formulated, is intractable. It is not possible to assign a net present value to a block until it is known when the block is to be mined but the time at which a block is mined is not known until the pit is designed.

The problem could be formulated as a constrained scheduling problem thereby avoiding the circular constraints referred to above. However, such a solution is yet to be formulated and/or achieved within reasonable computing time. The most common approach to the problem is to design an optimal pit/quarry shell using maximum (undiscounted) profit as the optimizing criterion and then to schedule the blocks within this shell in such a way as to maximize net present value. There is a growing number of publications in optimal scheduling of open pit mining (Onur 1992; Onur & Dowd 1993) but, in general, the problem has not attracted the same amount of interest as the optimal open pit design problem. Some authors have tackled the simpler but related problem of optimal overburden removal.

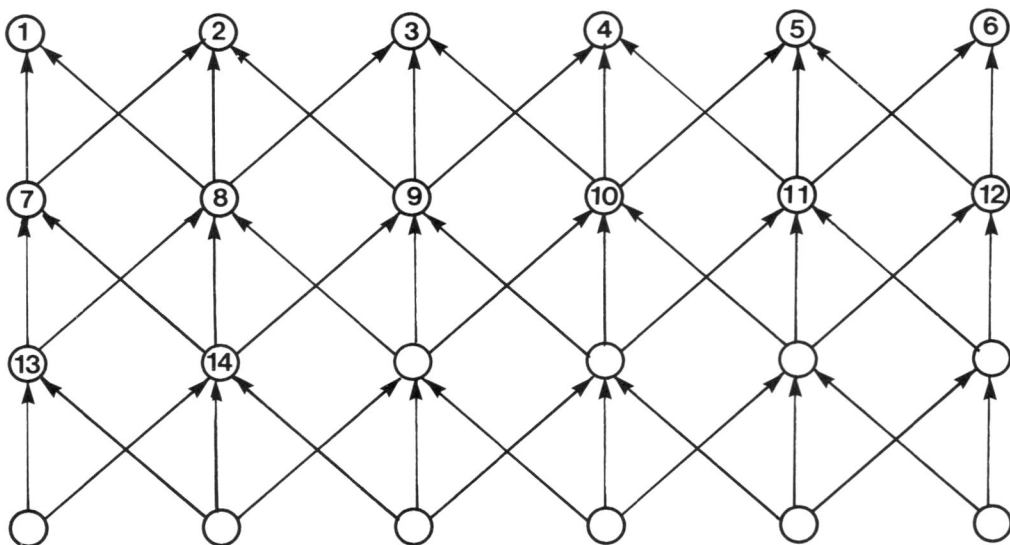

Fig. 7. Directed graph representing two-dimensional deposit model; nodes represent blocks and arcs define mining constraints.

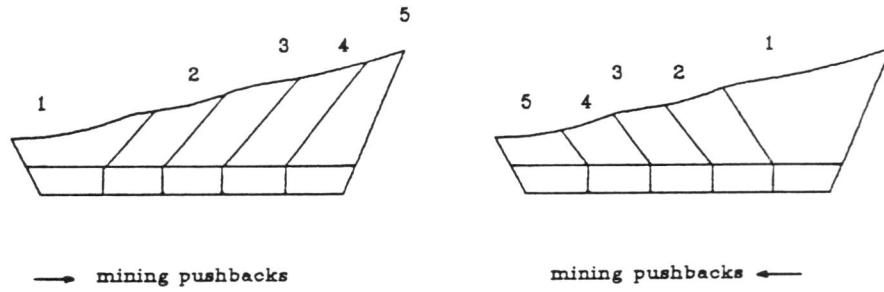

Fig. 8. Two possible mining sequences.

Table 4. *Discounted cash flows from two mining sequences*

Time period	Alternative 1 profit	Alternative 2 profit	Discount rate	Discounted alternative 1	Discounted alternative 2
1	100	60	0.909	90.0	54.5
2	90	70	0.826	74.3	57.8
3	80	80	0.751	60.1	60.1
4	70	90	0.683	47.8	61.5
5	60	100	0.621	37.3	62.1
Total	400	400		310.4	296.0

The simple example shown in Fig. 8 illustrates the importance of net present value in the determination of mining sequences. In this example the same section of an orebody is depleted in the same time period with a discount rate of 10% by using two different planning sequences. The results are shown in Table 4.

It can be seen from Table 4 that the best way of mining the orebody is to mine the most profitable parts during the early stages of the mining operation.

Open pit or quarry production scheduling is

the development of a sequence of depletion schedules leading from the initial conditions of the deposit, to the ultimate pit limits. Production scheduling may be long range or short range depending on the duration of the scheduling periods. Long range production scheduling is mainly concerned with such items as ore reserves, stripping ratios and major investment usually on a year by year basis. Short range scheduling, on the other hand, is the development of a sequence of depletion schedules on a daily, weekly or monthly basis, which complies with restrictions imposed by the long range plans, plant capacities, inventories, equipment availability and the exising mining operation.

Regardless of the type of ore or mineral product mined, there are certain basic data that are required in any production scheduling problem. These are:

(1) the tonnage and grade of each block;
(2) specific gravity of ore and waste;
(3) the revenue value of each block;
(4) deposit block model;
(5) present pit/quarry layout or overall optimum limits of the pit/quarry to be scheduled;
(6) mine life or production rate (depending on whether short term or long term scheduling is used);
(7) the maximum and minimum allowable grade to be mined or to be fed to the processing plant in any time period;
(8) the maximum and minimum allowable production rate of waste and ore;
(9) working slope angle in the pit;
(10) minimum pit bottom dimensions;
(11) discount rate;
(12) preproduction rate and period (if required).

Various methods are available to solve the scheduling problem including linear programming, goal programming and dynamic programming (Dowd 1976; Dowd & Elvan 1987; Onur 1992; Onur & Dowd 1993).

Sensitivity and risk analysis

The optimal pit/quarry is a function of a large number of variables, the most important of which are grades, costs, prices, geomechanics and wall slopes. Many of these variables are unknown and must be estimated.

In many cases the pit/quarry shape and the resulting profit will change dramatically when grades, costs and prices change. By using a computer based optimal design package the sensitivity of the pit design to changes in the values of these variables can be assessed (David et al. 1974; Dowd & David 1976; Dowd 1980). Such an assessment is called sensitivity analysis. The assessment of the sensitivity of the pit/quarry design to probabilistic changes in all of these variables is called risk analysis. One powerful form of risk analysis is to use geostatistical simulation (David et al. 1974; Dowd & David 1976; Dowd 1980) to provide a series of alternative deposit models based on different amounts and quality of data. Each alternative deposit model can be combined with different values of economic and financial parameters to design a new pit/quarry and calculate the resulting cash flow. An analysis of these cash flows will quantify the risk involved in the project.

Practical implementation of optimal design software

Commercial open pit design software packages can be purchased but these are generally very expensive. In-house development of such software can also be prohibitively expensive. In the interests of stimulating the use of computer aided optimal pit/quarry design a listing of a Fortran program for the Lerchs–Grossmann algorithm is given in the Appendix to this paper. This is a very elementary implementation of the algorithm and it is intended as a basis for further development rather than a final product. The program is written for fixed slopes which are governed by the block dimensions. In addition it is assumed that a block revenue model is available.

Advantages of optimum quarry design

The major advantages of implementing a computer based optimal pit/quarry design algorithm are:

- Economic   quarries are designed so as give maximum return on capital;
- Technical   quarries can be generated very rapidly for a whole range of changes in variables—costs, prices, geological conditions, mining equipment, mining method, *in situ* grades, wall slopes, processing assumptions, etc;
- Risk   risk and sensitivity analyses can be performed very rapidly;
- Cost saving   quarry/pit shapes can be generated very rapidly in a fraction of the time it takes to design pits manually.

Appendix: Fortran listing of Lerchs–Grossmann algorithm

```
***********************************************************************
*                                                                     *
*     Lerchs-Grossman method for determining optimal open pit limits  *
*                                                                     *
* The deposit is divided into a three-dimensional array of rectangular*
* blocks and a profit value is assigned to each.  These values are stored*
* in a three-dimensional matrix VAL(i,j,k) with dimensions :          *
*                 numx  -  number of rows                             *
*                 numy  -  number of columns                          *
*                 numz  -  number of levels                           *
* The maximum values of these dimensions are set in a parameter state-*
* ment to nx, ny and nz respectively                                  *
* LOG(i,j,k) is a matrix with the same dimensions as VAL and which is *
* used to indicate whether block (i,j,k) is inside (=1) or outside (=0)*
* the pit                                                             *
* IPLAN(i,j) has dimensions IK and JK respectively and defines contours*
* of the optimum pit by storing the pit level at horizontal location  *
* (i,j) :                                                             *
* If IPLAN(i,j) = 0 none of the blocks in cell (i,j) have been mined. *
* If IPLAN(i,j) = k the blocks in cell (i,j) on level k and above have*
* been mined.                                                         *
* Other working matrices have dimensions :                            *
* IROOT(lkm,2), ITREE(nem), IPATH(ipkm,3), ND(nem,2), D(nem), NORM(knm)*
* where lkm, nem, ipkm and knm are experimental; a sufficient value for*
* each of these is usually numx*numy*numz/20                          *
*                                                                     *
* Based on an original coding by S. Korobov.  Adapted by P.A. Dowd    *
*                                                                     *
***********************************************************************

      Parameter (lkmax=2000,nemax=2000,ipkmax=2000,knmax=2000)
      Parameter (nxmax=50, nymax=50, nzmax=10)

      Dimension val(nxmax,nymax,nzmax),log(nxmax,nymax,nzmax),d(nemax),
     1          iplan(nxmax,nymax),iroot(nemax,2),itree(nemax),
     2          nd(nemax,2),ipath(ipkmax,3),norm(knmax)

      data pi/3.14159263/

900   format(40f9.1)
905   Format(1x,40i2)
910   format(/20x,'Total value of blocks included in pit = ',
     1       E15.5,//)
920   Format(' Enter the number of blocks in the x, y and z directions',
     1       ' respectively'/)
925   Format(' Enter the x, y and z dimensions respectively of the',
     1       ' blocks'/)

      Read in data

      Write(6,920)
      read(5,*) numx,numy,numz
      write(6,925)
      read(5,*) ixdim,iydim,izdim
      do 5 k=1,numz
         do 5 i=1,numx
            read(4,*)   (val(i,j,k),j=1,numy)
5     continue

      Initialise arrays and variables

      Do 10 i=1,numx
```

```
      do 10 j=1,numy
        iplan(i,j)=0
10    continue
      do 15 i=1,numx
        do 15 j=1,numy
          do 15 k=1,numz
            log(i,j,k)=0
            if (val(i,j,k).eq.0.00) log(i,j,k)=1
15    continue
      s=0.
      lk=0
      ne=0
      nem=0
      ipkm=0
      lkm=0
      knm=0
```

Begin with uppermost level of blocks and remove all positive valued
blocks. These blocks belong to optimal open pit : add their values
to s, record their inclusion in the pit via log(i,j,1) and add them
to the contour array iplan(i,j)

```
      do 20 i=1,numx
        do 20 j=1,numy
          if (val(i,j,1).le.0) go to 20
          s=s+val(i,j,1)
          log(i,j,1)=1
          iplan(i,j)=1
20    continue
```

Increment the level counter (k) by 1 and add the blocks on the Kth
level

```
      k=1
25    if (k.ge.numz) go to 400
      k=k+1
30    if (lk.ge.0) go to 35
32    ks=k
      go to 380
35    ltr=1
```

Connect blocks on Kth level to the root and establish trees
itree   contains tree number
d   contains value of tree

```
40    nts=iroot(ltr,1)
      nds=itree(nts)
      if (d(nds).gt.0.) go to 55
41    if (ltr.lt.lk) then
          ltr=ltr+1
          go to 40
      endif

42    ltc=1
45    if (ltc.gt.lk) then
          ks=k
          go to 380
      endif
      nts=iroot(ltc,1)
      nds=itree(nts)
      if (d(nds).le.0.) go to 50
      lar=ltc
      lsw=4
      go to 340
```

```
    50  ltc=ltc+1
        go to 45

    55  lar=ltr
    60  lsw=1
        go to 340
    65  call coord(node,numx,numy,k1,j,i)
        if (k1.eq.1) go to 350
    70  ny=(lg-1)*numx*numy+(n-1)*numx+m
        do 80 l=1,lk
           lir=l
           ntw=iroot(l,1)
           ndw=itree(ntw)
           if (d(ndw).gt.0.) go to 80
           ntk=ntw+iroot(l,2)-1
           do 75 lt=ntw,ntk
              na=itree(lt)
              na1=nd(na,1)
              na2=nd(na,2)
              if (ny.eq.na1) go to 95
              if (ny.eq.na2) go to 95
    75     continue
    80  continue
        go to 350

    90  ny=(lg-1)*numx*numy+(n-1)*numx+m
        cpm=val(m,n,lg)
        log(m,n,lg)=2
        ne=ne+1
        nd(ne,1)=0
        nd(ne,2)=ny
        d(ne)=cpm
        itree(ne)=ne
        lk=lk+1
        iroot(lk,1)=ne
        iroot(lk,2)=1
        if (lkm.lt.Lk) lkm=lk
        if (nem.lt.ne) nem=ne
        lir=lk
    95  nd(nds,1)=node
        nd(nds,2)=ny
        mbw=iroot(lir,1)
        mew=iroot(lir,2)+mbw-1
        mbs=iroot(lar,1)
        mes=mbs+iroot(lar,2)-1
        iroot(lir,2)=iroot(lir,2)+iroot(lar,2)
        iroot(lar,1)=0
        iroot(lar,2)=0
        if (mew+1-mbs) 100,140,120

   100  ires=itree(mbs)
        n1=mew+1
        n2=mbs-1
        do 105 n=n1,n2
           nf=n2-n+n1
           itree(nf+1)=itree(nf)
   105  continue
        itree(mew+1)=ires
        do 110 l=1,lk
           if (iroot(l,1).eq.0) go to 110
           if (.not.(iroot(l,1).gt.mew.and.iroot(l,1).le.mbs)) go to 110
           iroot(l,1)=iroot(l,1)+1
   110  continue
        if (mbs.eq.mes) go to 140
```

```
        mbs=mbs+1
        mew=mew+1
        go to 100

120     do 135 m=mbs,mes
            ires=itree(mbs)
            n1=mbs+1
            n2=mew
            do 125 n=n1,n2
                itree(n-1)=itree(n)
125         continue
            itree(mew)=ires
            mbw=mbw-1
            do 130 l=1,lk
                if (iroot(l,1).eq.0) go to 130
                if (.not.(iroot(l,1).ge.mbs.and.iroot(l,1).le.mew))go to 130
                iroot(l,1)=iroot(l,1)-1
130         continue
135     continue

140     lcon=1
        go to 310
145     continue
        ipa=ip
150     n=ipath(ipa,1)
        if (n.eq.nds) go to 155
        d(n)=d(nds)-d(n)
        ipa=ipath(ipa,3)
        if (ipa.ne.0) go to 150
155     lar=lir
        lsw=3
        go to 340
160     if (node.ne.ny) go to 350
        ipa=ip
165     nn=ipath(ipa,1)
        d(nn)=d(nn)+d(nds)
        ipa=ipath(ipa,3)
        if (ipa.ne.0) go to 165
170     kn=1
        norm(kn)=lir
175     do 180 kt=1,kn
            if (norm(kt).eq.0) go to 180
            lar=norm(kt)
            lsw=2
            go to 340
180     continue
        go to 30
185     continue
        do 190 ip=1,ipk
            if (ip.eq.1) go to 190
            md=ipath(ip,1)
            nod=iabs(ipath(ip,2))
            if (ipath(ip,2).lt.0.and.d(md).le.0.) go to 195
            if (ipath(ip,2).gt.0.and.d(md).gt.0.) go to 195
190     continue
        norm(kt)=0
        go to 175
195     nd(md,1)=0
        nd(md,2)=nod
        nod1=nod
        ip1=ip
200     iq=ipath(ip,3)
        md1=ipath(iq,1)
        d(md1)=d(md1)-d(md)
```

```
          if (ipath(iq,3).eq.0) go to 205
          ip=iq
          go to 200
      205 do 230 iq=ip1,ipk
             mc=ipath(iq,1)
             ndc=iabs(ipath(iq,2))
             naf=ipath(iq,3)
             if (ndc.eq.nod1) go to 215
             ip=naf
      210    if (ip.eq.ip1) go to 215
             ip=ipath(ip,3)
             if (ip.lt.ip1) go to 230
             go to 210
      215    do 225 n=nit,nitk
                if (itree(n).ne.mc) go to 225
                if (n.eq.nitk) go to 225
                mem=itree(n)
                n1=n+1
                do 220 nz=n1,nitk
                   itree(nz-1)=itree(nz)
      220       continue
                itree(nitk)=mem
                go to 230
      225    continue
      230 continue
          do 235 n=nit,nitk
             if (itree(n).eq.md) go to 240
      235 continue
      240 iroot(lar,2)=n-nit
          lk=lk+1
          iroot(lk,1)=n
          iroot(lk,2)=nitk-n+1
          kn=kn+1
          norm(kn)=lk
          if (knm.lt.kn) knm=kn
          if (lkm.lt.lk) lkm=lk
          go to 175

      245 n=ipath(1,1)
          s=s+d(n)
          do 250 ip=1,ipk
             n=ipath(ip,1)
             nd(n,1)=0
             nd(n,2)=0
             d(n)=0.
             node=iabs(ipath(ip,2))
             call coord (node,numx,numy,k1,j,i)
             log(i,j,k1)=1
             if (iplan(i,j).lt.k1) iplan(i,j)=k1
      250 continue
          ne1=ne
          n=0
      255 n=n+1
      260 if (n.eq.ne) go to 280
          if (nd(n,2).eq.0) go to 265
          go to 255
      265 n1=n
          n2=ne-1
          do 270 na=n1,n2
             nd(na,1)=nd(na+1,1)
             nd(na,2)=nd(na+1,2)
             d(na)=d(na+1)
      270 continue
          ne=ne-1
```

```
      m1=ne1
      do 275 m=1,m1
         if (itree(m).gt.n) itree(m)=itree(m)-1
275   continue
      go to 260
280   if (nd(ne,2).eq.0) ne=ne-1
      do 285 n=nit,nitk
         itree(n)=0
285   continue
      iroot(lar,1)=0
      iroot(lar,2)=0
      lcon=2
      go to 310
290   continue
      if (nitk.eq.ne1) go to 300
      n1=nitk+1
      do 295 n=n1,ne1
         itree(nit+n-n1)=itree(n)
295   continue
300   do 305 l=1,lk
         if (iroot(l,1).lt.nit) go to 305
         iroot(l,1)=iroot(l,1)-nc
305   continue
      go to 42

310   l=0
315   l=l+1
320   if (l.eq.lk) go to 335
      if (iroot(l,1).eq.0) go to 325
      go to 315
325   l1=l
      l2=lk-1
      do 330 la=l1,l2
         iroot(la,1)=iroot(la+1,1)
         iroot(la,2)=iroot(la+1,2)
330   continue
      if (lir.gt.l1) lir=lir-1
      lk=lk-1
335   if (iroot(lk,1).eq.0) lk=lk-1
      if (lk.eq.0) go to 32
      go to (145,290), lcon
340   nit=iroot(lar,1)
      nc=iroot(lar,2)
      nitk=nit+nc-1
      ipk=1
      nnd=itree(nit)
      ipath(1,1)=nnd
      ipath(1,2)=nd(nnd,2)
      ipath(1,3)=0
      ip=1
345   node=iabs(ipath(ip,2))
      nn=ipath(ip,1)
      go to (65,350,160,350),lsw
350   do 360 n=nit,nitk
         nnd=itree(n)
         if (nnd.eq.nn) go to 360
         if (node.ne.nd(nnd,1)) go to 355
         ipk=ipk+1
         ipath(ipk,2)=nd(nnd,2)
         ipath(ipk,1)=nnd
         ipath(ipk,3)=ip
         if (ipkm.lt.ipk) ipkm=ipk
         go to 360
355      if (node.ne.nd(nnd,2)) go to 360
```

```
            ipk=ipk+1
            ipath(ipk,2)=-nd(nnd,1)
            ipath(ipk,1)=nnd
            ipath(ipk,3)=ip
            if (ipkm.lt.ipk) ipkm=ipk
 360    continue
        if (ip-ipk) 365,370,370
 365    ip=ip+1
        go to 345
 370    go to (41,185,450,245),lsw
 380    im=0
        jm=0
        sm=0
        do 385 i=2,numx-1
            do 385 j=2,numy-1
                if (log(i,j,ks).gt.0) go to 385
                if (val(i,j,ks).le.0) go to 385
                if (val(i,j,ks).le.sm) go to 385
                sm=val(i,j,ks)
                im=i
                jm=j
                km=ks
 385    continue
        if (sm.eq.0.) go to 25
        log(im,jm,km)=2
        ne=ne+1
        nd(ne,1)=0
        nd(ne,2)=(km-1)*numx*numy+(jm-1)*numx+im
        d(ne)=sm
        itree(ne)=ne
        lk=lk+1
        iroot(lk,1)=ne
        iroot(lk,2)=1
        if (nem.lt.ne) nem=ne
        if (lkm.lt.lk) lkm=lk
        nds=ne
        ltr=lk
        go to 55

        Print results

 400    write(6,910) s
        kollu=0
        kol=0
        kol1=0
        do 415 k=1,numz
            do 410 j=1,numy
                do 405 i=1,numx
                    if (log(i,j,k).eq.1) iplan(i,j)=k
                    if (log(i,j,k).eq.1.and.val(i,j,k).lt.0) kol=kol+1
                    if (log(i,j,k).eq.1.and.val(i,j,k).gt.0) kol1=kol1+1
                    if (log(i,j,k).eq.1.and.val(i,j,k).ne.0) kollu=kollu+1
 405            continue
 410        continue
 415    continue
        write(6,*)'total number of blocks in  pit',kollu
        write(6,*)'number of positive blocks in pit',kol1
        write(6,*)'number of negative blocks in pit',kol
        do 420 i=1,numx
            write(6,905) (iplan(i,j),j=1,numy)
 420    continue
        stop
 450    zz=1
        write(6,900) zz,nem,lkm,node
```

```
      stop
      end
      subroutine coord(n,numx,numy,k,j,i)

      Subroutine to determine the array index co-ordinates of node (block)
      n given that there are ik (x direction) x jk (y direction) nodes
      on each horizontal level
      Array index co-ordinates are returned as (i,j,k)

      kt=n/(numx*numy)
      k=kt+1
      if (n.eq.kt*numx*numy) k=k-1
del   jt=(n-numx*numy*(k-1))/numx
      j=jt+1
      if ((n-numx*numy*(k-1)).eq.ik*jt) j=j-1
      i=n-numx*numy*(k-1)-numx*(j-1)
      return
      end
```

References

DAVID, M., DOWD, P. A. & KOROBOV, S. 1974. Forecasting departure from planning in open pit design and grade control. *12th Symposium on the application of computers and operations research in the mineral industries (APCOM)*. Colorado School of Mines, 2, F131–F142.

DOWD, P. A. 1976. Application of dynamic and stochastic programming to optimize cutoff grades and production rates. *Transactions of the Institution of Mining and Metallurgy*, 85, A22–A31.

—— 1980. The role of certain taxation systems in the management of mineral resources. *In:* JONES, M. J. (ed.) *National and international management of mineral resources*. IMM, London, 329–335.

—— & ELVAN, L. 1987. Dynamic programming applied to grade control in sub-level open stoping. *Transactions of the Institution of Mining and Metallurgy*, 96, A171–A178.

—— & DAVID, M. 1976. Planning from estimates: sensitivity of mine production schedules to estimation methods. *In:* GUARASCIO, M., DAVID, M. & HUIJBREGTS, C. D. (eds) *Advanced geostatistics in the mining industry*. NATO ASI Series C: Mathematical and Physical Sciences, Reidel Pub. Co. Dordrecht, Netherlands, 163–183.

—— & ONUR, A. H. 1992. Optimizing open pit design and sequencing. *23rd Symposium on the application of computers and operations research in the mineral industries* (APCOM). AIME, Littleton, Co., 411–422.

—— & —— 1993. Open pit optimization: I Optimal pit design. *Transactions of the Institution of Mining and Metallurgy*, 102, A95–A104.

LERCHS, H. & GROSSMANN, I. F. 1965. Optimum design of open pit mines. *CIM Bulletin*, 58, 47–54.

ONUR, A. H. 1992. *Optimal open pit design and planning*. PhD thesis, Department of Mining and Mineral Engineering, University of Leeds, UK.

—— & DOWD, P. A. 1993. Optimization of open pit mine design: II Production scheduling and the inclusion of roadways. *Transactions of the Institution of Mining and Metallurgy*, 102, A105–A113.

WHITTLE, J. 1989. *The facts and fallacies of open pit optimization*. Whittle Programming Pty Ltd, North Balwyn, Victoria, Australia.

Geostatistical estimation of manganese oxide resources at the Nsuta Mine

S. AL-HASSAN & A. E. ANNELS

Mineral Resource Evaluation Research Unit, Department of Geology, University of Wales College of Cardiff; PO Box 914, Cardiff CF1 3YE, UK

The Nsuta manganese deposit is located about 6.5 km south of Tarkwa in the southwestern region of Ghana. It is associated with a carbonate horizon within a thick series of interbedded grey tuffs and thin argillaceous horizons of Precambrian, Upper Birimian age. The manganese oxide mineralization, produced by supergene enrichment of the stratiform carbonate horizon, occurs as massive, patchy or bedded accumulations of psilomelane, with numerous veinlets and cavity fillings of pyrolusite. Due to the impact of folding and faulting and past mining activity (open-pit), the remaining resources have a complex spatial distribution. The study described is based on a portion of this deposit referred to as Hill D (South Crest). Initial statistical studies revealed the existence of two assay populations which appear to correspond to two geologically identifiable ore types that cannot be differentiated at the present scale of mining. Two geostatistical methods (ordinary kriging and indicator kriging) were used to evaluate the resource and the estimates so produced were then compared. A wireframe model was used to constrain the three dimensional geostatistical block model.

Manganese occurrences are known in almost all the Regions of Ghana, but the main deposits are found in the Western Region. However, the Nsuta manganese deposit is economically the most important. The other deposits are either too small or their grades are so low that they are currently uneconomic to exploit.

The Nsuta manganese deposit is situated alongside the Sekondi–Kumasi railway line. It is only about 6.5 km south of Tarkwa (Fig. 1), one of the centres of extensive small scale gold mining in the country. The deposit occupies five hills, designated A, B, C, D, and E, which are oriented along two lines with bearings of N15°E–N20°E and which have been traced continuously for about 4 km. They are interconnected by saddles and some of them are divided into two parts—north and south crests.

The deposit was discovered in 1914 by the then Gold Coast Geological Survey. Open pit mining of the deposit started in 1916 and has been in operation since then.

The manganese oxide ore is generally classified according to grade as:

(i) R-grade (or 'battery-grade') $\geq 52\%$
(ii) standard metallurgical grade 48–52%
(iii) B–grade 45–48%
(iv) C-grade 40–45%

For a long time manganese resources have been estimated at the Nsuta mine using the cross-sectional method. The high grade ore, which contained up to 60% Mn, is now almost depleted and this has led to the beneficiation of lower grade material. The low grade of the mineralization, coupled with escalating operating costs and the volatility of metal prices, demands more careful short and long term planning and, in particular, more accurate grade predictions.

The objective of this paper is to present and discuss the results obtained from the application of geostatistical techniques to the evaluation of the manganese oxides within Hill D (South Crest) using samples from drill holes which were logged between 1955 and 1960.

In any geostatistical estimation exercise it is essential that a thorough understanding of the geology exists. In this case, there is a serious lack of geological information on the deposit and thus the authors were forced to rely on the construction of a 3D wireframe model, based on the down-hole grade distribution, to control the geostatistical block modelling of the deposit.

Geological setting

The Nsuta deposit occurs at the eastern edge of a large Precambrian cratonic block extending from Ghana westwards through the neighbouring countries of West Africa (Dixon 1979). The

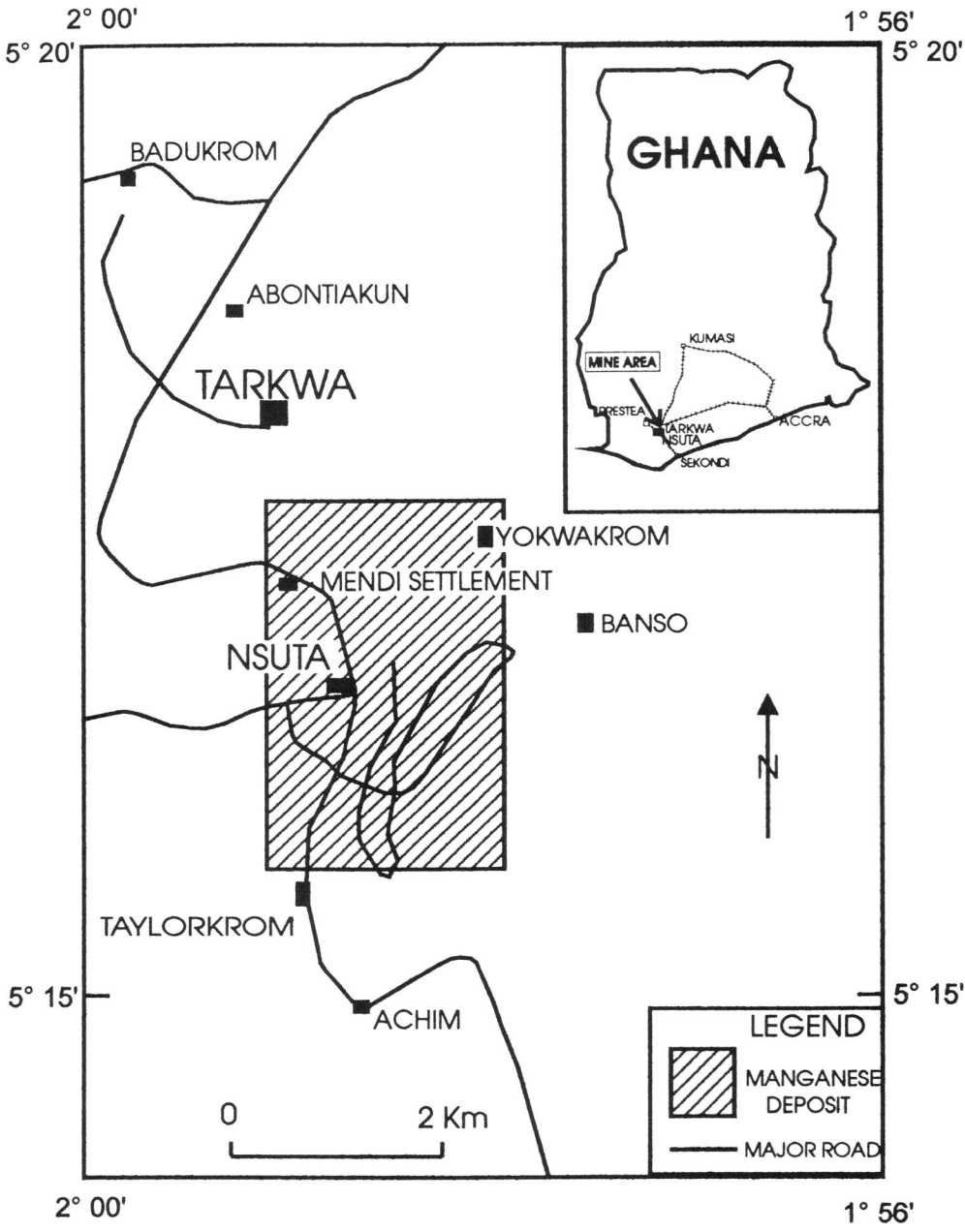

Fig. 1. Location of the Nsuta manganese deposit (shaded block) and map of Ghana showing the location of Nsuta (insert). (Reproduced from Kesse 1985.)

supercrustal rocks of this area are subdivided into two main groups, namely, the older Birimian Group and the younger Tarkwaian Group. The Birimian is subdivided into Lower and Upper Groups. Manganese occurs as a definite horizon within the Upper Birimian (Soper 1979).

Manganese carbonate forms the primary source from which the manganese oxide has been derived by weathering processes involving laterization and supergene enrichment. The chief manganese oxide minerals are pyrolusite and psilomelane.

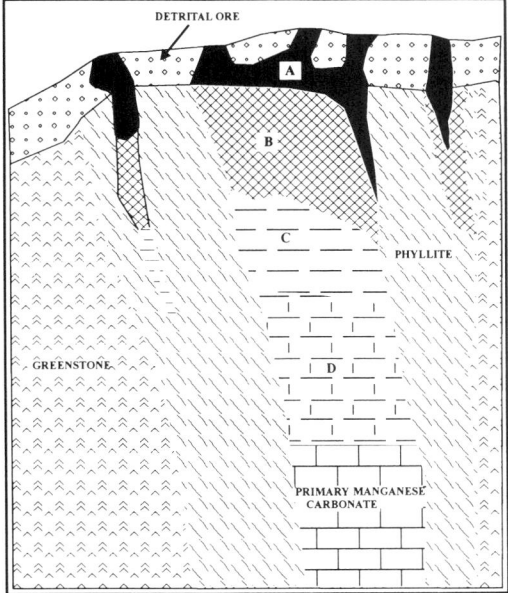

Fig. 2. Diagrammatic section through a typical orebody (modified after Service 1943). A, hard lateritic ore; B, black ore leached of quartz; C, porous ore cut by quartz veins; D, compact ore with bands of phyllite and tuff. (Not to scale.)

The *in situ* oxides usually conformably overlie the carbonate horizon but may be cross-cutting and lenticular when associated with the meta-tuffs. The areal distribution is closely linked to the structure of the underlying Birimian (Kesse 1976). A diagrammatic section through the deposit is illustrated in Fig. 2. At the top there is an indurated and reconstituted lateritic cap of manganese oxide (A) which overlies a high grade, porous black ore (B) from which nearly all traces of quartz veining have been removed by leaching and in which little argillaceous material remains. This grades downwards into a lower grade, porous ore (C) containing remnant quartz vein material. There is also a general downward increase in the amount of argillaceous impurities in the form of clays and halloysite. At the lower levels the oxide is interbedded with weathered beds of phyllite and tuff (D) (Service 1943).

The overlying detrital ore is the result of more recent weathering and erosion of the manganese oxide bodies exposed at the surface. It consists of rounded boulders and small fragments of manganese oxide mixed with lateritic clay and covered by a layer of lateritic clayey soil.

The distribution of oxide mineralization is complex due to folding and faulting of the host Birimian rocks. The principal faults dip steeply (60–90°) and trend in three main directions, viz. (a) east or slightly north of east; (b) north to NNW, (c) NNE to NE. These directions conform with the main directions of faulting in the area underlain by the Birimian rocks (Service 1943). There are also post-Birimian intrusives which cut through the manganese bearing horizon and add to the complexity of the horizon. These include metabasalts, metadiorites and granite porphyries. Quartz veins are abundant throughout the mine area where they vary in thickness from 0.03–1.3 m.

Data organization

The data used in this study consist of samples taken at an average interval of 1.5 m (5 feet) from 152 vertical boreholes drilled by both churn drills and diamond drills. The holes were drilled on a random stratified grid whose grid dimensions are approximately 24 m × 24 m. NX, BX and AX diamond drill bits were used giving an average core diameter of about 43 mm. The drilling strategy used in the mine usually involves the use of churn drills (150 mm) until hard formations are encountered when the hole is continued using diamond drilling. The logs from the 152 holes contain the coordinates and elevations of the borehole collars, a brief description of the geological units from which the samples were taken and the percent manganese (Mn).

The initial data organization involved the production of five files: collar, lithological, assay, survey and drill files, using Lotus 123.

To facilitate the creation of section plots and orebody modelling, the local coordinate system was transformed into a new coordinate system by a clockwise rotation through 38° so that the new eastings are parallel to the average direction of the drill lines (section lines).

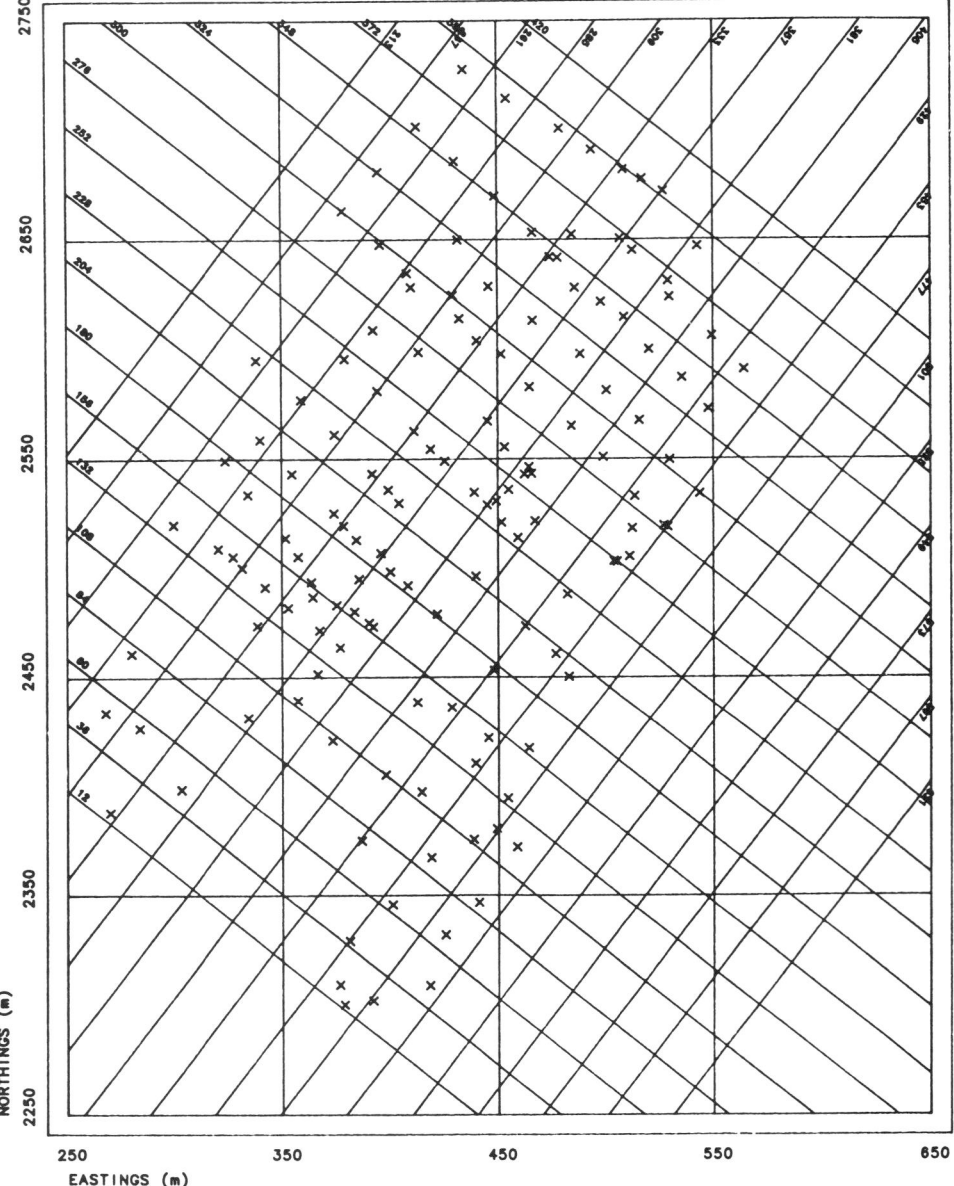

Fig. 3. Location of drill holes (in local coordinate system) and superimposed grid (in rotated coordinate system).

Figure 3 shows the drillhole layout (in the local coordinate system) together with the superimposed mesh (in the rotated coordinate system).

Variable length samples were composited in each lithological unit over 1.5 m depth intervals form the collar in each borehole and for each drill type. Attempts to composite the samples over 6 m, corresponding to the bench height, were abandoned since they failed to produce sufficient number of samples to allow the production of good semi-variograms.

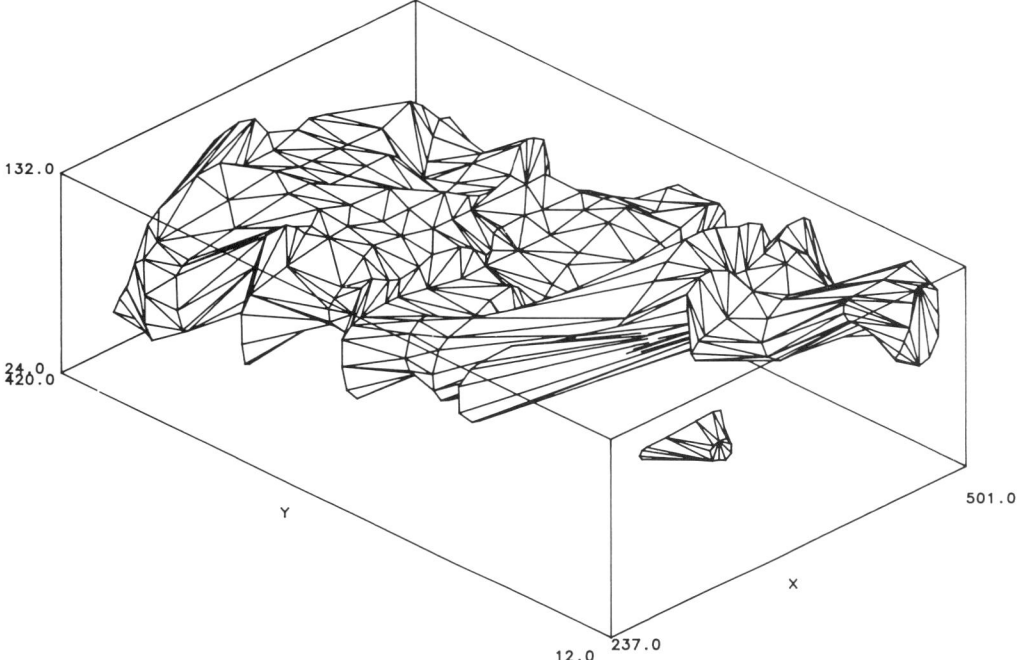

Fig. 4. Wireframe model of the manganese oxide mineralization at Hill D (South Crest).

Morphology of the mineralized zone

The shape of the ore body is controlled by the folding and faulting of the host rocks and thus marked changes can occur. Service (1943) states that the thickness of ore-grade material can vary in the study area from about 21 m to 1.8 m over a lateral distance of just 146 m.

Good geological data are necessary in order to delineate precisely the morphology of the mineralized zone. However, as was mentioned earlier, these data are not available and therefore the downhole assay profile had to be utilized to delimit the ore body in section.

Envelopes (also called perimeters) of the mineralized zone were digitized from drill hole sections. These were then linked together by triangulation to form a three dimensional wireframe model (Fig. 4). Since the envelopes were drawn in order to incorporate all of the mineralization in the drill holes without the application of a cut-off grade, this model is the best possible approximation of the morphology of the mineralized zone in the absence of structural data. The obvious advantage of such a model is that it can be used to constrain the resource estimates. Unconstrained linear estimators have the tendency to assign grades to blocks that fall well outside the mineralized zone. The selection of blocks within the model is achieved by appending the wireframe model (using Datamines's 'ADDMOD' facility), in which each block is coded, to the block grade model and selecting only those blocks from the combined model which have been so coded.

Statistical analysis

Mixing of diamond drill and churn drill samples

A relationship exists between sample volume and shape (support) and the statistical distribution of samples. Increasing the support has the effect of reducing the dispersion variance due to the 'smoothing effect', and the distribution becomes more symmetrical. The only statistic which does not change is the mean (Isaaks & Srivastava 1989; Dowd & Milton 1987).

If groups of samples are to be combined for statistical analysis, then they should belong to statistically similar distributions. Thus, to investigate whether the data produced by diamond and the churn drilling belong to statistically similar populations, F- and t- tests were carried out on these two populations. These tests indicate that the diamond drill and churn drill samples can be combined. The samples from the total database were then composited over 1.5 m intervals from the collar (regardless of their geological unit or drill type) in preparation for further statistical analysis.

that it is a complex population with a major mode at about 55% Mn and a minor mode close to the 20% Mn. The presence of bimodality is confirmed by the marked inflexion (Marsal 1987) at approximately 50% Mn in the normal probability plot (Fig. 5b).

This bimodality is probably closely related to the genesis of the mineralization in that the major mode reflects high grade supergene enriched oxide (hereafter referred to as Type 1 oxide) while the minor mode represents material from the less altered oxide with clay and the reconstituted oxide in the tuffs (hereafter referred to as Type II oxide).

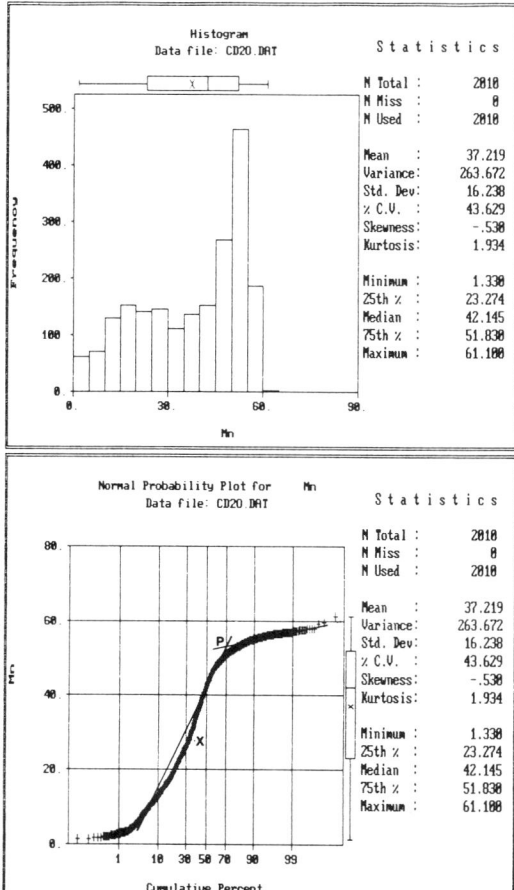

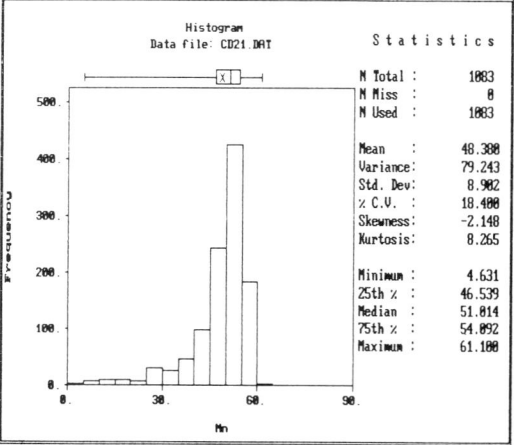

Fig. 6. Histogram of Type I oxide sample values.

Fig. 5. Statistics for total oxide population. (a) Histogram; (b) normal probability plot.

Frequency distributions

The histogram of the grade of the 1.5 m composite samples is shown in Fig. 5a. It is apparently negatively skewed but with evidence of bimodality. This is probably due to the fact

To ensure that sample values come from a homogeneous distribution, the two populations were separated by retrieval from the original database according to their lithological code. Figure 6 shows the histogram of the type I oxide which is negatively skewed indicating an inverse log-normal distribution. The histogram of the type II oxide is illustrated in Fig. 7a. This has a coefficient of skewness very close to zero (0.4) and the near-straight curve of the normal probability plot (Fig. 7b) indicates that the distribution may be accepted as normal. The negative skewness exhibited by the global histogram is due to the dominance of the Type I oxide.

The negative skewness of the Type I oxide is typical of the grade of highly concentrated elements such as iron (Journel & Huijbregts 1991). According to Koch & Link (1970), this type of distribution has a natural limit, Z, called the 'chemical barrier', and the frequency curve

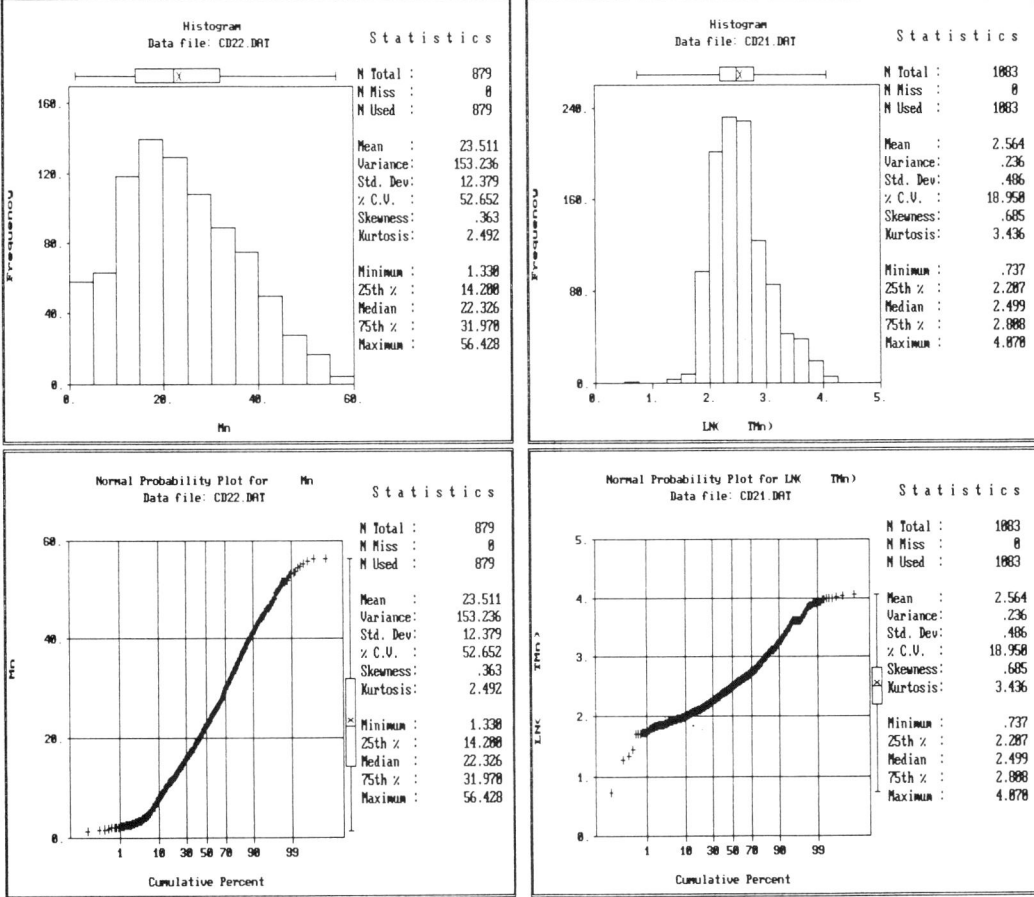

Fig. 7. Statistics for Type II oxide sample values. (a) Histogram; (b) normal probability plot.

Fig. 8. Statistics for the logarithm of transformed values of Type I oxide sample values. (a) Histogram; (b) normal probability plot.

of the logarithm of their transform $(Z-x)$ is normal; where x is the raw data variate whose distribution is inverse log-normal. The chemical barrier, Z, of Mn in pure pyrolusite (MnO_2) is 63.19%.

Figure 8a shows the histogram of the natural logarithm of the transform T_{Mn} ($=63.19-Mn$) where Mn is the raw manganese grade of 1.5 m composite samples. It has coefficients of kurtosis and skewness of 3.4 and 0.7 respectively which are close to those of a perfect normal distribution. Also, the log-probability plot of T_{Mn} (Fig. 8b) approximates to a straight line. It is not a three-parameter log-normal distribution since the calculated additive constant is -2.2 and it produces some negative values whose logarithms cannot be obtained. Thus, T_{Mn} is assumed to be a two-parameter log normal variate.

Geostatistics

Geostatistically, the two apparently different component populations should be treated separately. One of the criteria for being able to infer the characteristics of the mineralization from the characteristics of the samples is that there must be a sufficient degree of homogeneity in the mineralization as measured by the samples (Dowd 1988). Thus, geostatistical modelling should be done with homogeneous sample populations. However, the two types of mineralization are mixed in many places, and are therefore not easily separable at the present scale of mining (6 m benches). Journel & Huijbregts (1991) suggest that the most practical approach is to combine them and assume that, though the global distribution is bimodal, the ore is homogeneous at the scale of a 6 m bench.

The estimates resulting from ordinary kriging (OK) of a bimodal distribution are suspect as this is a parametric (distribution dependent) estimation method. Systematic over-, or under-estimation of local grade estimates may thus result. Indicator kriging (IK), which is nonparametric (distribution independent) was thus used to evaluate the *in situ* manganese oxide resources so that the results from these two methods could be compared.

The three-dimensional structure of the deposit was investigated by calculating directional semi-variograms on the horizontal plane and also in the vertical direction. The four horizontal directions considered were: N–S (0°), NE–SW (45°), E–W (90°), and NW–SE (135°). These directions refer to the rotated grid system. A lag spacing of 24 m, equal to the average drillhole spacing, and an angular tolerance of 22.5° were used for the horizontal semi-variograms while the mean vertical semi-variogram was calculated using a lag spacing of 1.5 m (composite sample length) and a horizontal angular tolerance of 90°. All the geostatistical studies were done using the DATAMINE software package. In each direction, the program automatically calculates three types of semi-variograms simultaneously, viz, absolute, pairwise-relative, and logarithmic semi-variograms.

In all the cases, some directional anistropy was evident in the horizontal semi-variograms and the maximum anisotropy ratio was about 1.6. According to Royle & Hosgit (1974), anisotropism may be considered insignificant if the anisotropy ratio is less than 1.8; and hence, structural isotropy was assumed in the horizontal regionalization throughout this study.

Ordinary kriging

Variography. Semi-variograms were calculated for the Mn and T_{Mn} varieties. Though their semi-variograms were very similar, the absolute semi-variogram of Mn displayed the best structure. The failure of the logarithmic semi-variogram of T_{Mn} to produce the best structure as expected is probably due to the fact that the grade distribution is not truly inverse lognormal (cf. Fig. 8). Thus, the statistical modelling of the oxide grade was done using the Mn variate.

The mean experimental horizontal semi-variogram, with the simple spherical model fitted to it, is shown in Fig. 9. The model parameters are:

$C_0 = 5(\%)^2$
$C = 258.67(\%)^2$
$a = 65\,\text{m}$

The experimental absolute semi-variogram for the vertical direction was also modelled with a simple spherical model, with zero nugget variance (cf. Fig. 10). The model parameters are:

$C = 171(\%)^2$
$a = 20\,\text{m}$

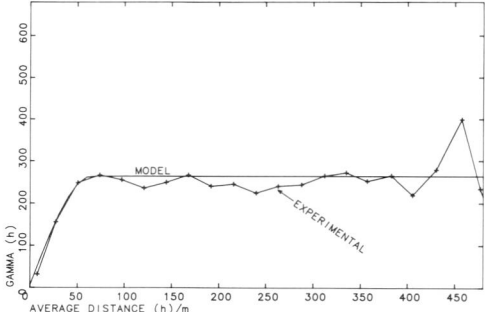

Fig. 9. Mean horizontal semi-variogram of oxide sample values (total data set).

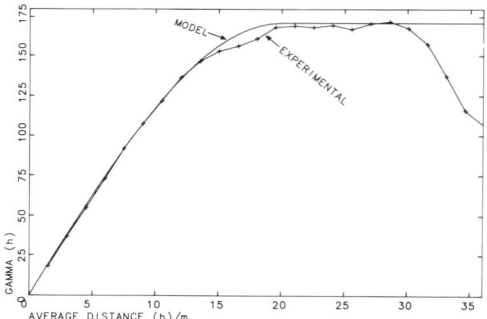

Fig. 10. Vertical directional semi-variogram of samples values (total data set).

Deregularization. The available data are not defined on point support and so it is necessary to deduce a point model from the regularized models by the process of 'deregularization' (also called 'deconvolution'.)

Since the diameter of the samples is very small with respect to the range of the horizontal mean semi-variogram, the experimental regularized mean horizontal semi-variogram and its theoretical model can thus be treated as point semi-variograms.

In the case of the vertical semi-variogram,

however, the sample length of 1.5 m is quite a substantial proportion of the range (8%) and so it cannot be considered as a quasi-point semi-variogram. Using the 1D regularization graphs, the parameters of the point semi-variogram were determined as:

$C_0 = 5.46(\%)^2$
$C = 171.7(\%)^2$
$a = 18.5\,m$

Readers interested in the practice of deregularization are referred to Annels (1991) and Journel & Huijbregts (1991). It is evident that the zero nugget variance observed in the regularized vertical semi-variogram model is due to the obscuring of micro-structures by the 'smoothing effect' resulting from 'compositing' of the point values over the sample length.

Three-dimensional model. Although the point vertical model has practically the same nugget variance as the mean horizontal quasi-point semi-variogram, it has a smaller spatial variance. This suggests that zonal anisotropism exists (the direction of zonality is in the horizontal plane). The three-dimensional point support model adopted consists of an isotropic structure with a nugget variance and a longer range anisotropic structure which is only relevant in the horizontal plane. The general equation of the three-dimensional model (Journel & Huijbregts 1991) is:

$$\gamma(h) = C_0 + C_1\gamma_1(r) + C_2\gamma_2(h_D) \quad (1)$$

where,
$r = |h| = (h_x^2 + h_y^2 + h_z^2)^{1/2}$
C_0 = isotropic nugget variance which appears on the point semi-variograms
$C_1\gamma_1(r)$ = first isotropic structure with spatial variance, C_1 and range, a_1
$C_2\gamma_2(h_D)$ = second isotropic structure with spatial variance, C_2 and range, a_2, depending only on the distance h_D $((h_x^2 + h_y^2)^{1/2})$ in the isotropic horizontal regionalization.

The parameters of the three dimensional model (adjusted after cross-validation) are shown below:

$C_0 = 5.46(\%)^2$
$C_1 = 171.7(\%)^2 \quad C_2 = 87.67(\%)^2$
$a_1 = 18.5\,m \quad a_2 = 65\,m$

Geological interpretation of semi-variograms. Geological interpretation of the above semi-variograms is important. The apparent zonal anisotropy in the mineralization is probably due to supergene processes. The amount of manganese being concentrated by these processes is highly variable because of differences in the underlying bed-rock and by local factors such as faulting. The deposition and hence, the concentration of manganese oxide, is more uniform in the vertical direction than in the horizontal direction as transport of the element in percolating water was mainly in the vertical direction in the weathering profile. Hence the vertical semi-variogram has a lower sill value than the mean horizontal semi-variogram.

Local resource evaluation. Local block (24× 24×6 m³) estimates were calculated using the DATAMINE software. The calculation was done using the total data population and the parameters of the three-dimensional model. The axes of the search ellipsoid used were 65 m, 65 m, and 18.5 m in the X, Y and Z directions respectively. The output from this program contains the coordinates (XC, YC, and ZC) of the centre of each block, the kriged estimate, the kriging variance and the block identifer (IJK).

Indicator kriging

This method involves the use of the indicator values obtained from the total samples and the raw sample values of the two sub-population separated on the basis of a threshold value.

The indicator values are obtained by applying the threshold value (COG) as:

$$i(x_i) = \begin{cases} 1 \text{ if } Z(x_i) \geq COG \\ 0 \text{ if } Z(x_i) < COG \end{cases} \quad (2)$$

where, $i(x_i)$ is the indicator variable and $Z(x_i)$ is the raw sample value from the global sample population.

The selection of the threshold is somewhat arbitrary (Parker 1991) but statistical techniques have been adopted to provide a more logical basis. In some cases, the cumulative coefficient of variation curve has been used (Parker 1991), whilst in others the cumulative frequency curve has been used (Dowd 1992).

In this study, a normal probability plot of the grade of composite samples was used (Fig. 5b) which shows that, apart from the lower 12% of the data, two straight lines can be constructed through the curve. These meet at a point, P, between 50% and 55% Mn. The 52% point was taken as the threshold as this is also the economic cut-off grade for 'R-grade' ore in the mine. However, variography of the raw values above 52% Mn showed very high nugget effects,

ε, for the vertical and mean horizontal directional semi-variograms of 70 and 130% respectively. This makes the use of linear kriging inappropriate. Experience has shown that linear kriging generally gives superior results if ε is less than 50% (Annels 1988).

Following this failure, the threshold was set to 35%—the operational cut-off grade for the mine. This value is also close to the overlap, X, between the two populations (cf. Fig. 5b). Hereafter, those samples with values greater, or equal to, 35% Mn are referred to as 'high grade' samples and those below it are referred to as 'low grade' samples.

Table 1. *Statistics of 1.5 m composites*

	Total	High grade	Low grade
Number of samples	2010	1202	808
Mean (%)	37.2	49.1	19.5
Median (%)	42.1	50.6	19.9
Variance (%)^2	263.7	35.1	76.9

Statistics for the 1.5 m composite sample grades for the overall population and for the high and low grade populations are shown in Table 1. From this table, it may be observed that the low grade samples have a disproportionate influence on the variability. 35% appears to be an effective and practical threshold point at which the two sub-populations may be discriminated and separately treated.

The characteristics of each sub-population are separately quantified and then the proportion of the high grade ($\geq$COG) value in any block is obtained from kriging the indicator values. The average grade of a block, V_i, is defined as:

$$Z^*(V_i) = \{1 - p^*(V_i)\} Z_{LG}^*(V_i) + p^*(V_i) Z_{HG}^*(V_i) \quad (3)$$

where

$Z_{LG}^*(V_i)$ is the kriged grade of block, V_i, using only the low grade ($<$COG) values;

$Z_{HG}^*(V_i)$ is the kriged grade of block, V_i, using only the high grade values;

$p^*(V_i)$ is the estimated proportion of the high grade material in the block, V_i, using indicator values

Variography. The raw values of each sub-population and the indicator values of the global population were used to calculate the semi-variograms in the horizontal plane and also in the vertical direction. A three-dimensional model was obtained for each case and checked by cross-validation techniques.

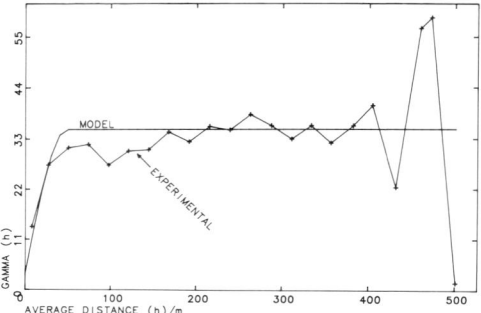

Fig. 11. Mean horizontal directional semi-variogram for high grade sample values (Mn $\geq$ 35%).

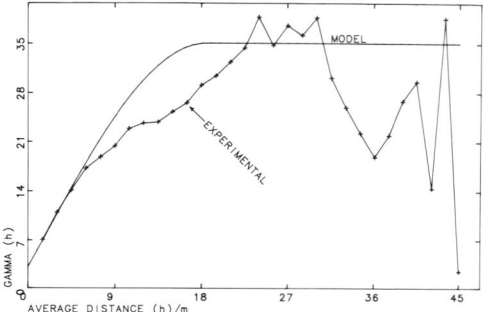

Fig. 12. Vertical directional semi-variogram for high grade samples values (Mn $\geq$ 35%).

(i) *High grade oxide.* The mean experimental horizontal and vertical semi-variograms are illustrated in Figs 11 and 12. In both cases, a simple spherical model was fitted. The quasi-point horizontal semi-variogram model differs only from the deregularized vertical semi-variogram model in the range. This indicates the presence of geometric anisotropy (anisotropy ratio equal to 2.8) in the three-dimensional model of the high grade values. The parameters of this model adopted (after cross-validation) are shown below, where a_x, a_y and a_z represent the ranges in the X and Y directions (horizontal plane) and the vertical direction respectively.

$C_0 = 3.61(\%)^2$
$C = 32(\%)^2$
$a_x = a_y = 48\,\text{m}$
$a_z = 17\,\text{m}$

(ii) *Low grade oxide.* Semi-variograms for the low grade population are similar to those obtained for the high grade population. They are, however, more erratic, which may be attributable to the low number of pairs used to produce some of the points. They exhibit simple structures, with the vertical semi-variogram having a shorter range than the mean horizontal semi-variogram. The parameters of the three-dimensional point model adopted are:

$C_0 = 12.73(\%)^2$
$C = 66.55(\%)_2$
$a_x = a_y = 80\,\text{m}$
$a_z = 34.5\,\text{m}$

(iii) *Indicator values.* Semi-variograms from the indicator values show similar, but better, structures to those of the global sample population (used for ordinary kriging). The point vertical semi-variogram of the indicator values has the same nugget variance but a lower sill value than the quasi-point horizontal semi-variogram model. A zonal anisotropic model has, therefore, been adopted. The second structure is not relevant in the vertical regionalization. The parameters of the three-dimensional model are shown below:

$C_0 = 0.025$
$C_1 = 0.181$ $C_2 = 0.035$
$a_1 = 19.5\,\text{m}$ $a_2 = 75\,\text{m}$

Local resource evaluation. The *in situ* block grade was estimated by combining the three estimates from the above three stages. For each block, the kriged grade is obtained using equation (3).

The kriging variance cannot, however, be combined in the usual way since the IK estimates are not independent of each other. There is no direct way to obtain a single kriging variance of a block estimate. Dowd (pers. comm.) suggests that an approximation could be achieved by using the kriging variance obtained from the high grade estimator. This approach was adopted, though where blocks had no high grade estimates the kriging variance obtained from the low grade estimator was taken as the estimator of the block kriging variance, $\sigma_k^2(V_i)$.

Model validation

Using the back-estimation technique, which has been clearly explained by Dowd & Milton (1987), the three-dimensional models were cross-validated. Figure 13 shows an example representing the scatter plot of the actual against the estimated value using the OK. The linear regression parameters, shown in this figure, are close to those expected for perfect correlation, and thus the model satisfactorily characterizes the spatial variability of the oxide grade.

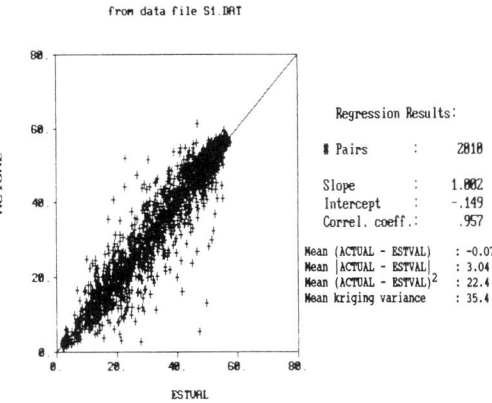

Fig. 13. Cross-validation of semi-variogram parameters for the total data set using OK.

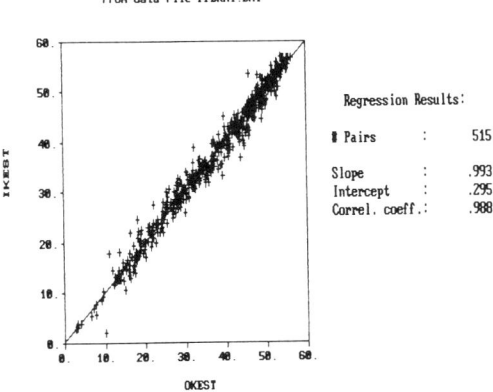

Fig. 14. Scatter plot of IK block grade estimates on OK block grade estimates.

Comparison of the OK and IK estimates

The summary statistics of the block estimates obtained from both OK and IK are shown in Table 2. It is evident that there is little difference in the grade estimates. This observation is supported by the near perfect correlation observed in the scatter plot, shown in Fig. 14, of IK estimates (IKEST) on OK estimates

(OKEST).

However, when the block kriging variances are compared a significant difference emerges (see Table 2). The kriging variance associated with each block represents the accuracy of its grade estimate. Low kriging variance implies high reliability and vice versa.

Table 2. *Statistics of OK and IK block estimates*

Statistic	OK	IK
Mean (%)	35.9	36
Varaince (%)^2	167.2	168.8
C.V. (%)	36	36.1
Skewness	−0.5	−0.4
Minimum (%)	2.7	2
Maximum (%)	53.7	56.7
Mean σ_k^2 (%)^2	59	11*
Minimum σ_k^2 (%)^2	18.6	1.7*
Maximum σ_k^2 (%)^2	137.9	36.5*

* Value is an approximation.

On average, the kriging variance obtained from OK is about five times that from IK. Whereas the minimum and maximum kriging variances from IK are 10.9(%)^2 and 36.5(%)^2 respectively, for OK they are 59(%)^2 and 137.9(%)^2 respectively.

This big difference between IK and OK is due to the ability of the three-stage IK to reduce the variability of the bimodal global population by using its sub-populations which are more or less unimodal (homogeneous). Effectively, this can be likened to modelling the three-dimensional distribution of the global sample values with a lower sill (re-scaling). According to Isaaks & Srivastava (1989), re-scaling the semi-variogram does not change the block grade estimates, only the kriging variance.

This difference can be of practical significance. Kriging variance, which is a measure of the deviation of estimated value from the true value, is particularly important to mine planners in categorizing resources, and in the determination of the optimum drill-hole spacing. Kriging variance values above a pre-determined limit may call for in-filling drilling to lower the kriging variance and ensure greater reliability of the estimates. It is also very important in verifying how well the 'experimental' grade-tonnage curves estimate the 'true' grade-tonnage curves.

Conclusion

The operational cut-off grade of 35% appears to be a good threshold value at which to separate the manganese oxide database into two quasi-homogeneous sub-populations.

The results obtained from this study indicate that, for the manganese oxide, there is practically no significant difference in the grade estimates produced by the application of OK and IK techniques. However, the kriging variance of block estimates obtained from OK is, on the average, about five times higher than those produced by IK. This is because IK is able to effectively re-scale the semi-variogram parameters used in OK through the use of statistically identified quasi-homogeneous sub-populations.

Geostatistical studies require a knowledge of the spatial variability of the mineralization through semi-variograms and these depend on the degree of homogeneity of the distribution of the sample population. How well the kriging technique will perform depends on the reliability of the semi-variogram model parameters as, among other factors, they influence the kriged grade and kriging variance. Although geostatistics provides a means of predicting estimation variances, they are dependent on assumptions made during the estimation process and if incorrect assumptions are made such measures of uncertainty may be misleading. The results from this study indicate that values of kriging variance should be treated with caution unless the nature of the sample distribution is known and the kriging technique used specified.

The difference between these two methods is very important in mine planning. Where only a global estimate is of interest, for instance at the exploration stage, OK may be preferable to IK since it is simpler to apply and the semi-variograms are easier to model having a more clearly defined structure.

On the other hand, where local block estimates are required with the best possible precision, for example in production planning, the need for low kriging variances cannot be over-emphasized. In such cases, it is preferable to use IK despite the large amount of time and higher level of expertise that it requires. Great economic benefits can be gained which may offset the additional cost that may accompany the use of IK. For example it may be possible to minimize any additional drilling that high kriging variances, obtained from OK, might suggest is needed.

The authors would like to thank the Ghana Government for granting the first author a scholarship to pursue a PhD course at the University of Wales, Cardiff. Without this support it would have been almost impossible to present this paper.

We would also like to thank the Ghana National Manganese Corporation, Nsuta, for their technical support and permission to use this data. The support of the geology staff at the mine is gratefully acknowledged.

References

ANNELS, A. E. 1988. *An Introductory Course in Geostatistics—Lecture Notes.* University College, Cardiff (unpublished).

—— 1991. *Mineral Deposit Evaluation—A Practical Approach.* Chapman & Hall.

DIXON, C. J. 1979. *Atlas of Economic Mineral Deposits.* Fletcher & Sons Ltd, UK.

DOWD, P. A. 1988. *Basic Geostatistics for the Mining Industry.* MSc Course notes, Leeds University (unpublished).

—— 1992. Geostatistical ore reserve estimation: a case study in disseminated nickel deposit. In: ANNELS, A. E. (ed.) *Case Histories and Methods in Mineral Resource Evaluation.* Geological Society, London, Special Publications, **63**, 243–255.

—— & MILTON, D. W. 1987. Estimation of a section of the Perseverance Nickel deposit. *In:* MATHERON, G. & ARMSTRONG, M. (eds) *Geostatistical Case Studies.* D. Reidel Publishing Company, Dordrecht, Holland, 39–67.

ISAAKS, E. H. & SRIVASTAVA, R. M. 1989. *An introduction to applied geostatistics.* Oxford University Press, Oxford.

JOURNEL, A. G. & HUIJBREGTS, C. 1991. *Mining Geostatistics.* Academic Press, London.

KESSE, G. O. 1976. Manganese Ore Deposits of Ghana. *Ghana Geological Survey Bulletin*, **44**, 13–35.

—— 1985. *The Mineral and Rock Resources of Ghana.* A. A. Balkema/Rotterdam/Boston.

KOCH, S. G. & LINK, R. F. 1970. *Statistical Analysis of Geological Data.* John Wiley & Sons, Inc.

MARSAL, D. 1987. *Statistics for geoscientists.* Pergamon Press.

PARKER, H. M. 1991. Statistical Treatment of Outlier Data in Epithermal Deposit Reserve Estimation. *Mathematical Geology*, **23**, 175–199.

ROYLE, A. G. & HOSGIT, E. 1974. Local estimation of sand and gravel by geostatistical methods. *Transactions of the Institution of Mining and Metallurgy (Section A: Mining industry)*, **82**, A53–A62.

SERVICE, H. 1943. *The Geology of Nsuta Manganese Ore Deposits.* Gold Coast Geological Survey Memoirs **5**, 5–28.

SOPER, M. A. R. 1979. *The Geology of Nsuta Manganese Deposit and surrounding Districts* (Draft for Geological Society of Ghana Book 'The Geology of Ghana'). Unpublished.

Structural reconstruction and mineral resource evaluation at Zinkgruvan Mine, Sweden

A. E. ANNELS,[1] S. INGRAM[1] & L. MALMSTROM[2]

[1]*Department of Geology University of Wales, PO Box 914, Cardiff CF1 3YE, UK*
[2]*Vieille-Montagne Sverige, Centrumvägen, S-690 42 Zinkgruvan, Sweden*

Abstract: The use of geostatistics in ore evaluation studies is based on the relationship between point values and their spatial distribution. When an orebody is deformed the syn-depositional spatial relationship between control points is altered and it is extremely difficult to represent these points on a single plan suitable for ore reserve estimation. To overcome this problem, deformed orebodies could be dissected into a number of individual units and analysed separately, but this results in a reduction of control points available for analysis. An alternative to orebody dissection is to unroll the deposit so as to return control points back as close as possible to their original relative spatial locations.
  This paper describes attempts to reconstruct the Nygruvan orebody structurally at Knalla mine, Zinkgruvan, Sweden, which is operated by Vieille-Montagne Sverige. The deposit is located in the Bergslagen geological province of south-central Sweden, and has features in common with stratiform volcanogenic massive Zn–Pb sulphides and sediment hosted exhalative Zn–Pb deposits. Nygruvan is essentially a thin (5–20 m) tabular deposit of large lateral extent, dipping at 60–80° to the northeast, which was folded and faulted during the Svecofennian orogeny.
  Current ore resource evaluation methods employ the polygon method for global resources using diamond drill information, and the extrapolation of chip sample data between mine levels to produce reserves of mining blocks. These methods do not take account of orebody structure, assuming it to be a planar structure of variable thickness.
  Structural reconstructions of thrust belts and deformed basins are usually achieved by drawing sections in the plane containing the movement vector. Sectional reconstruction methods do not, however, give consideration to transverse movements. If strain is triaxial, a two dimensional restoration using cross sections is not appropriate, instead a three dimensional method should be used. To date little work has been done on the development of a suitable method applicable to deformed orebodies
  A small area of the mine has been selected for preliminary studies and a variety of manual methods applied in order to develop a suitable computerized restoration algorithm (these include two dimensional sectional and triaxial methods). A new Datamine unrolling package, which will be tested on this deposit, is described. A geostatistical study will be undertaken to generate a block reserve model once the deposit has been unrolled satisfactorily.

The computation of the ore reserves of highly deformed and faulted stratiform or stratabound orebodies has always presented a problem for the mining geologist. Attempts to introduce geostatistical techniques present a further problem in that variography is difficult, if not impossible, in a situation where structural continuity is broken by faults and folding and thus it is impossible to undertake 2D variography and block modelling either on plan or on vertical longitudinal projection (VLP). In these situations, the geologist has to resort to the production of palinspastic maps (structural unrolling); a process which is often tedious in the extreme. Prior to this exercise the effects of faulting have also to be removed. Once accomplished, the resultant plan, although possibly reflecting the shape of the orebody prior to tectonism, presents a highly distorted representation of the orebody and associated development as they are today. Such plans are thus often found unacceptable or incomprehensible by mining personnel.
  This paper presents progress to date in the development of a computer-based method for the structural unrolling and reconstruction of deformed orebodies which is aimed largely at resource evaluation and not necessarily at geological or palaeogeographical or sedimentological reconstructions. The methods under

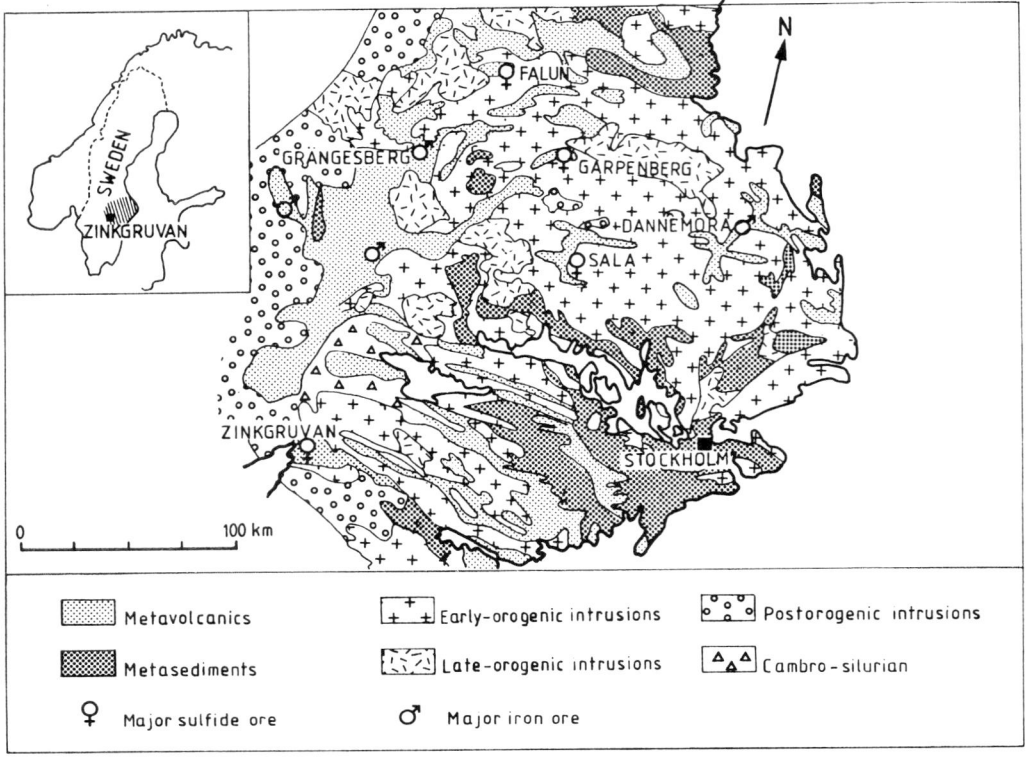

Fig. 1. Geological map of the Bergslagen ore province.

development are designed to cope with a range of degrees of deformation from relatively simple open folds to complex structures which are also highly faulted. The aim is to preserve the areas of ore-blocks at the expense of some distortion so that tonnage estimates are unaffected. At the same time, blocks defined on these unrolled plans for local resource estimation will be easily identifiable in that they will be bounded, where possible, by mine levels and mining block boundaries.

The Zinkgruvan deposit in south-central Sweden was chosen as a suitable subject for study in that comprehensive assay databases are available and good geological control on the orebody has been achieved by detailed underground mapping. In particular, it was decided to concentrate on the eastern or Nygruvan section of the orebody which shows the necessary range of styles and intensities of deformation on which the various methods could be tested. As well as utilizing 'in-house' semi-computerized techniques, the ability of Datamine International's new 'Unfold' facility to achieve the aims of the project will be assessed (Datamine 1992).

It is hoped that this exercise will allow the production of new databases containing the assay information from diamond drill core and chip samples whose 3D coordinates have been transformed by the unrolling and reconstruction process. This will allow more accurate directional variography of the metal accumulations and thicknesses because, hopefully, the samples will be in the same relative positions as they were immediately prior to deformation with only minor angular changes in directional vectors between pairs of samples. Such changes are acceptable given the tolerance angles usually applied during variography (e.g. 10 to 22.5°). From this study, kriging of irregular polygons should be possible to estimate the resources of individual mining blocks or stoping areas in the mine and to determine the global reserves of the Nygruvan deposit. More accurate information will also be available to assess the efficiency of current sampling practices and their ability to produce ore reserve estimates to the required level of precision.

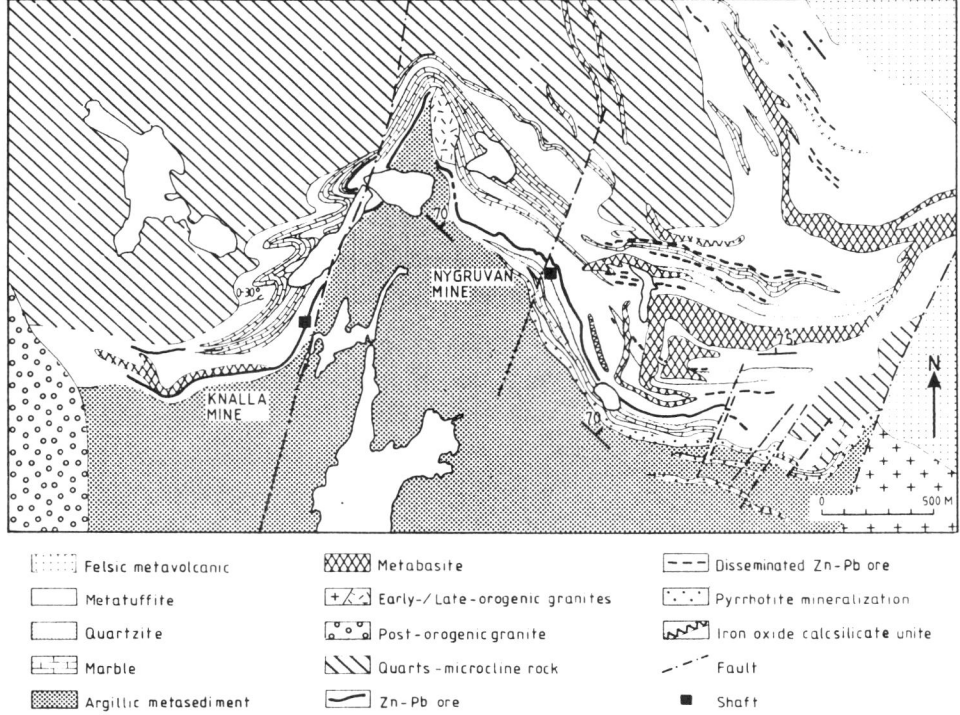

Fig. 2. General geological map of the Zinkgruvan mine area (after Hedström *et al.* 1989).

Introduction to the Zinkgruvan Deposit

Zinkgruvan is a stratiform massive Zn–Pb–Ag deposit which is situated in the southernmost part of the Bergslagen ore province in south-central Sweden (Fig. 1). It has a strike length of approximately 4.6 km and has been proven to depths of 1300 m. It is estimated that the original *in situ* tonnage was close to 40 million tonnes. In 1857 the Vieille Montagne Company acquired the deposit and started mining on a large scale. Today, the deposit is exploited in two underground workings, the Nygruvan mine and the Knalla mine (Fig. 2). Two mining methods are applied, the first of which is drift and benching and the second, cut and fill. Production in 1992 reached 650 000 metric tonnes of ore grading 10.1% Zn, 2.4% Pb and 59 ppm Ag.

Geology of the deposit

The Bergslagen ore province is in an area of supracrustal and plutonic rocks formed during the Sveco-fennian orogeny. The predominantly rhyolitic volcanic rocks are dated at 1.88 Ga (Welin *et al.* 1980). The supracrustal rocks of the Zinkgruvan area consist of three lithostratigraphic groups (Hedstrom *et al.* 1989); the Metavolcanic Group is in the lowest stratigraphic position, followed by the Metavolcano-Sedimentary Group, and finally at the top of the stratigraphic succession is the Meta-Sedimentary Group. This succession is, however, structurally inverted as the orebodies lie within the overturned northern limb of a syncline whose axial plane dips in a northerly direction. The Zinkgruvan Zn–Pb–Ag orebodies are tabular, sheet-like and have thicknesses which range from a few metres to a maximum of 25 m. A NNE-trending fault system exists in the region and two major fault zones with this trend can be observed in the Zinkgruvan area. One of these subdivides the ore deposit into two sections; the Knalla mine in the west and the Nygruvan mine in the east. The other fault zone interrupts the continuity of the ore in the Nygruvan mine (Fig. 2). In this mine, the dip averages close to 70°, while at the Knalla mine and in the western parts of the Nygruvan mine, the variation is much larger due to intense folding.

The ore zone occurs in the Metavolcano-Sedimentary Group, in which the dominant rock

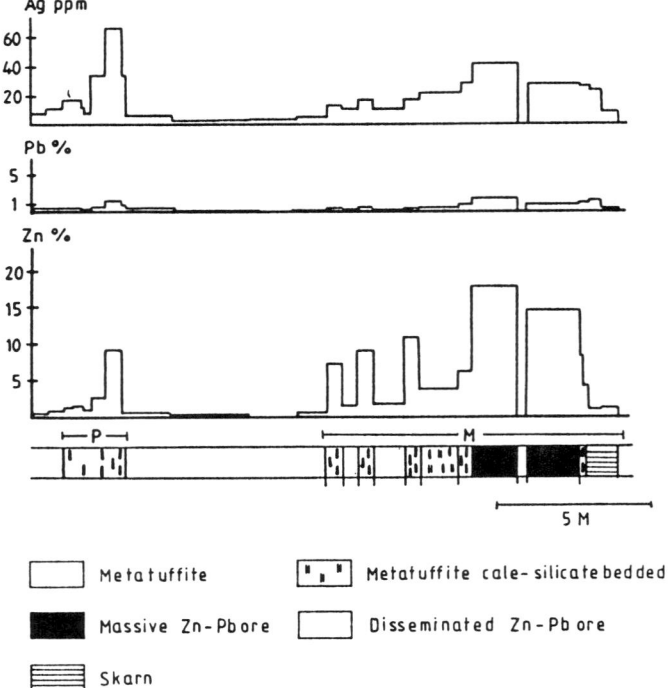

Fig. 3. Variation across the ore zone in the eastern part of Nygruvan mine. M, main ore bed; P, parallel ore bed (after Hedström *et al.* 1989).

type is a metatuffite. Intercalated with the metatuffite are calc-silicate rocks, marbles, quartzites and tuffaceous metasediments. The ore minerals, sphalerite and galena, are hosted by a quartzitic metatuffite.

The ore zone itself can be divided into two units, the Parallel Ore and the Main Ore (Fig. 3). The Parallel Ore, which is situated about 5–10 m stratigraphically below the Main Ore, is not mined at the present time. The Main Ore horizon is more or less continuous along the whole deposit. However, there are some differences between the eastern and the western parts which are described below.

A typical stratigraphic section across the Main Ore in the middle and eastern part of the Nygruvan mine (Fig. 3) consists, close to the stratigraphic footwall, of two to three layers of disseminated sulphide. Varying amounts of sulphides give rise to a pronounced stratification which is consistent along strike and down dip. The layers have a thickness of around 1 m and a combined Zn plus Pb grade of up to 15%. They are separated by a barren metatuffite. A unit of calc-silicate and carbonate bedded, quartzitic metatuffite appears at a higher stratigraphic level with less than 6% combined Zn

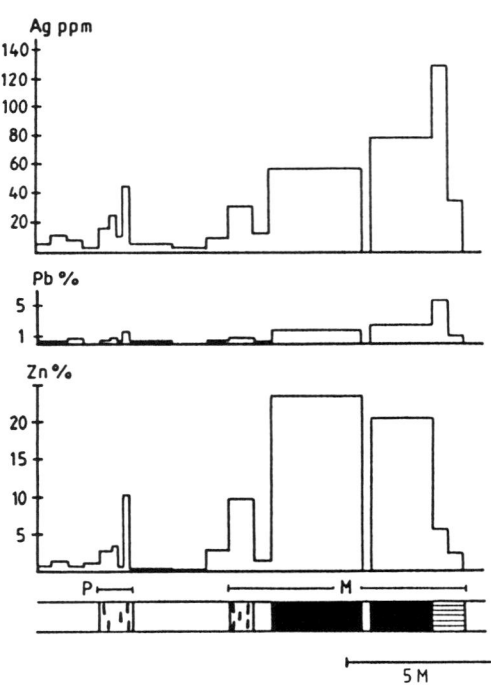

Fig. 4. Variations across the ore zone in the western part of the Nygruvan mine. For legend see Fig. 3. (after Hedström *et al.* 1989).

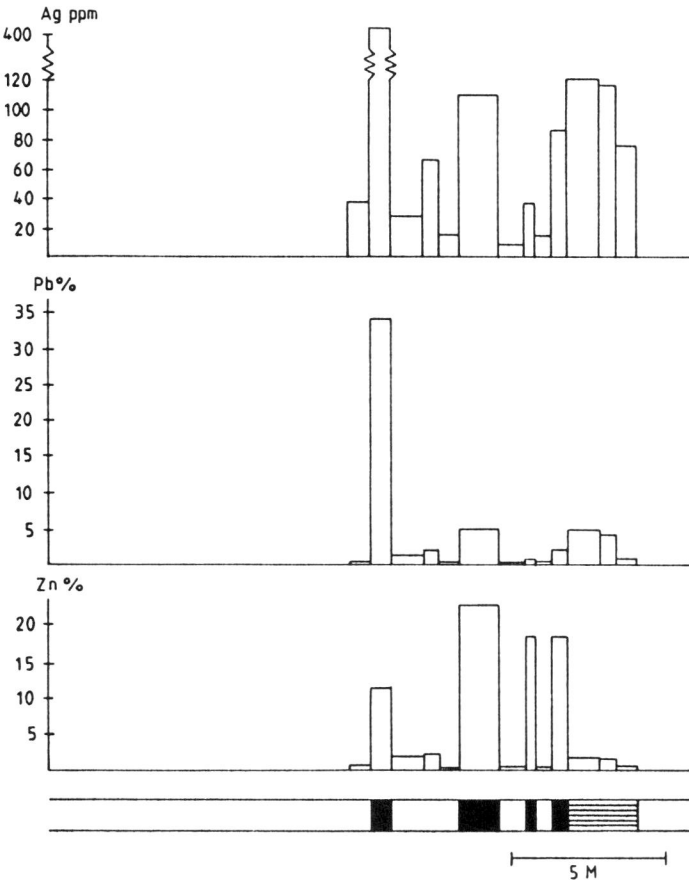

Fig. 5. Variation across the ore zone in the Knalla mine. For legend see Fig. 3.

plus Pb and a width of 1–2 m. At the top of the Main Ore unit there is an up to 8 m thick, massive, Zn–Pb bed which is overlain by a skarn layer. The massive ore has a combined grade of up to 55%. The skarn, whose main constituent is diopside, is around 1.5 m thick but it contains only a few percent of Zn and Pb though the grade does increase towards the contact with the massive ore (Fig. 3). Disseminated pyrrhotite is abundant in the skarn. The skarn is also more or less continuous throughout the whole deposit which makes it an ideal marker for unrolling purposes. Individual beds and layers are generally continuous and consistent in grade over distances of several hundred metres in both the strike and dip directions.

A stratigraphic section through the ore zone in the western part of the Nygruvan mine is shown in Fig. 4. The Parallel Ore is not so well developed here and the calc-silicate and carbonate bedded, quartzitic metatuffite are missing and disseminated sulphides are now found in only one layer or are totally missing. The massive ore is also more variable in thickness ranging from less than 1 m up to 15 m. A slight increase in galena can also be observed.

Further to the west, in the Knalla mine, the Parallel Ore has disappeared and the layers stratigraphically below the massive ore are also missing (Fig. 5). The Main Ore horizon here consists of layers of massive sphalerite and galena ore intercalated with metatuffite and it varies between 0 and 20 m in thickness with the most rapid changes occurring along strike. A further increase in galena is also evident. The skarn is even more variable in width and encloses a calcite layer with disseminations of galena. Current drilling in the deeper parts of the

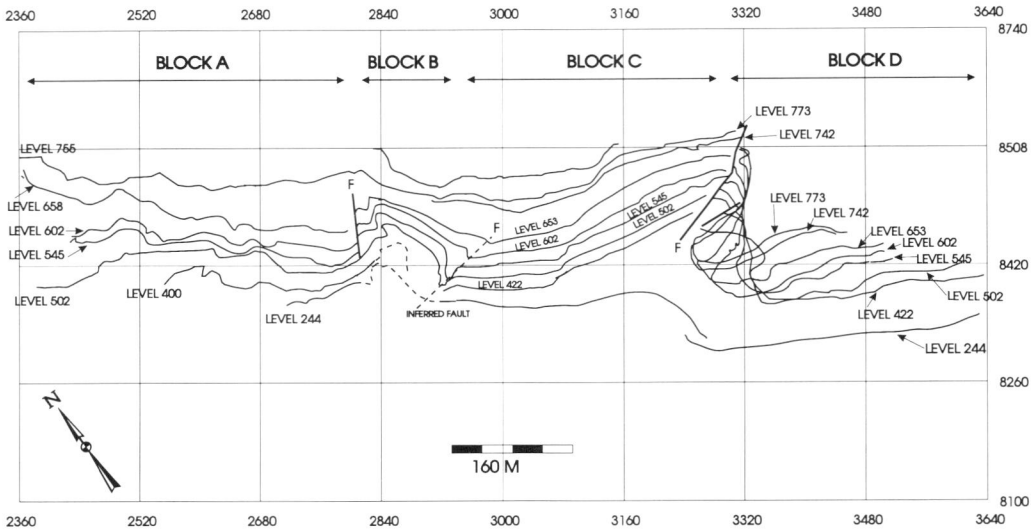

Fig. 6. Structure contour plan of the ore–skarn contact for the Nygruvan Mine, constructed using level and sub-level geological plans.

Knalla mine indicates the existence of a more complex style of mineralization.

Structure of the Nygruvan Orebody

In the Nygruvan section of the mine, the orebody has a strike length of about 1.8 km and, in areas which have not undergone intense folding, has dips which generally lie between 68 and 70° to the northeast. Where folding is intense, dips can become sub-vertical. Below the 400 m mine level, the ore–skarn contact has a high degree of continuity. Unfortunately, geological mapping in the upper mine levels was poor and there is thus a low level of confidence in any interpretation of the geological structure at these levels. It was thus decided to perform the unrolling exercise between the 502 m level and the 800 m level, the deepest mined level in this section of the mine. A clearly distinguishable tuff layer is intercalated within the ore zone which shows intense small scale isoclinal folding, indicating that a significant contraction has occurred in the spatial dimensions of the ore zone over and above that caused by the large scale folds. This shortening of the ore zone by this small scale isoclinal folding will not have any significant impact on the validity of the geostatistical study of the *in situ* resource but would have to be taken into account if unrolling were contemplated for the purposes of a palaeogeographic interpretation or a determination of the original dimensions of the deposit as laid down on the sea floor.

A structure contour plan of the ore–skarn contact has been constructed using level and sub-level geological plans for the whole of the Nygruvan ore body and the result is summarized in Fig. 6. This was only possible due to the high degree of lateral continuity of the orebody below the 400 m level. The Parallel Ore, although present in the Nygruvan section, has not been included in this study as it would complicate the unrolling exercise and, there are no plans to mine it at the present time.

There are two main areas of deformation within the Nygruvan section, each associated with a set of closely spaced faults which cause considerable dislocation of the ore zones. Before reconstruction can be achieved, it is necessary to determine whether the faults are the cause of the folding or vice versa, and also the magnitude of the throw on the faults. The amplitude of the folding increases to the southeast with most deformation and shortening occurring in the strike direction. The most complex fold occurs between, $X = 3240$ and $X = 3400$, in fault block D (see Fig. 6) where the ore horizon is folded back on itself in the strike direction in addition to being faulted. This area will require careful treatment during unrolling to ensure that distortion is minimized. A second area, which lies between $X = 2760$ and $X = 2920$ in fault block B, contains a much less complex fold structure with associated faults, and should pose less of a problem during unrolling. The adjacent areas are relatively undeformed and can be used to represent the situation where open folding exists.

Data collection and ore evaluation methods at Zinkgruvan

The purpose for which samples are collected, and the sampling technique and density used, depends on the stage that different areas of the mine have reached in their development and evaluation.

In the exploration phase, information is obtained by core drilling and the analysis of core samples. The spacing of the drill holes depends mainly on the complexity of the geology. In the western part of the ore deposit, where the folding is much more intense, the spacing can go down to 40 m while in the eastern part the spacing is up to 100 m. The ore reserves calculated from this drilling are done by manual polygon methods on a vertical longitudinal projection.

In the exploitation phase, when more exact information is needed on the location of assay hanging-walls and footwalls and on the thicknesses and grades of the potential ore zone, chip sampling, sludge drilling, geological mapping and 'down-the-hole' geophysical surveys (natural gamma, gamma–gamma and magnetic susceptibility logging) may be used. One of the most important reasons for this work is grade control, i.e., to ensure that overbreak during stoping is kept to a minimum and that the grades of the ore remain above the economic threshold. Chip sampling is undertaken by marking a line across the width of the ore zone in the back of the drive or stope; this is then divided into lengths depending on the lithology and mineralization. These lines are spaced at intervals of 15–20 m along strike so that every stope on every main level is sampled. The samples are collected using a hand-held air-pick. A series of 15 mm diameter circular holes are drilled to a depth of 10 mm over the sample length. The cuttings are sucked into the bag of a vacuum cleaner which is changed for every new sample. Sampling is undertaken from the footwall to the hanging-wall following the pre-marked line. Footwall and hanging-wall formations are also sampled to the maximum width of the excavation. The samples are then analysed for Zn, Pb, Ag, Cu and Fe. This sampling strategy means that the spacing of lines of chip samples between mine levels can be up to 150 m in the more uniform parts of the orebody if no significant grade changes have been observed. When the reserves are being estimated for a stope, the grade is determined by weighting the length weighted grade of a line of chip samples by its 'volume of influence' which is determined by the distance between mine levels. A minimum stoping width, which is determined by the mining method to be employed, is also taken into account. For example 'cut and fill' stopes require a minimum width of 3.0 to 4.1 m. A stope overbreak of 20–27% is also applied depending on the mining method. The end product of the calculation is the value of the ore expressed in Swedish Krona per tonne.

The total tonnage of a stope is calculated by using the distance between the levels on VLP and the horizontal cross sectional areas of the orebody on the two bounding mine levels. This method is, however, only possible in areas where the ore thickness and the grade are relatively uniform and where structural complexity is not too great. Global mine reserves can then be calculated by volume weighting these stope grades and accumulating the tonnages

Mapping of the stope backs is done on a regular basis at a scale of 1:400 but if more detailed information is needed it is undertaken at a scale of 1:200. In addition to the core drilling, systematic chip sampling and mapping are also used for ore reserve calculation.

Structural unrolling: previous work

To date there has been little published in the scientific literature regarding the unrolling of stratabound and stratiform deposits prior to the geostatistical evaluation of their *in situ* resource. However, an important paper by Sides (1987) describes an unrolling technique which was developed during reserve estimation at Rio Tinto Minera's Bama and Brandelos copper mines in north west Spain. These deposits consist of stratabound mineralized stockworks within a metamorphozed ophiolite complex which has been subjected to a relatively open style of folding.

In order to unfold the mineralized horizon, an interpreted median plane was selected as the structural reference plane (see Fig. 7). To account for lateral as well as vertical displacements, the reference plane was interpreted from two sets of sections oriented N–S and E–W. Two sets of control points were created (N–S, E–W), with points being marked at 25 m intervals in both directions from two reference lines. To calculate the unrolled position of a point (A) not located on the reference plane, a line is projected

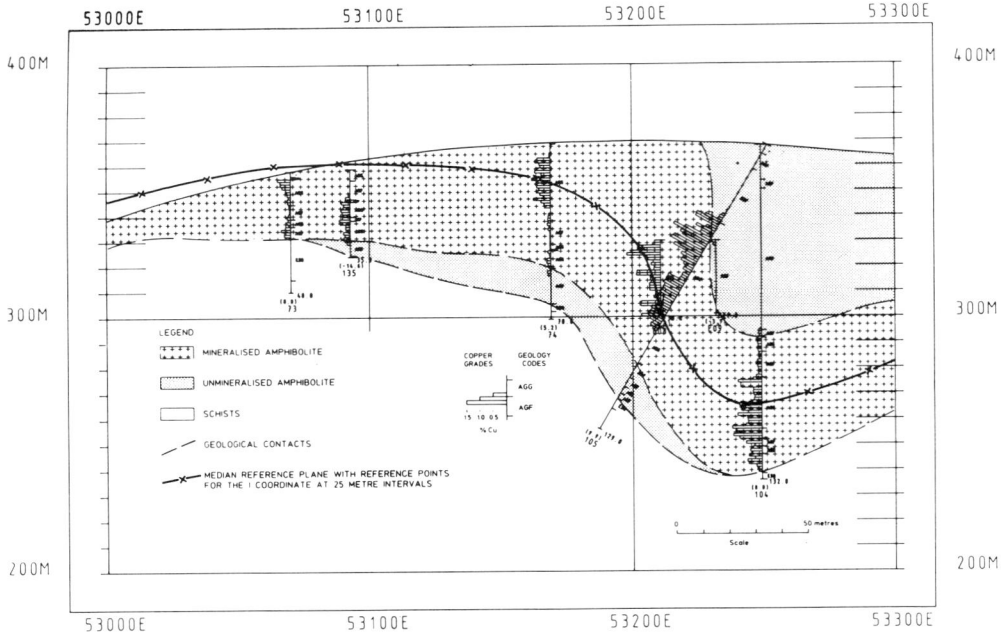

Fig. 7. Median plane used as the reference plane for unrolling (from Sides 1987).

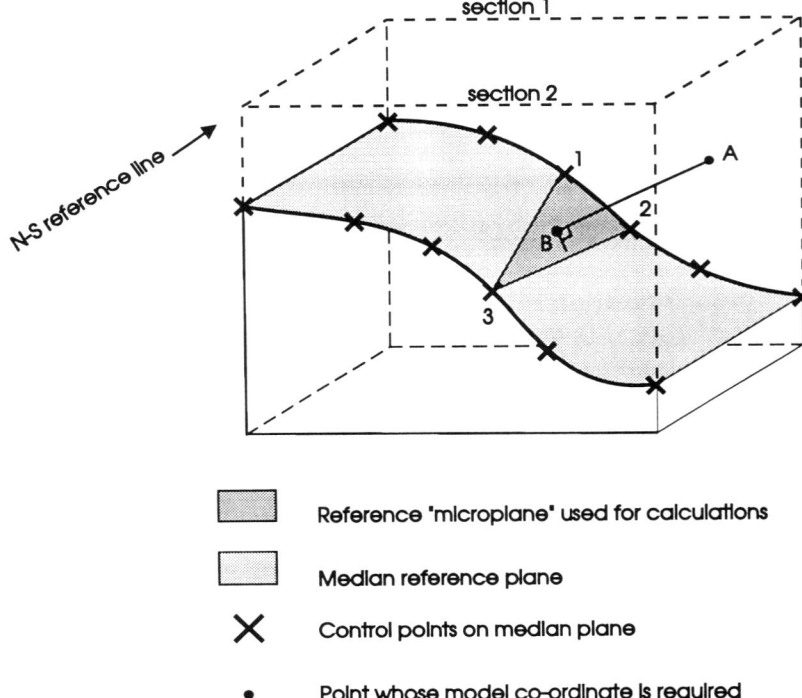

Fig. 8. Sketch of method used to calculate model k co-ordinates (based on Sides 1987, Fig. 5).

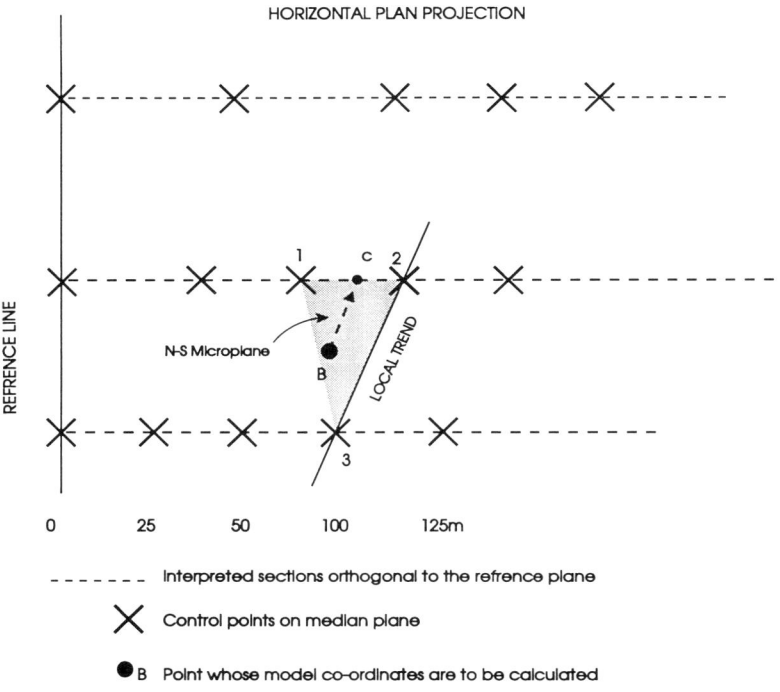

Fig. 9. Sketch of method used to determine control point coordinates (i,j) on median plane (based on Sides 1987).

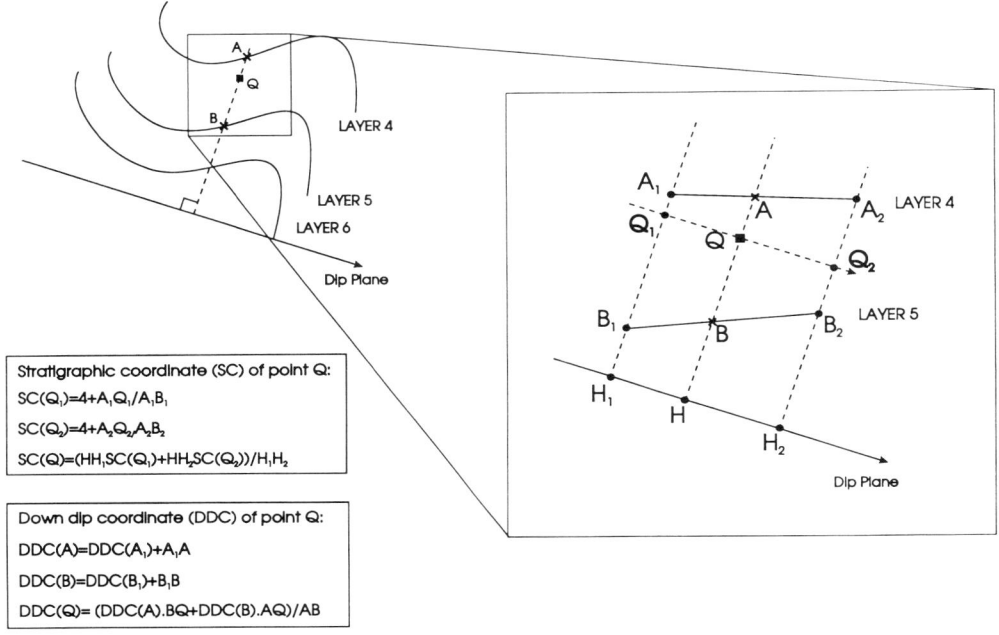

Fig. 10. Method to determine stratigraphic and down dip co-ordinates of a data point (based on David 1988).

orthogonally onto the reference plane from the point so as to intersect it at point B (see Fig. 8). A triangular 'microplane' is then defined from the closest three control points, on the two bounding section lines, first using the control points for the N–S direction (j), and then using the control points for the E–W direction (i). Two of the control points (i or j) will lie on the same i or j coordinate line while the third will have a different unrolled i or j coordinate. The local trend of the microplane is determined using the two common control points, (e.g., 2 and 3 on Figs 8 and 9), so that the projected point (B) may be extrapolated parallel to this local trend to point C on Fig. 9, which lies between control points 1 and 2 on the same section line. The unrolled coordinate (i or j) for point C is then determined by linear interpolation between the values of the adjacent control points on the section line. The new unfolded Z coordinate (k) is the projection distance (A–B) which is calculated and then assigned a positive or negative value depending whether the point is above or below the median reference plane.

Transformed unfolded coordinates (i, j and k) were then used to calculate semi-variograms for copper % in the usual way. As a result, Sides (1987) found that a significant lowering of the nugget effect was achieved for horizontal directional semi-variograms indicating that the error introduced by not unfolding the control points was not insignificant.

An additional method to unroll deformed ore bodies is described by David (1988). This uses the interpolation of distances between sections, and may be used only on cylindrical simple folds containing geologically distinguishable stratigraphic horizons. Three axes are defined; a strike axis, and two axes within sections perpendicular to the strike direction. One of the latter is the dip direction of an average dip plane for the whole orebody, while the other is perpendicular to this plane.

The first stage in the unrolling procedure is the construction of a perpendicular line from the reference plane through each data point (Fig. 10). To determine the stratigraphic coordinate (SC) of a data point (Q), its position on the perpendicular line drawn from the reference dip plane relative to the closest overlying stratigraphic layer must be determined. The actual value calculated, by linear interpretation, is the distance below this layer to the data point expressed as a ratio of the distance between this layer and the immediately underlying layer (e.g. a point half way between layers 4 and 5 would be allocated a stratigraphic coordinate value of 4.5).

To determine the down dip unrolled coordinate of data points the intercepts of this perpendicular line within the two bounding stratigraphic horizons are calculated relative to adjacent control points on each horizon (again by linear interpretation) (Fig. 10). As the geological distances around the folds from a reference line to the control points on each horizon are known, this allows the unrolled coordinates of the data points to be calculated. If a point lies between sections then linear interpolation is undertaken between sections.

No account is given by David (1988) as to the separation, or frequency of points used to control the unrolling (control points) or as to whether their selection is based on a user defined or rigid grid.

Structural unrolling at Nygruvan

Figure 6 shows that the Nygruvan orebody can be subdivided along strike into four blocks, each separated by a fault. The deformation intensity and style in each differs significantly and thus an ideal situation presents itself to test various unrolling methods. At this preliminary stage of the project, two areas were selected for further study, namely fault blocks B and C on Fig. 6. At a later stage, fault block A, which is intermediate between B and C in terms of intensity of deformation, and fault block D, which is an example of extreme deformation, will be unrolled.

Unrolling method 1: minor folding

This method of unrolling is suitable for areas in which only minor deformation has taken place, and in which open folds exist (or the limbs of such folds). Fault block C, between $X=2950$ and $X=3300$ m in Fig. 6, is ideal in this respect. The orebody thickness, or strained thickness resulting from deformation, is not considered in this planar geometric unrolling procedure, which uses the ore–skarn contact as a marker horizon. This contact lies on the stratigraphic hangingwall of the orebody, and was chosen as it is a clearly defined contact which can be easily located in underground exposures and in drill core. Structure contours of the ore–skarn contact shown on Fig. 6 are based on geological mapping along mine levels and sub levels.

This unrolling method is based on a grid of fixed points which are unrolled from baselines selected perpendicular to the unrolling directions (Fig. 11). A set of grid lines, approximately normal to the strike, are superimposed over a plan of ore–skarn structure contours (labelled

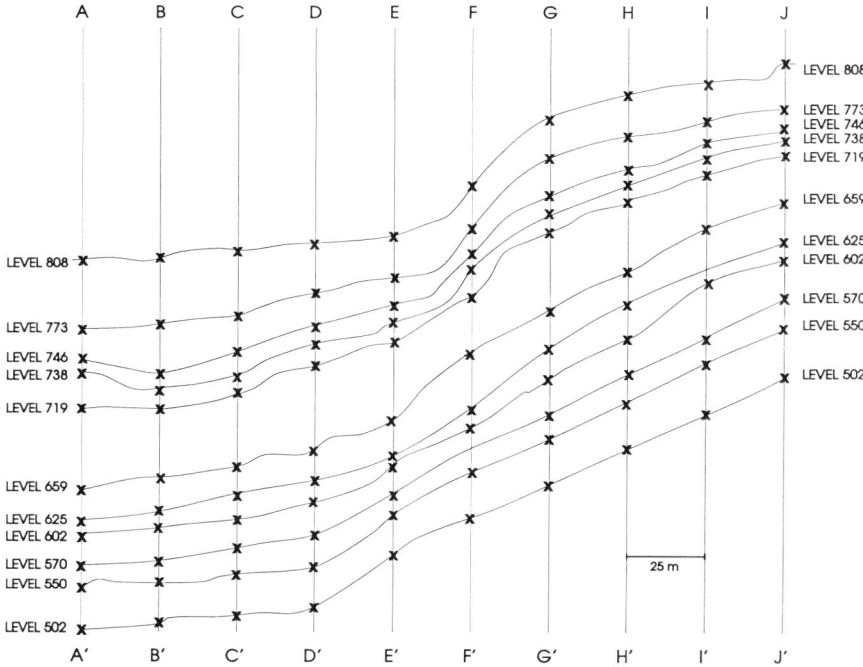

Fig. 11. Control points in fault block C for unrolling method.

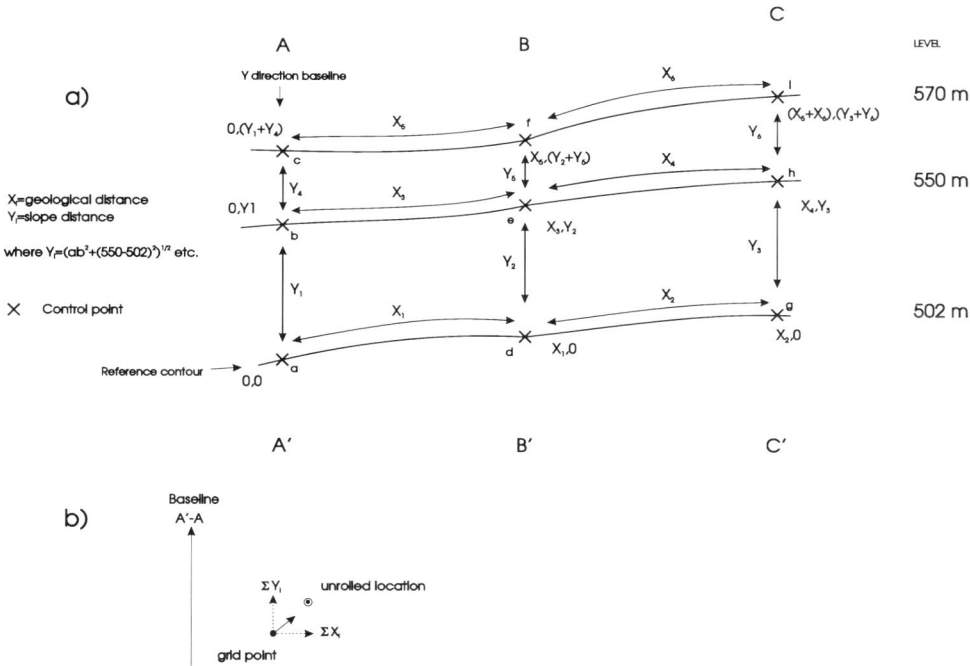

Fig. 12. (*a*) Explanation of algorithim used in unrolling method 1. (*b*) Accumulation of *X* and *Y* geological distances to calculate unrolled coordinates of an unrolled point.

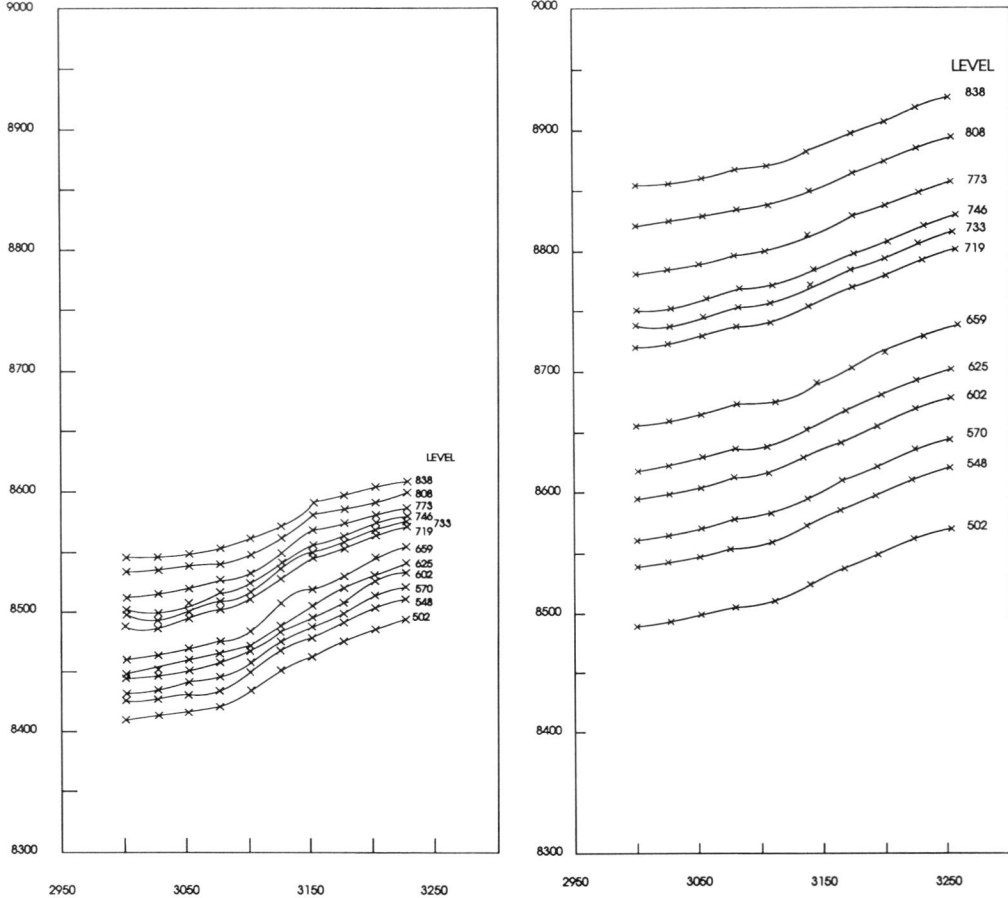

Fig. 13. Pre-unrolled plan of control points in Nygruvan fault block C.

Fig. 14. Unrolled plan of control points using unrolling method 1 in Nygruvan fault block C (Z elevation = 502 m).

A'–A through to J'–J in Fig. 11). Each grid line and mine level contour intersect at a point whose three dimensional coordinates (x, y, z) can be determined; the z coordinate being determined from the elevation of each mine level. A grid of control points is thus established whose spacing is regular in only one direction, i.e., in the strike direction of the orebody. In the down-dip direction along grid lines, e.g., A'–A, the grid spacing is irregular being controlled by the contour separation, which is ultimately controlled by the mine level and sub-level separation and dip. Two baselines were selected so that the control points could be unrolled in the x and y planes, relative to a set of datum points. The first of these baselines was the highest mine level contour which was used for unrolling in the y direction (i.e. mine level −502 m) and hereafter

referred to as the reference contour. The first control point (now a datum point) on each mine level contour was used to construct the baseline for unrolling in the x direction (in this case line A'–A).

The geological distance is the true distance between two stratigraphically equivalent points measured around the folded structure. The distance between each control point along a mine level contour is measured in a direction away from the y direction baseline (commencing from datum points along line A'–A in Figs 11 and 12a). For processing purposes, this distance is associated with the control point to which it was measured. The geological distance is thus greater than, or equal to, the direct point to point distance. All geological distances between control points along each mine level contour are

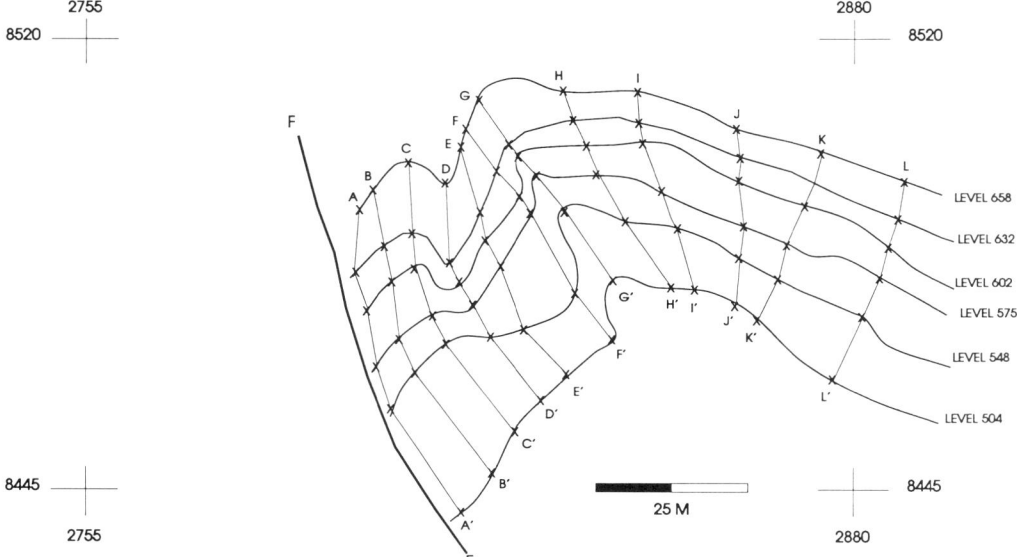

Fig. 15. Unrolling method 2 for more complex structures using linking of user defined geologically equivalent control points. Fault block B Nygruvan.

thus accumulated in the x direction, thus unrolling the control points along the strike of the orebody.

To unroll the control points in the down-dip direction onto a horizontal plane in the y direction, distances are measured from the first control point on a mine level contour to the corresponding first control point in the next mine level contour below (moving down grid line A′–A, in Figs 11 and 12a). The first distance that is measured is that between level −502 m and −550 m (e.g. a–b in Fig. 12a), then the exercise is repeated between −550 m and −570 m (e.g. b–c in Fig. 12a) and so on. This is then repeated for each grid line. Because the orebody is inclined, the slope distance (Y_i) must be calculated, using the measured horizontal distance, and vertical mine level separation. It is assumed that the segment of ore between each pair of control points is planar. These distances are then accumulated from the reference contour (−502 m), producing an unrolled grid in the y direction. The unrolled x, y coordinates of each control point are thus the accumulated x and accumulated y increments as in Fig. 12b. A plan can now be constructed of the unrolled control points in the x–y plane with the z coordinate being equal to the elevation of the reference contour. All the data acquired were compiled and manipulated on an EXCEL spreadsheet from which an ASCII file was produced of the unrolled coordinates. This file was then input to SURFER to produce a graphical output. The original and unrolled plots produced are presented in Figs 13 and 14.

Limitations of the method

This method of unrolling has, however, a number of limitations which restrict its use to areas of minor deformation. For example, once a grid orientation has been selected on a geological basis, measurements of distances between control points are in a set direction so that when a change in orebody orientation occurs the distances between the control points are not those in the strike and dip direction of the deposit. Unrolling then takes place in directions across the dip and strike. The problem is particularly serious when the orebody orientation becomes sub-parallel to the grid lines, so that large strike lengths of ore occur between control points on adjacent grid lines, reducing the resolution of the structure. Also, the unrolling of overturned fold limbs is not easily achieved by this method.

The selection of grid dimensions is critical in that a large grid dimension will only provide a small number of control points on the structure, while a small grid dimension will provide many more control points and thus much greater resolution of orebody structure.

Geologically equivalent points are not selected between mine level contours due to the rigid nature of the grid system, and the unrolling of a structure becomes largely a geometrical rather than a geological exercise. The control points are not selected on a geological basis, and thus the best use is not made of the data available. Unrolling in one direction only results in an accumulation of errors in this direction

This technique requires all the data to be in a very rigid format for processing which does not make it easily applicable to all types of deformation.

One advantage of this method is, however, that near regular ore blocks may be defined prior to the unrolling, and unrolled blocks can be easily related to the original blocks. These blocks may in fact be mining blocks defined by mine level and mine structure section lines (mining block boundaries). Reserve estimates made on this basis will then be of direct benefit / interest to mining personnel.

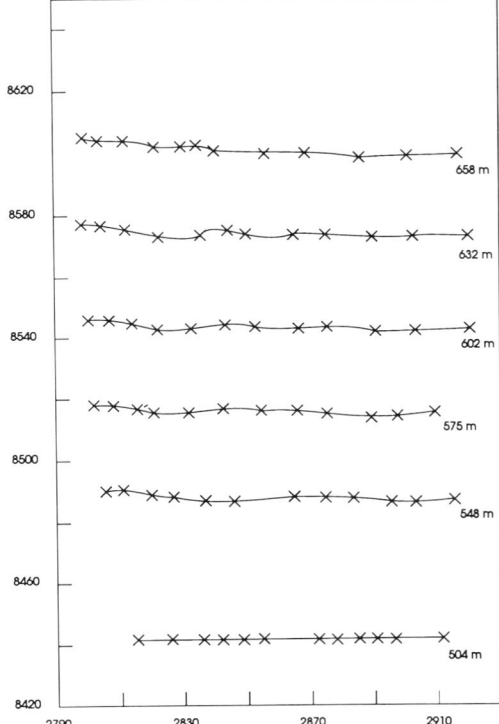

Fig. 17. Unrolled plan of control points in Nygruvan fault block B using unrolling method 2 employing a fixed baseline of datum control points in the Y direction only and a straight baseline in the strike direction.

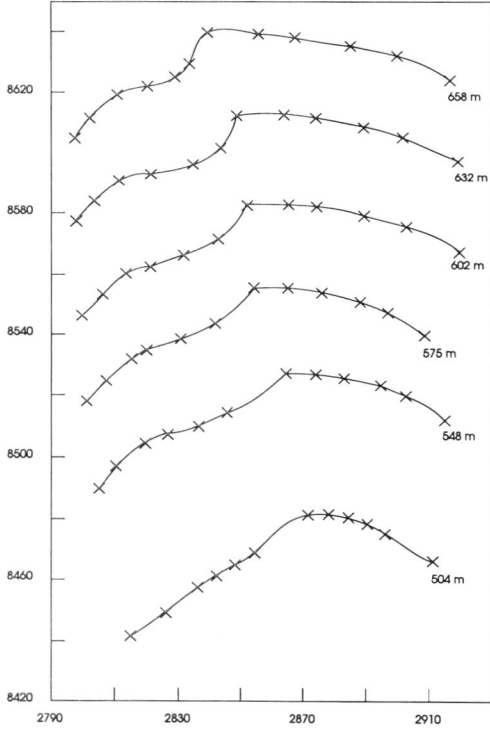

Fig. 16. Unrolled plan of control points in Nygruvan fault block B using unrolling method 2 and employing fixed baselines of datum control points in the X and Y directions.

Unrolling method 2: complex folding

This modification is a development of the first method to enable the unrolling of more complex structures. Fault block B from the Nygruvan orebody (Fig. 6) was deemed a suitable fold structure on which to test this method. In this case, the user defines control points on each mine level contour, moving down-dip to each successive mine level contour, and linking each control point with structural tie-lines (Fig. 15). The control points which are linked along lines A′–A through to L′–L, are defined by the geologist and are not restricted to a rigid grid system. It is now possible to link geologically equivalent control points together down the dip of the orebody which later will be unrolled together. A random grid of control points is thus created defining irregularly shaped grid blocks. Geological distances are measured between the control points along mine level contours and slope distances are calculated in the down-dip direction between linked control points as in method 1. These slope distances are accumulated along lines A′–A to L′–L.

Two separate attempts were made to unroll

the fault block B fold after the control points had been linked and geological distances measured. It soon became apparent that the choice of baseline, and control points to use as datum points, is critical. Initially, all control points along line A′–A and the −504 m mine level contour were used as datum points (Fig. 15). Unrolling was achieved by accumulating the measured geological distances away from both baselines of datum points, in the x and y directions. The resulting unrolled plan (Fig. 16) can be compared with the pre-unrolled structural plan of control points on mine level contours (Fig. 15). As can be seen, the two plans remain basically similar, except that the unrolled plan is stretched in the x and y directions. The form of the unrolled levels is influenced by the shape of the baseline datum (−504 m level). A second attempt was then made to unroll this particular fold using only datum points along line A′–A. The −504 m contour was thus unrolled in a straight line by the accumulation of measured geological distances in the x direction. Each successive lower contour was then unrolled relative to the datum points along A′–A and the now straight −504 m contour, again by accumulation of measured and calculated geological distances. The resulting unrolled structure can be seen in Fig. 17. The wide spacing between contours is a function of the orebody dip (approximately 70°) when rotated onto the horizontal −504 m reference plane.

Limitations of the method

The initial control grid is irregular in both directions and thus each unrolled grid block also has an irregular shape. As a result, it will be difficult to relate these blocks to mining blocks on the original pre-unrolled grid. A second grid may have to be overlaid on top of the irregular control grid to define mining blocks which can then be unrolled using vectors associated with the original unrolling grid.

This method has successfully unrolled a thin folded tabular orebody but thought must be given to the effect that the thickness of the ore horizon and the type or classification of the fold structure has on the unrolling process.

Unrolling of borehole intersection points

The methods described above have enabled the structure of the orebody to be unrolled, but what is actually required is the production of unrolled coordinates for drill hole or face samples which may lie structurally above the ore–skarn contact. This exercise has not yet been undertaken but is the next phase of the project. Initially, it will be necessary to project the mid-points of these samples orthogonally (or parallel to the dip of the axial plane depending on the mechanism of fold formation) onto the the pre-unrolled reference plane. The four control points at the corners of the quadrilateral containing each of

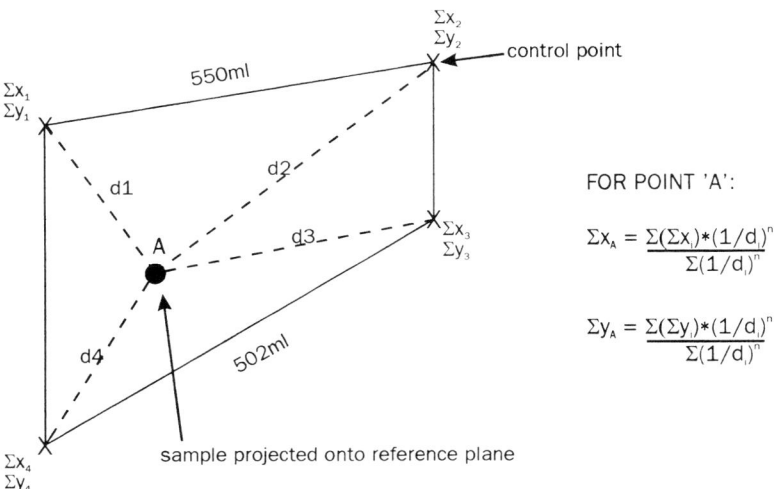

UNROLLED COORDINATES FOR POINT 'A' FROM UNROLLED CONTROL POINTS

Fig. 18. Method by which unrolled coordinate of a drill hole or sample point may be calculated using inverse distance weighting methods.

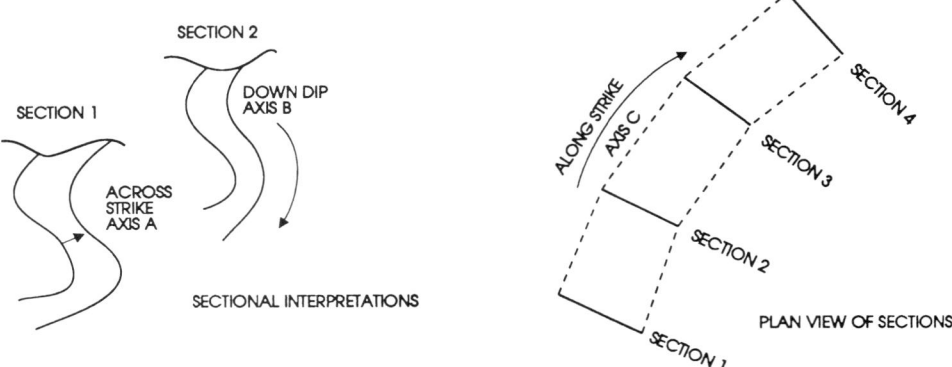

Fig. 19. Coordinate axis in the unfolded coordinate system (UCS).

the projected points will then be located and the distances to each calculated (Fig. 18). Inverse distance weighting methods will then be applied to the incremental vectors in both the x and y directions at each of the four points. This will then allow the accumulated x and y increments at the sample point to be estimated giving the unrolled coordinates of the point relative to the reference baselines.

Datamine 'unfold' methology

The semi-computerized unrolling techniques described above are being developed in order to make direct comparisons with a new, as yet unreleased DATAMINE facility for the unfolding of deformed tabular orebodies. This new module, 'Unfold' (Datamine 1992), has reached the 'beta-test' stage, undergoing site testing. Unfolding of folded structures is achieved by determination of geological distances on sections. The program calculates an unfolded coordinate system (UCS) for the down hole 3D coordinate of sample mid points drill hole data so that semi-variogram analysis may be undertaken. A kriged or inverse distance weighted block model may then be be calculated in the unfolded coordinate system and the results back-transformed to the world coordinate system within the folded stratified unit using standard DATAMINE methods. The following description is based on a more detailed analysis of the method in the Datamine 'Unfold' reference manual (Datamine 1992), on which Figs 18 to 21 are based.

A standard x, y, z grid is transformed to UCS whose axes are neither orthogonal nor straight lines (Fig. 19). UCS axis A is oriented across strike of the orebody, axis B down-dip on a centre line between the hanging-wall and foot-wall of the orebody and axis C is oriented along strike of orebody.

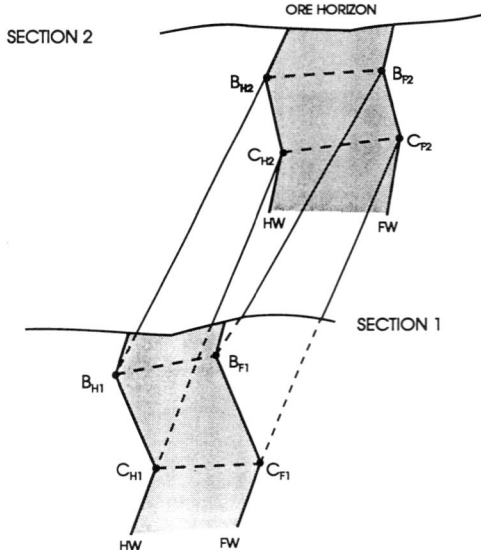

Fig. 20. Creation of control points between sections where: B_{H1}, B_{F1}, C_{F1}, C_{H1} and B_{H2}, B_{F2}, C_{F2}, C_{H2} have the same along strike coordinate, i.e. UCS-C coordinate; B_{H1}, B_{H2}, C_{H2}, C_{H1} and B_{F1}, B_{F2}, C_{F2}, C_{F1} have the same across strike coordinate, i.e. UCS-A coordinate; B_{H2}, B_{F2}, B_{H1}, B_{F1} and C_{H1}, C_{H2}, C_{F2}, C_{F1} have the same down dip coordinate, i.e. UCS-B coordinate.

A series of hexahedrons is created within the defined stratified unit whose sides are oriented parallel to the three unrolling axes A, B and C (Fig. 20). To create a hexahedron, links are

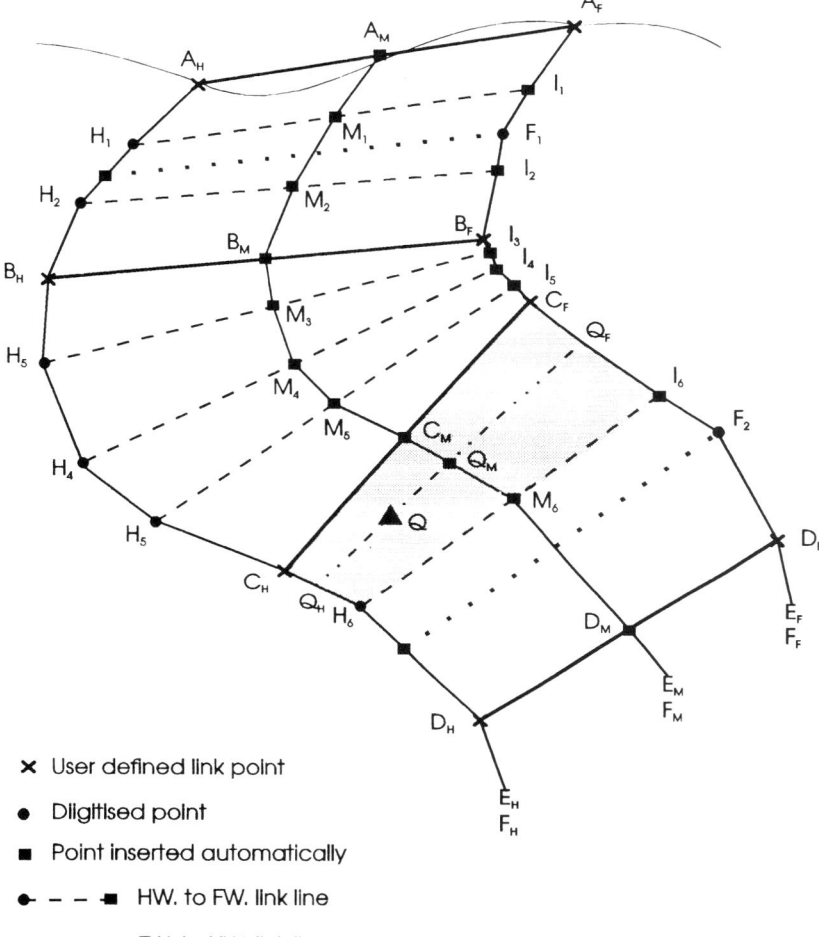

Fig. 21. Linking of control points between hanging-wall and footwall in DATAMINE 'Unfold'.

made between hanging-wall and footwall strings, which are user defined during digitization in exactly the same way as when tagging a wire frame model. Points are linked which are considered to have the same down-dip UCS coordinates and to be geologically equivalent (Fig. 21), though linking can also be done automatically using a DATAMINE algorithm. Any point within the stratified unit will lie within one of these hexahedrons, and its coordinates can thus be transformed into the UCS, which may be scaled in one of four ways:

(1) 'normalized'—unrolled distances are calculated as a ratio of the reasured distance (along each axis) to the total axis length giving a value between 1 and 0;

(2) 'adjusted'—the normalized value is multiplied by the appropriate average axis length;

(3) 'true length'—length measured from the UCS origin;

(4) 'world co-ordinate system'—in standard x, y, z coordinates.

To calculate the coordinates of any point Q, based on the A, B and C axes, DATAMINE 'Unfold' takes every digitized point on the hanging-wall and creates a corresponding point on the footwall, taking account of the links defined by the user (see Fig. 21). This is achieved by measuring the digitized distance from the first linked point on the hanging-wall to the second (e.g. A_H H_1 H_2 B_H in Fig. 21) and calculating the ratio between the distance from the first linked

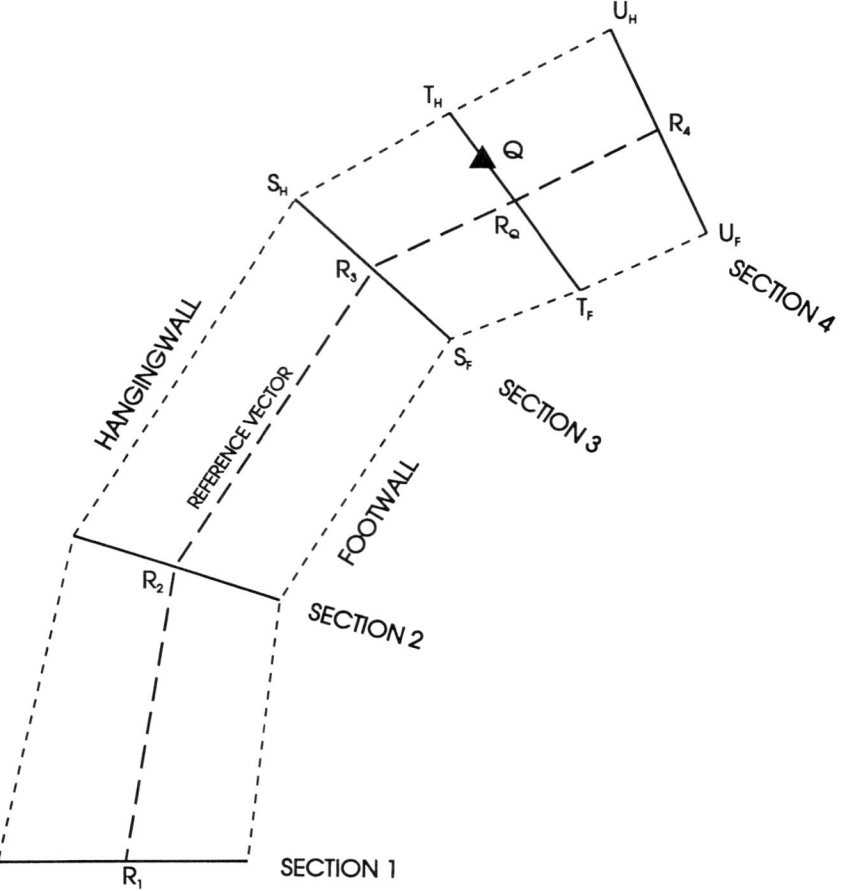

Fig. 22. Linking of control points between adjacent sections to calculate the UCS-C coordinate in DATAMINE 'Unfold'.

point to the first digitized point (e.g. $A_H H_1$) as below:

$$R = A_H H_1 / A_H H_1 H_2 B_H$$

This same ratio is then applied to the footwall so that a new point can be created on this string (e.g. I_1). In this case:

$$A_F I_1 / A_F F_1 B_F = R$$

All digitized points on the hanging-wall are then treated in the same way to create a new set of equivalent points on the footwall. This process is then repeated by creating new points on the hanging-wall from digitized points on the footwall.

A series of quadrilaterals is thus created in section, and it is now possible to calculate the UCS A and B coordinates, for point Q. A median plane containing points M1, M2, M3 etc, is created down the section of the stratified unit as in Fig. 21. Coordinate A (in the direction of the across strike axis) is then calculated for point Q by first identifying the quadrilateral within which the point falls, i.e., C_H, C_F, I_6, H_6 (shaded area). A straight line is drawn through Q intersecting the hanging-wall and footwall such that the ratio of each end length $C_H Q_H / C_H H_6$ is the same as $C_M Q_M / C_M M_6$. The normalized value of the A coordinate for point Q is:

$$A_N = Q_H Q / Q_H Q_F$$

and the normalized value of the B coordinate is:

$$B_N = A_M B_M C_M Q_M / A_M B_M C_M D_M E_M F_M$$

This only deals with a point lying on a section, but for a point lying between sections, the method must be extended to three dimensions. Points are linked between sections in a similar manner to that used on individual sections. Links are made from link points in the hanging-wall of one section to the hanging-wall of the adjacent section, and also between each digitized point. The point Q now lies within a wire frame hexahedron which, if sliced through Q such that the intersections divide the hanging-wall and footwall equally, a quadrilateral will be created. The A and B coordinates can now be calculated as described earlier. The user only needs to define links between the hanging-wall link points on adjacent sections, for the footwall to footwall links are done automatically by the software.

If Fig. 22 is taken to represent the maximum horizontal extent of the hanging-wall and footwall of the orebody in plan then the C coordinate can be calculated as follows. A reference plane is drawn through the centre of the deposit and a line is drawn through point Q such that the hanging-wall and footwall distances between sections are divided in the same proportions such that,

$$S_H T_H / S_H U_H = S_F T_F / S_F U_F.$$

All points on any given section will have the same along strike UCS coordinate. The total distance along strike is calculated by creating a control point on each section and calculating the length of line joining these control points. This reference line then defines the average unrolled strike length of the orebody (RL).

The normalized UCS coordinate C for point Q is calculated as below:

$$C_N = R_1 R_2 R_3 R_Q / R_L$$

The end product of this exercise is thus the production of an unrolled normalized coordinate (A_N, B_N, C_N) for a sample mid-point in the orebody. Repetition of this process for all samples in the database allows the production of a new transformed database on which variography can be undertaken. At the time of writing, this package has not been tested on the Nygruvan orebody.

Conclusion and future work

The three methods described for the structural unrolling of ore deposits show considerable promise and development work and testing will continue so that a preferred methodology can be selected leading to a geostatistical study at some time in the future. Though the work to date has concentrated on unrolling control points on a reference plane, there appears to be no particular problem in extending the method to cope with either chip or core samples which may lie above or below the reference plane. It was suggested earlier that after projection of the sample mid-points normally onto this plane, a weighting technique, such as inverse distance weighting, could be used on the four unrolled control points at the corners of the quadrilateral containing the projected point. This will provide the necessary coordinates for the sample mid-points. Other methods of calculating these coordinates from the adjacent control points will also be investigated.

The authors would like to acknowledge the invaluable logistical and financial support of Vieille-Montagne Sverige for this project and also the award of a NERC–Extractive Industries Partnership Scheme (EIPS) studentship to S. Ingram. They would also like to thank Datamine International for suggesting that the work be undertaken at Zinkgruvan and providing an advance copy of 'Unfold' for test work, and Dr Edmund Sides for reviewing and commenting on the manuscript.

References

DAVID, M. 1988. *Handbook of Applied Advanced Geostatistical Ore Reserve Methods*. Elsevier.
DATAMINE AUSTRALIA 1992. *Datamine Unfold Reference manual*.
HEDSTRÖM, P., SIMEONOV, A. & MALMSTRÖM, L. 1989. The Zinkgruvan Ore Deposit, South Central Sweden: A Proterozoic, proximal Zn-Pb deposit in distral volcanic facies, *Economic Geology*, 84, 1235–1261.
SIDES, E. 1987. An alternative approach to the modelling of deformed stratified and stratabound deposits. *12th APCOM proceedings 1987*.
WELIN, E., WIKLANDER, U. & KÄHR, A-M. 1980. Radiometric dating of quartz-porphyritic rhyolite at Hällefors, South Central Sweden *Geologiske. Föreningens Stockholm Förhandlinger*, 102, 269–272.

The application of geostatistical techniques to *in situ* resource estimation in the sand and gravel industry

J. ARTHUR & A. E. ANNELS

Mineral Resource Evaluation Research Unit, Department of Geology, University of Wales, College of Cardiff, PO Box 914, Cardiff CF1 3YE, UK

Abstract: In June 1990, a research project was commenced at a site in Southern England, in which sands and gravels of the Kesgrave Formation are overlain by Anglian boulder clay. This site was, at the time, under evaluation by RMC (UK) Ltd as a potential new source of natural aggregate.

The primary aim of the research was to ascertain whether standard geostatistical techniques could be applied to Quaternary fluvio-glacial sand and gravel deposits. Comparisons would then be possible with the results of evaluations carried out using inverse distance weighting and digital terrain modelling techniques.

The data used in the project were collected by shell and auger drilling on a localized detailed drilling grid (25 m spacing), and a less regular grid (100 m hole spacing) covering the whole site. The latter was carried out as part of a standard site evaluation. The variables used in the study included mineral and overburden thickness, size gradings as defined by BS:812, and petrography.

The work has highlighted the need for improved methods of drilling and sampling in unconsolidated deposits. Loss of fine ($<75 \mu$m) material and poor sampling of the coarser (>40 mm) fraction had resulted in excessively high nugget variances which caused difficulties in the modelling of semi-variograms for these size fractions.

This paper presents the results of the geostatistical modelling of the above parameters using a range of methods including indicator variography and kriging. In particular, the deviation of the kriged block values from specific British Standards was assessed in order to determine the potential end use of the *in situ* material, before or after selective blending/processing.

The research described in this paper has been carried out at a site in south-central Essex which is characterized by Quaternary fluvio-glacial braided river gravels and sands overlain by a chalky boulder clay. The gravels are Beestonian in age and the overlying till is the Anglian Lowestoft lodgement till which covers large areas of East Anglia and the East Midlands. The area is a major producer of aggregate for the south east of England, mostly from the unconsolidated Quaternary gravels of Essex, Suffolk and Norfolk. A relatively small amount is also produced from offshore areas around the coasts of these counties. The original aim of the research was to assess drilling and sampling practices within the sand and gravel industry and the application of standard geostatistical *in situ* evaluation techniques to deposits of loose aggregate.

The gravels are predominantly composed of flint and quartzite with minor amounts of sandstone and volcanic rock. The principal end use is in concrete production and for the purpose of this research, the British standards relevant for concrete specifications were used (i.e. BS:812 and BS:882).

From a total drillhole database of 527 boreholes, three separate datasets were eventually produced to analyse the effect of various drilling patterns and densities in different areas of the site. The drill hole location map (Fig. 1) covers an area of 3.5 km by 1.5 km. The close spaced drilling in the west of the area is the site of the 'research drilling'. The remainder of the boreholes were drilled on an approximately 100 m grid spacing and represent a standard evaluation drill grid for the company undertaking the evaluation.

The 'research drilling' database consists of 197 boreholes, with spacings down to 25 m together with an additional 30 redrills which were sited 1 m from the original hole. The 100 m database consists of 382 boreholes including holes in the research area left after removal of those which did not correspond to a 100 m random stratified grid. The resulting dataset was regarded as a typical evaluation drilling grid.

Data from borehole records were used to

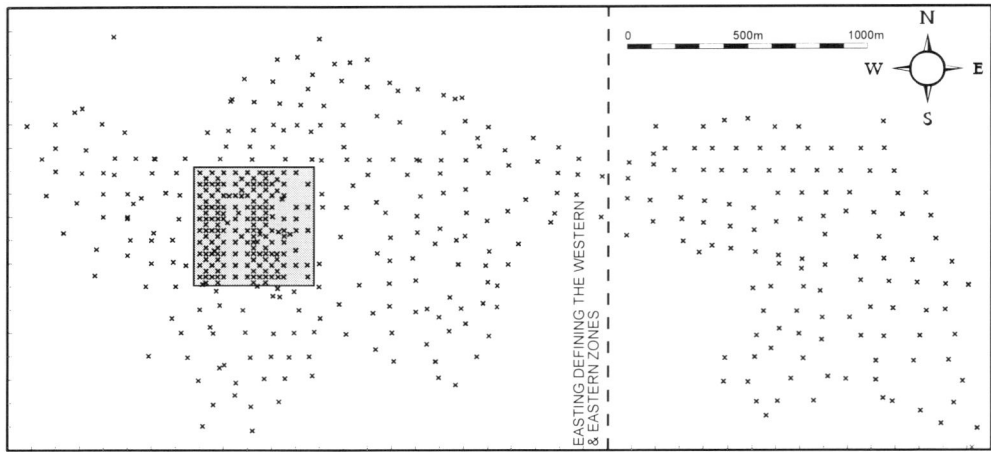

Fig. 1. Borehole location plan.

produce cross-sections and contour plots of elevation and horizon thicknesses. These plots provided the first steps towards the production of a geological model for the site. Statistical and geostatistical analyses were then carried out on thickness values for gravel and overburden and the stripping ratio. Gradings values were produced according to the specifications laid out in BS:812 and geostatistical analysis was then carried out in an attempt to correlate the results with those produced by geological modelling of the site.

Drilling and sampling methods

Drilling was carried out using shell and auger cable percussion rigs with an 8 inch (200 mm) bailer. A 30 kg sample was bagged for every 1 m, or part thereof, drilled in mineral. In the research area 30 boreholes were redrilled approximately 1 m from the original site to provide data on the precision of the drilling and sampling methods employed.

The 1 m samples from each hole were stored on site, logged and a hole composite was produced for the mineral horizon. If a layer of clay interburden appeared in the gravel or other clear evidence of a natural break, then this was reflected in the compositing of samples from the hole. In these cases it was necessary to produce more than one composite. Thirty complete sets of 1 m samples were split before compositing and two duplicate composites produced for each of these holes. Thirty hole composites were also split to form duplicate samples. All these additional composite samples were bagged and sent for analysis with a false line number in order to disguise their origin.

The drill hole grid in the research area was designed in order to obtain the minimum requirement of at least 30 sample pairs in each of the four principal directions up to a lag value of 7 with a 25 m lag spacing north–south and east–west and a 36 m spacing in the NE–SW and NW–SE directions. This was regarded as the minimum required to obtain an accurate model for the semi-variograms.

Precision of drilling and sampling

Use of the same drilling method and up to six drilling crews allowed a measure of continuity to the data. There are, however, serious problems with the shell and auger method of drilling as detailed by Barrett (1989), who states that loss of fines ($<75\,\mu$m) can be considerable during the drilling process. The shell and auger method normally relies on the presence of water within the hole and the gravel is fluidized by the upward motion of the bailer within the water column. The fluidized sample is then collected during the downward stroke. However, to prevent the bailer filling with water there are several drainage holes which allow the water to escape; this water invariably carries with it a proportion of the finer material from the gravel. Additional losses of fines also occur when they are driven into the side walls and base of the hole by the surging action of the bailer. When the sample is brought to the surface it is decanted onto a sampling board (commonly a sheet of corrugated iron) which allows excess water to

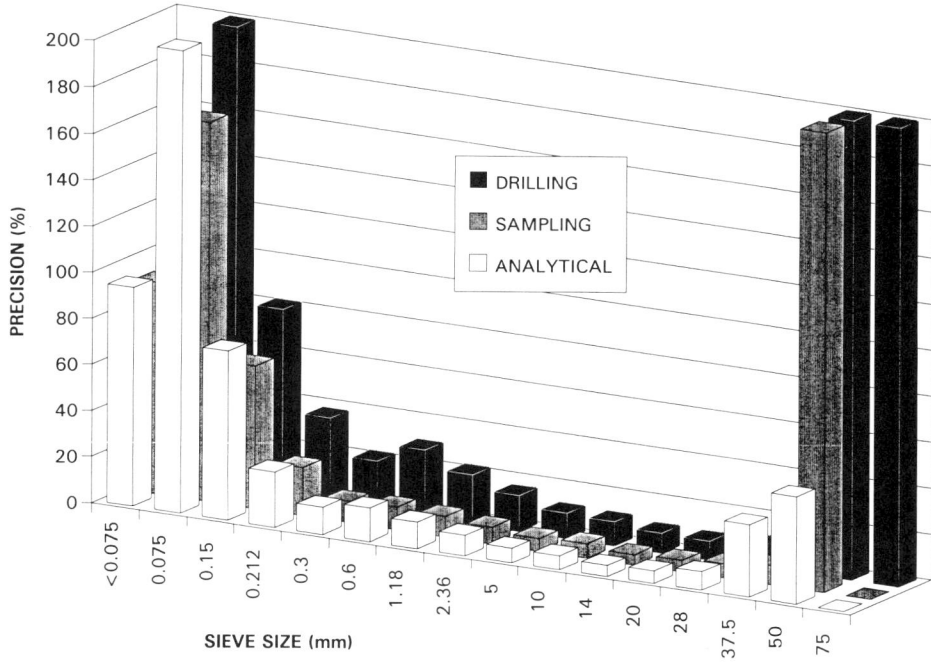

Fig. 2. Histogram of precision values from the three duplicate sets for sieve sizes between 75 mm and 76 μm.

Table 1. *Error combinations present in the three duplicate sample populations*

Duplicate population	1	2	3	4	5	6	7
Redrills	√	√	×	×	√	√	√
Sample duplicates	√	√	√	×	√	√	×
Composite duplicates	√	√	×	√	√	√	×

drain off, again carrying with it fine material. During this process large cobbles tend to roll off the sample pile into the field. Contamination from the previous sample can occur if care is not taken in cleaning the sampling board. Settling of the sample occurs within the sample bags which can result in sample bias unless thorough mixing is undertaken before production of composites. Drainage from burst sample bags also results in further loss of fine material from the sample.

To assess the loss of material during the drilling process, dry holes were drilled at seven locations adjacent to previously drilled holes to allow a direct comparison. The results showed that, on average, there was a loss of 75% of the fines (<75 μm) using standard wet drilling techniques. This figure may also underestimate the total loss due to the possible loss of fines during the dry drilling itself. The thick overburden (average 9 m) prevented the digging of trial pits on the site to assess directly the *in situ* fines content. Only seven boreholes were redrilled dry so the average values are not statistically valid; however, they gave consistently higher values for the content of silt and fines in all the dry holes as was expected.

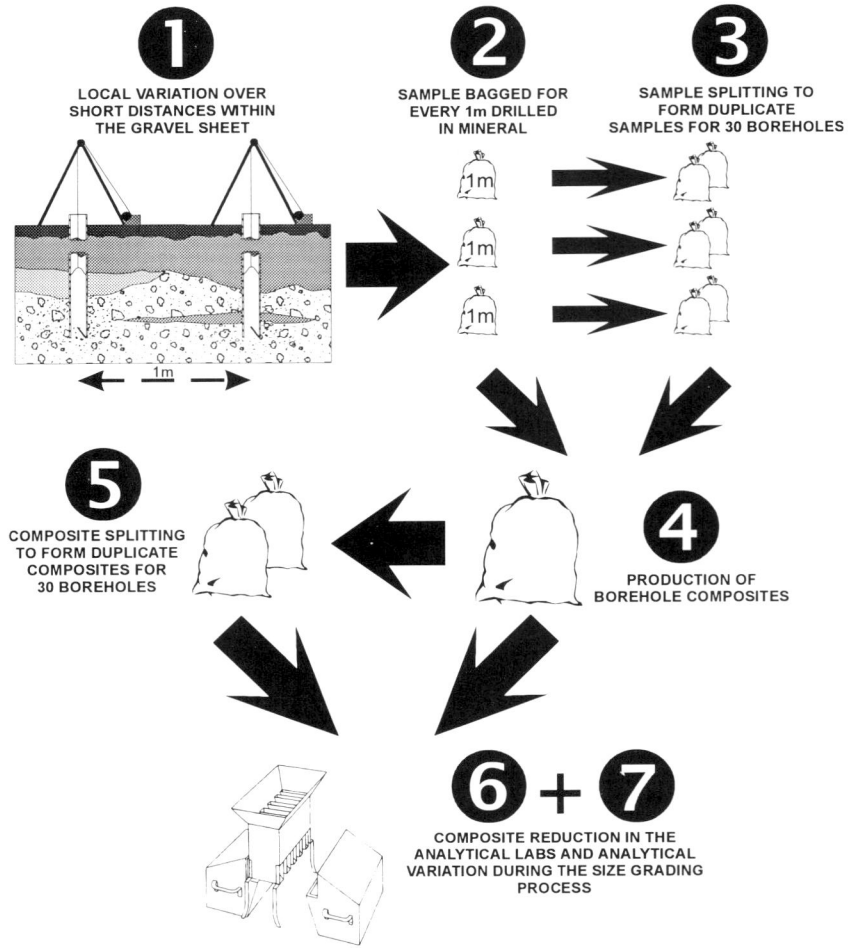

Fig. 3. Flow chart highlighting areas in the drilling and sampling process where errors occur.

Inaccuracies in the values for the fine-grained fractions will have a knock on effect for values of the coarser grain size fractions. Values are measured as a percentage passing a certain sieve size so if a fine size fraction is recorded as 2% passing but has suffered a 75% loss, the remaining 6% will have to be assigned to the coarser fractions. If several size fractions are being undersampled the effect on the coarser fractions will be even more pronounced.

The precision and associated errors were calculated using the equations presented by Garrett (1969) (Fig. 2). The flow diagram (Fig. 3) shows that seven potential sources of error are possible in the sampling procedure while Table 1 shows the combination of individual errors likely within each of the three sets of duplicate samples. The duplicates studied represented redrills, original sample duplicates and composite duplicates. The individual errors are summarized below:

(1) sample collection at the rig (including drilling);
(2) compositing of original samples;
(3) reduction of original samples to produce duplicates;
(4) reduction of composites to produce duplicates;
(5) reduction of composites at analytical laboratories;
(6) analytical variation;
(7) local variation between redrill sites (at 1 m spacing).

The precision is generally good for samples from the 75 mm fraction down to the 600 μm fraction with a value below 10%. However, as

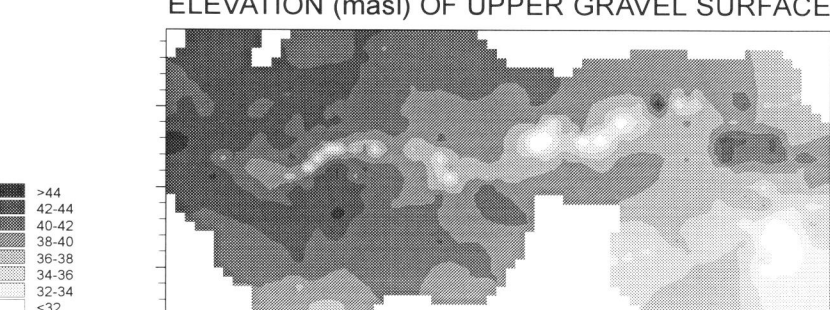

Fig. 4. Topographic plots showing features on the upper surfaces of the basement and gravel sheet.

the grain size diminishes from 600 μm to 75 μm the precision deteriorates dramatically to around 100%. The precision of thickness values from the redrills is good at 8.53%; however, this gives no indication of the true magnitude of the errors (accuracy). The poor precision results for the fine fraction samples confirm the presence of errors in the size grading data due to problems induced by sampling procedures.

Geological analysis

All boreholes were completed in the London Clay basement, and the collars were surveyed to provide an accurate elevation for each of the intersected horizons. Two-dimensional contour plots of gravel and overburden thickness were produced along with 3D plots showing the basement and gravel surface topography. Borehole sections were then produced using SURPAC.

The upper surface of the London Clay shows a marked change in elevation in the centre of the drilled area. The average elevation changes from 35.2 m in the west to 30.6 m in the east (a.m.s.l.). A marked break in slope forms a distinctive feature running NNW across the centre of the area (Fig. 4) which may indicate the presence of a palaeo-river terrace or it may even be a natural settling feature due to the presence of a buried channel in the underlying Chalk (McGregor & Green 1983).

The orientation of linear depressions in the east of the area lie in a roughly NNE to NE direction corresponding with palaeocurrent directions measured from foresets in surround-

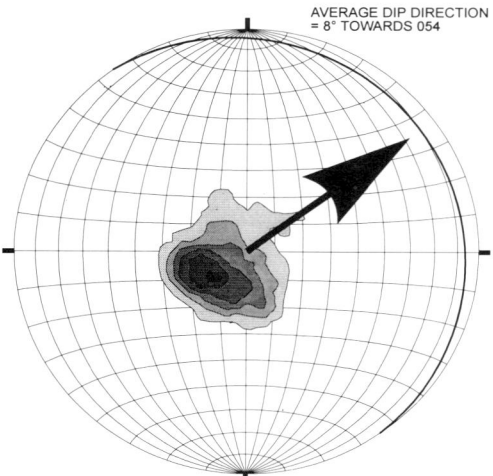

Fig. 5. Pole concentrations from dip directions measured on foresets within the Kesgrave gravel sheet.

ing workings (Fig. 5) and from published literature on the Kesgrave Gravels (Rose et al. 1976; Rose & Allen 1977).

The greatest thickness of gravel occurs in the east of the area and coincides with a depression in the basement topography. However, the upper surface of the gravel shows only a slight drop in elevation from west to east and is generally flat lying with few, obvious, topographic anomalies. The exception to this is in the west where a linear zone of thinner gravel occurs. The thinning of the gravel sheet coincides with a thickening of the overlying boulder clay and is unrelated to features of the basement topography. The zone is narrow in the west (50–100 m wide), but widens towards the east to in excess of 500 m. In the 3D surface plots this zone has been identified as an erosional scour. In places the gravel sheet has been completely removed and the underlying basement surface also shows evidence of scouring and erosion.

The erosive channel probably formed between the closing stages of the deposition of the Kesgrave Gravel sheet and the onset of the Anglian Glaciation. Reworking of the upper layers of the Kesgrave Gravels has been recognized elsewhere and the deposits formed have been classified as the Barham Sands and Gravels. These deposits only occur locally and are not laterally extensive, generally occupying hollows in the surface of the Kesgrave Gravels and occasionally interbedded with the lower layers of the Anglian Boulder Clays (Rose & Allen 1977; Hey 1980). It is possible that the scour channel seen here is the source for some of

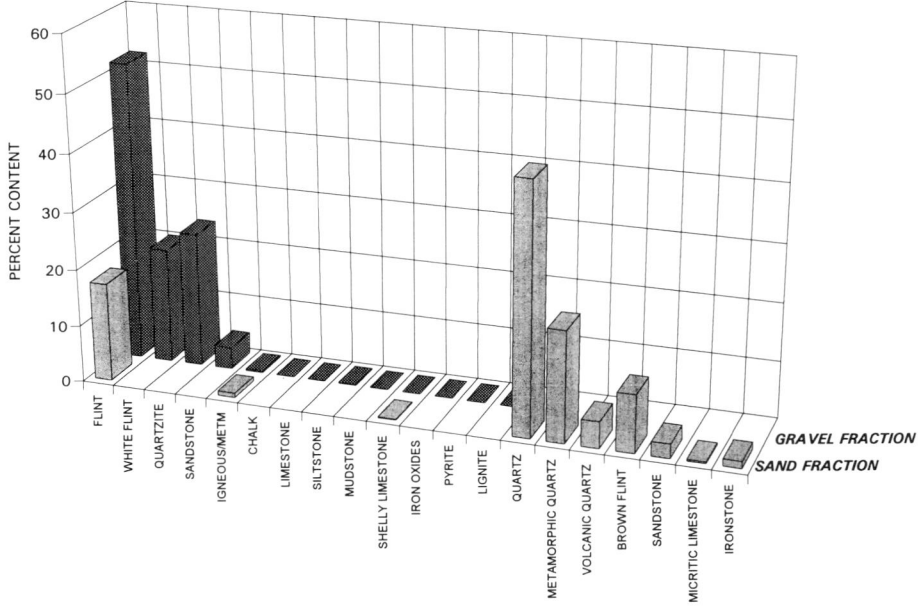

Fig. 6. Histogram showing mean concentrations for 20 lithological groups present within the gravel (20–5 mm) and sand (5 mm–75 μm) fractions.

the reworked gravels located in the overlying boulder clay, the lower layers of which contain a high proportion of gravel and sand indicating some reworking immediately prior to the deposition of the glacial tills. In the east, the gravel sheet also shows signs of erosion over a wide, NE-trending zone (Fig. 4) in which a thin gravel sheet of possibly Barham age occurs intermixed with the boulder clay. Petrographic analysis was carried out on all composites from the 25 m spacing drill grid, and two separate datasets constructed for the gravel fraction (20–5 mm) and sand fraction (5 mm–75 μm). Twenty lithological groups were identified and the mean % contents are summarized in Fig. 6. Lithologies dominant in the gravel fraction include flint, massive quartz (quartzite) and sandstone. In the sand fraction the dominant mineralogies are quartz, metaquartzite and flint (Reading University 1992).

The high flint content of the mineral reflects the path of the drainage system from the west where it would have been draining off the Cretaceous chalk escarpments around the London Basin. Igneous and metamorphic clasts occur in both size fractions. Hey & Brenchley (1977) conclude that the most likely source for this material is the Ordovician volcanic sequences of North Wales. Glacial drainage systems may have transported this material either incorporated into an ice sheet or as a bedload in glacial meltwater drainage systems. However, clasts of similar rock types are found in conglomerates of the Sherwood Sandstone Group (formerly known as Bunter Pebble Beds) in the East Midlands and the erosion of these sequences may be the direct source of the igneous material present in the Kesgrave Gravels.

Reserve estimation

The calculation of univariate statistics is a necessary first step in any evaluation of the *in situ* resources of a deposit, whether geostatistics is to be used or not. In some cases, it may become apparent at an early stage that geostatistical evaluation is not feasible and that statistical methods would provide the best estimator. The statistical study allows the distribution of the data within the deposit to be observed. The presence of sub-populations within the data may indicate geologically distinct zones which may have to be treated separately in the final evaluation.

The main advantage of geostatistical methods for reserve estimation is their ability to calculate confidence limits and precision values for the final estimates. Kriging variances, block errors and extension variance calculations can all indicate the magnitude of the influence of the surrounding sample points on the final block estimate.

The use of geostatistics allowed the spatial characteristics of the deposit to be taken into consideration during the estimation excercise. By the very nature of a fluvial gravel deposit, a high degree of variability in thickness would be expected throughout the site. In any fluvial system there are different zones corresponding to erosion, deposition, areas of slack water, abandoned meanders, flood plain deposits, bars, channels, etc. (Reading 1986; Selley 1976). In all cases, where the semi-variograms could be modelled, a spherical model was used.

It was decided to approach the resource evaluation from two directions. The first was the calculation of the *in situ* tonnage for the gravel sheet. The quality of the gravel is defined by the size grading analysis and, to a lesser extent, by the petrographic analysis of the material. This data was treated separately at the initial stage of the evaluation.

Part of the geostatistical evaluation involved the production of a structural model for the data (directional semi-variograms) which, it was hoped, would identify geological trends and so allow verification of the earlier interpretation of the borehole sections. Most of the work was carried out using the mining software package SURPAC, allowing the production of borehole cross-sections, contour plots of various horizons, extraction of data from specific zones of the site and the production of statistics and semi-variogram models.

Thickness data

The mean gravel thickness in the research area is 5.09 m compared to 5.98 m from the 100 m grid (Figs 7 & 8) and the frequency histograms show a near normal population distribution. However, there is a sub-population of low thickness values which is evident in both histograms and which is an indication of the zone of erosion and reworking which runs through the drilled area. The presence of these sub-populations was clearly seen in the probability plots. From these plots, the zone of erosion appears to contain gravel thicknesses of less than 4.3 m, although some of these values do occur in the main body of the gravel sheet away from the erosive zone. Another inflexion point appears on the probability plots at 2.8 m, below which it appears that gravel thicknesses are entirely controlled by the erosionary event as all these values lie within the zone of erosion. The 100 m spaced data

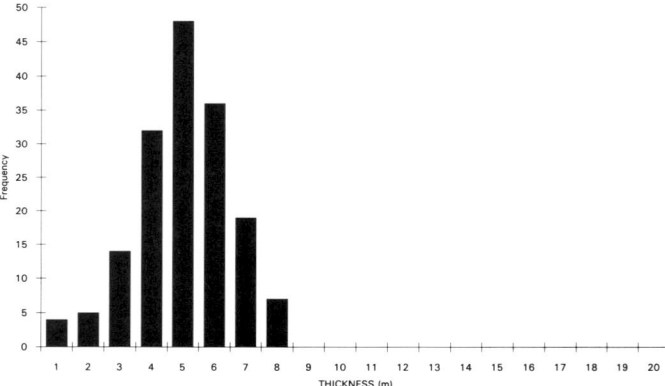

Fig. 7. Histogram of gravel thickness values from all boreholes drilled in the research area.

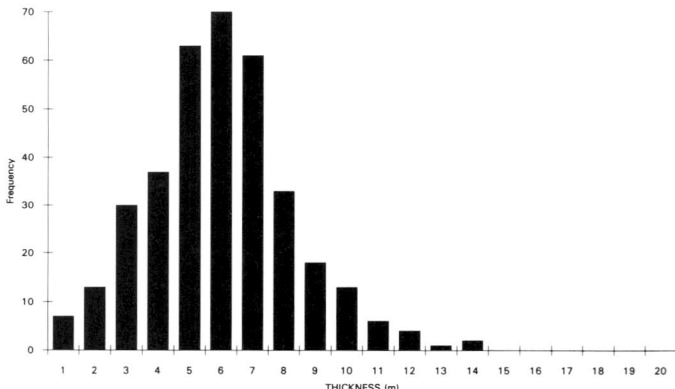

Fig. 8. Histogram of gravel thickness values from all boreholes lying on a 100 m spacing.

shows a greater range in thickness values with a maximum of over 14 m occurring in the north east of the site; this fact is reflected by the higher data variance.

Research area

The deposit was originally split into various zones defined on the basis of thickness and the borehole data in these zones was analysed as individual populations. In the research area, however, it was decided that, as the gravel sheet thickness was fairly uniform, only the erosive channel should be treated separately. The limits of the eroded zone were set on a statistical basis at the 4.3 m contour (Fig. 9). The average thickness for the 48 boreholes in this central zone was 2.04 m. The northern gravel sheet had an average thickness of 5.42 m from 68 boreholes and the southern sheet had a thickness of 5.95 m from 80 boreholes. These compare with an average thickness for the whole of the research area of 5.09 m.

Geostatistical analysis. The effect of sub-dividing the research area was to reduce the data variance for each of the resultant populations and this in turn had a marked effect on the semi-variograms produced for each of the datasets. Omni-directional semi-variograms were first produced before the calculation of directional semi-variograms in each of the four principal directions. Annels (1991) states that semi-variogram values should not be regarded as accurate beyond a lag distance equivalent to half the total length or width of the sampled area; in

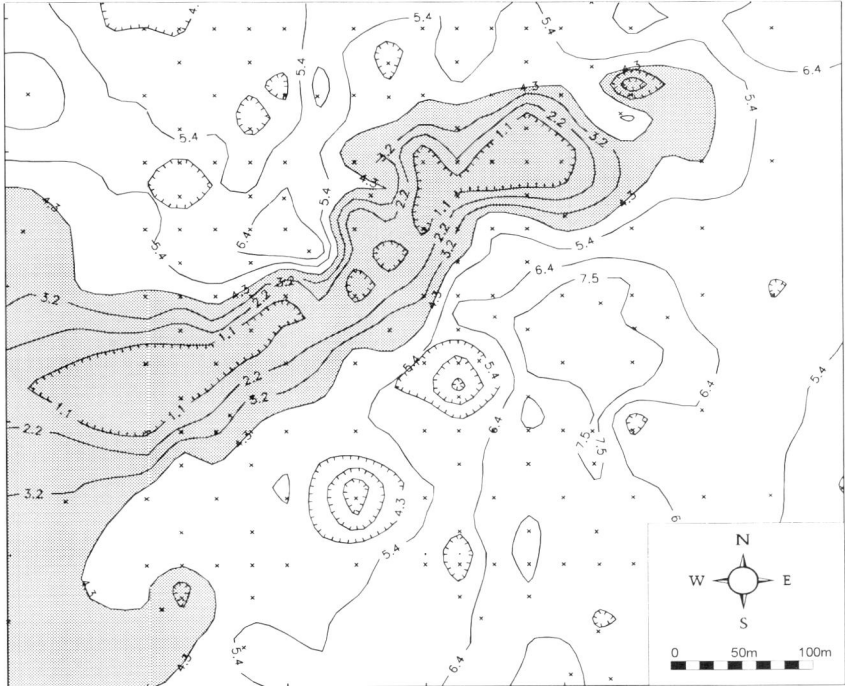

Fig. 9. Contour values for gravel thickness in the research area. Shaded area corresponds to eroded zone with thickness values of less than 4.3 m.

this case a maximum lag distance of 300 m would be indicated.

The semi-variogram for the total data from the research area shows a drop in the sill beyond a distance of about 250 m but this is not evident in any of the other plots from the sub-populations. This may be the result of 'hole effect' which is described by Hohn (1988) as 'an oscillation of the semi-variogram that reflects pseudoperiodicity of the variable'. In the case of the research area, this could be a reflection of the scour channel which passes through its centre. The semi-variogram, based on data with values less than 4.3 m excluded, has no apparent hole effect and the data variance is also much reduced.

The directional semi-variograms were produced for the total dataset in an attempt to define any structural features within the gravel sheet. The two plots for 045° and 090°, which run in the general direction of the channel, exhibited the longest ranges at 170 m and 210 m respectively. These two plots also had a generally lower variance than the omni-directional semi-variogram. The plots for the directions of 000° and 135°, which cut across the scour channels at high angles, exhibited a shorter range of around 140 m. The variance of these plots was higher than that for the omni-directional semi-variogram and they exhibited a high degree of scatter around the sill value reflecting the presence of hole-effect in these directions (Fig. 10).

There were not enough data points within the eroded channel to produce accurate directional semi-variograms. Also, the shape of the northerly and southerly portions of the research area, only allowed semi-variograms in the 045° direction to be accurately defined.

Cross-validation of the semi-variogram model produced from the total dataset produced acceptable results with a kriging variance of 2.67 (data mean = 4.99 m), all the data points lying in the zone of erosion were overestimated producing a slight positive skew to the error histogram. However, when the semi-variogram model for the data from which the eroded zone samples had been extracted was cross-validated the kriging variance was found to have dropped to 0.66 (data mean = 5.98 m) giving a much more accurate estimation for the data.

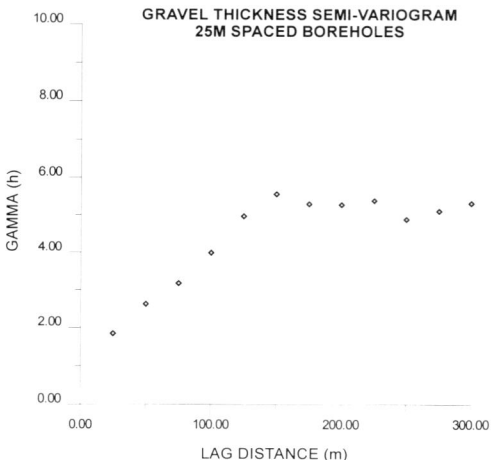

Fig. 10. Research area thickness omni-directional semi-variogram.

100 m data

The data from the 100 m spaced boreholes displayed a normal distribution with a slight negative skew due to the presence of a large sub-population of data points with a value of zero (approximately 7% of the total). As in the research area, this relates to a zone of secondary erosion which runs through the site in an east–west direction (Fig. 11). The spread of data values is greater than that in the research area because of the presence of a zone of thick gravel which occurs in the north east of the site. Figure 11 shows both of these features together with the distribution of boreholes lying on the 100 m grid. The 100 m data was split into east and west zones but the western subset is by far the larger of the two. The maximum thickness in the west is 11.9 m compared to 14.65 m in the east. The data variance is also lower than that in the east but in both cases, when the data from the eroded zone (4.3 m) is removed, the data variance decreases. This has the effect of producing a strong positive skew in the eastern dataset unless the high values are also deleted. Neither of the resultant data populations exhibited a log-normal distribution and it was decided to carry out the geostatistical study using untransformed data.

Geostatistical analysis. Once again both omni-directional and directional semi-variograms were produced for each of the datasets and the results are summarized in Table 2. In this case, seven different populations were used, namely, those representing the whole western and eastern area populations, and those with the eroded zone data extracted, and finally the eastern area with both the eroded and thicker zones extracted. Both of the eastern zone datasets with data extracted produced pure nugget effect semi-variograms. Though the total data from the eastern area produced an omni-directional semi-variogram with a nugget effect (ε) of 0.43 and a nugget variance of 3 there were insufficient data to produce reliable directional semi-variograms.

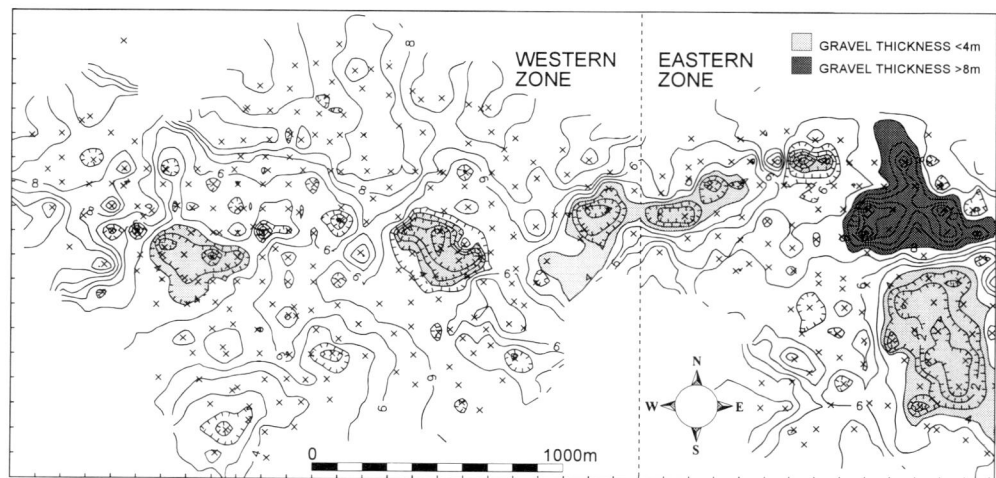

Fig. 11. Contour values for gravel thickness across the total area showing 100 m spaced boreholes. Shaded zones correspond to areas with thickness values below 4 m and greater than 8 m.

Table 2. *Summary of statistical and geostatistical properties of gravel thickness data for each of the subareas identified on the basis of geological interpretation of the borehole data (see Fig. 11).*

Data Population (no. of samples)	Statistical data				Geostatistical data (omni-directional semi-variograms)			
	Mean	Variance	Skewness	Kurtosis	Nugget variance (C0)	Co-variance (C1)	Range (a1)	Nugget effect
Total dataset (386)	5.98	7.57	−0.15	3.37	3.2	4.0	480	0.8
Total dataset >4.3 m (323)	6.73	4.75	0.366	4.38	1.8	1.6	250	1.125
Western zone (263)	6.04	6.67	−0.42	3.14	1.8	4.0	280	0.45
Eastern zone (123)	5.85	9.54	0.227	3.49	3.0	7.0	400	0.43
Western zone >4.3 m (230)	6.63	4.25	−0.06	3.51	1.0	1.8	230	0.55
Eastern zone >4.3 m (91)	7.07	5.43	1.284	4.82	Pure nugget effect			
Eastern zone >4.3 <8 m (8)	6.37	1.77	−0.24	3.06	Pure nugget effect			

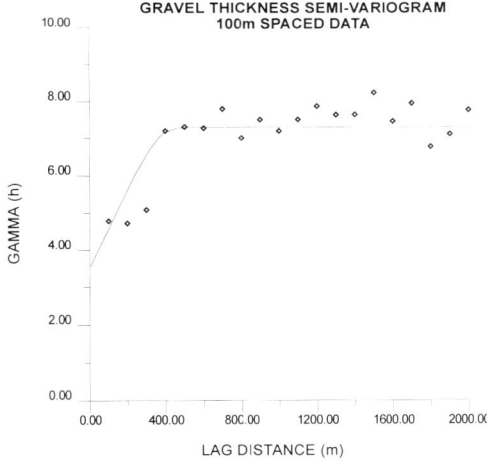

Fig. 12. 100 m spaced borehole semi-variograms for gravel thickness values.

The western area contained significantly more data points and good directional semi-variograms could thus be produced (Fig. 12). Those produced from the uncut dataset had an average nugget effect of $\varepsilon = 0.45$, a nugget variance of 1.8 and an omni-directional range of $a = 280$ m. The directional semi-variogram for 135° exhibited the longest range at 380 m while that for 045° exhibited a nested structure with ranges of 260 m and 380 m.

The semi-variogram for the western area, with the data from the eroded zone removed, were almost isotropic. The omni-directional semi-variogram had a range of 230 m whilst directional ranges varied from 210 m to 220 m. Sill values, in each case, were relatively constant and nugget effect values showed a limited range from $\varepsilon = 0.35$ to 0.44.

Cross-validation of the parameters determined for both data populations from the western area indicated that they were satisfactory in that their mean error values are close to zero and the mean kriging variances are similar to the data variances (Fig. 13). The conclusion is that, although the presence of the eroded zone is having an effect on the quality of the semi-variograms, those produced from the uncut data produce good cross-validation results. The reason for this is that the number of data points falling within the eroded zone is relatively small compared to the total data population. However, it should be noted that all boreholes with a gravel thickness of zero were overestimated and in some cases were given values in excess of 6 m. Ideally, eroded zones should be defined at an early stage in the evaluation and treated separately but in this case there were insufficient data points within the zone to allow estimation of its geostatistical parameters. Extraction of this data produced uniform variance values for the main body of the gravel sheet indicating its uniformity in the western

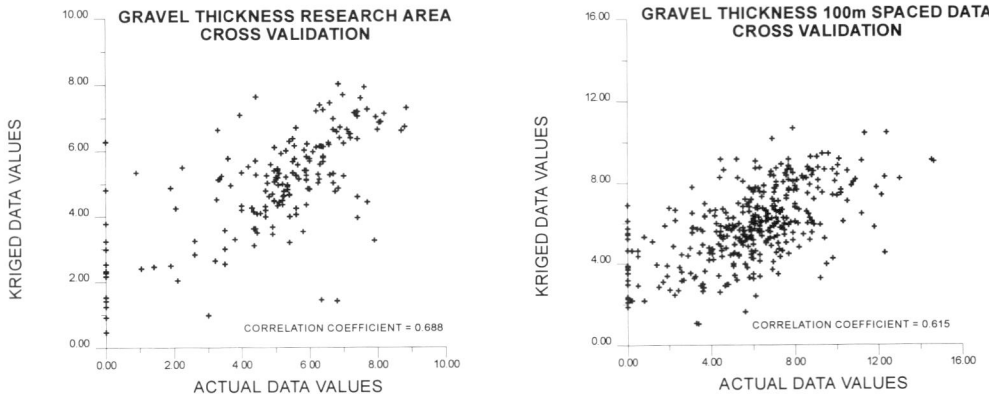

Fig. 13. Cross-validation plots for gravel thickness omni-directional semi-variograms from the research area boreholes and 100 m spaced boreholes.

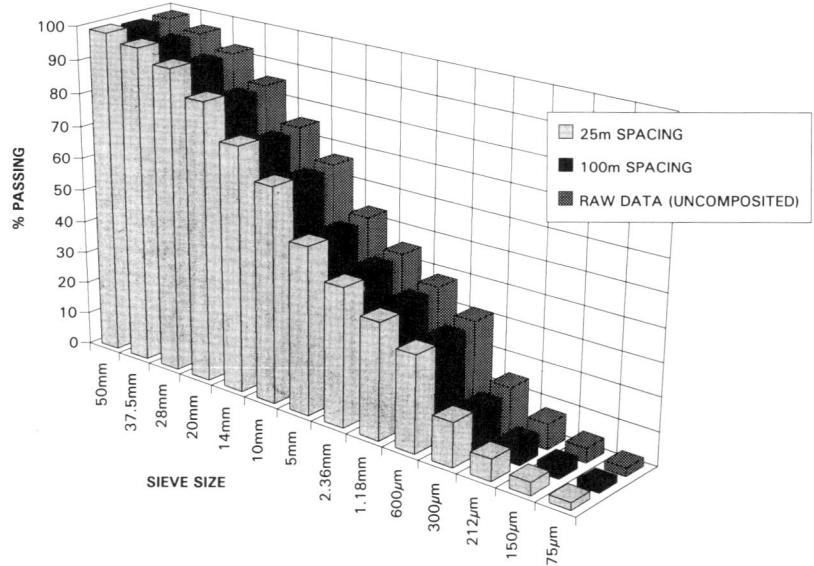

Fig. 14. Histograms for mean grading data from the research area, 100 m area and the uncomposited raw data values.

area. This again provides evidence for the best depositional origin of the thinning of the gravel sheet along the central axis of the area.

Size-grading data analysis

Size-grading analysis of the aggregate was carried out to British Standard BS:882:1983. Sieve sizes corresponded to BS:812: part 1:1975 and all sieves met the requirements of BS:410.

A total of 529 samples were graded from 405 boreholes, from both the research area and the surrounding RMC exploration area. Once again two datasets were constructed corresponding to the 187 boreholes sampled in the research area and the 251 boreholes from the 100 m spaced grid over the whole site.

Fifteen individual size fractions were calcu-

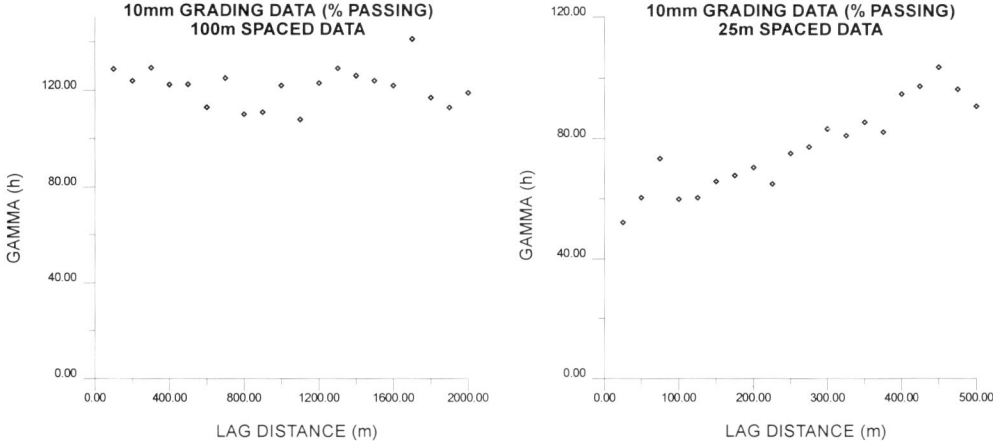

Fig. 15. Omni-directional semi-variograms for 10 mm size fraction (% passing) from the 25 m spaced borehole datasets.

lated and reported as the percentage of the sample passing the relevant sieve size. The maximum sieve size was 75 mm and the minimum was 75 μm. Figure 14 shows the mean values for percent passing each size fraction in the three datasets. Approximately 50% of the deposit lies within the sand fraction (5 mm–75 μm) and the mean value for silt content (<75 μm) varied between 2.07% and 2.5%. The distribution of data in the two datasets is very similar and a 't test' showed that only the data for the 50 mm and 37.5 mm populations failed the hypothesis that there was a relationship. It was thus assumed that the research area was a reliable subset of the total area. However, the spatial characteristics of the grading data cannot be shown by these basic statistics.

The statistical distribution of data within each size fraction also showed a correlation between datasets. Data for the 5 mm fraction, and all finer fractions, showed a log-normal distribution. Between 20 mm and 10 mm, however, the data were normally distributed, and above 20 mm the data exhibited a strong negative skew.

Semi-variograms were produced in eight directions together with an omni-directional semi-variogram for each size fraction between 20 mm and 75 μm. It was decided not to deal with the coarser fractions as they represented waste material or material that would be reprocessed by crushing. Each of the directional semi-variograms had an angular tolerance of 12.5 on either side of the direction vector giving a total coverage of 25° in each direction. It was hoped that the directional plots would enable the production of an ellipse of anisotropy for each size fraction.

Research area data

The 20 mm semi-variograms showed a maximum range of 109 m on a bearing of 157.5°. In all the other size fractions the major range was consistently found in one of two directions. Data for the fractions 14 mm to 1.18 mm exhibited a major range on a bearing of 067.5°, and data for the 600 μm to 75 μm fractions had a major range on a bearing of 022.5°. Most of the coarser fractions also exhibited a second long range structure in the north–south direction. The average range for the first structure was 140 m and for the second, when present, 234 m (Fig. 15).

100 m data

All semi-variograms produced from these data exhibited a pure nugget effect and could not, therefore, be modelled. The omni-directional semi-variograms for the fractions between 2.36 mm to 300 μm showed a structure with a very high nugget variance, but no directional semi-variograms could be modelled and pure nugget effect was again assumed (Fig. 15).

Cross-validation

Cross-validation was carried out on the semi-variogram models from the research area. Figure 16 shows the cross-validation graphical output for the 10 mm fraction data which is a good example of the distribution of results obtained from most of the size fractions. In general, the mean kriging standard deviation and the standard deviation of the error estimate were significantly close to each other, the mean error was close to zero, and the error histograms showed a normal distribution. However, a large number of data points are either under- or overestimated resulting in a high variance of the error distribution. Regression analysis also reveals a very poor correlation between measured values and kriged estimates.

Cross-validation was carried out on data from several subsets including the western area defined earlier and the northern part of the research area, north of the zone of erosion. In all cases, the cross-validation continued to give poor results with little or no correlation between estimates and real values.

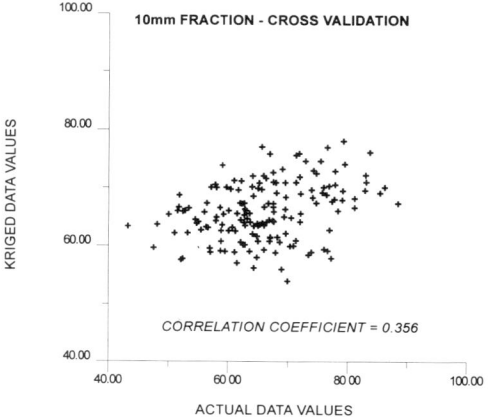

Fig. 16. Cross-validation correlation from 10 mm size fraction semi-variograms in the research area (25 m spacing).

Conclusions

(1) The coarse fraction (20–5 mm) exhibits a normal distribution while the fine fraction (5 mm–75 μm) exhibits a log-normal distribution.
(2) 100 m spaced data semi-variograms show no structure indicating that boreholes are too far apart to produce reliable estimates of the areas between them.
(3) Cross-validation of 25 m data semi-variogram models gave very poor correlation results between measured data values and kriged estimates.
(4) The use of a poor sampling method is probably causing 'geostatistical noise' obscuring the true nature of the semi-variogram structures for the size grading data. There is also a probable breakdown in stationarity within the deposit causing excessive drift in the semi-variogram. This breakdown is caused by the complex geological nature of the deposit although the geological cause has not yet been identified.
(5) Contour plots of horizon thickness, palaeotopography and size grading distribution indicate possible zones of channel flow, erosion, deposition and secondary reworking.
(6) Petrographic studies show a high percentage of far-travelled material including volcanic rocks indicating the probable influence of glacial action in the production of the bedload for the channel system.
(7) Better geological control is needed before further geostatistical work can be carried out on the size grading data. Subdivision of the deposit into both lateral and vertical zones should be carried out to allow production of vertical semi-variograms.
(8) Thickness data can be modelled effectively using geostatistical methods and the quality of the depth presentation data produced by the shell and auger rig is considered acceptable. However, grading data is very inaccurate due to the sampling methods employed by this type of drilling. Even at the close spacing of 25 m, the variation in data quality is too high to be able to predict values for specific blocks of ground.

Thanks are due to I. Williams and R. Fox at RMC (UK) Ltd for their continued support of this project; to RMC (UK) Ltd and NERC for the financial backing; the staff at M & B Geotechnical Laboratories for carrying out the size grading analysis; B. Blackwell at the Building Research Establishment and the staff of the Sedimentological Research Unit at Reading University for the petrographic work; and finally to the drillers without whom none of this work would have been possible.

References

ANNELS, A. E. 1991. *Mineral Deposit Evaluation: A Practical Approach*. Chapman & Hall.
BARRETT, W. L. 1989. Detailed site investigation procedures for aggregate resources. *Bulletin of the Institution of Mining and Metallurgy, Minerals Industry International*, **988**, 21–25.
BRITISH STANDARDS INSTITUTION BS 410. *Specifications for test sieves.*
—— BS 812:1975. *Methods for sampling and testing of mineral aggregates, sand and fillers.*
—— BS 882:1983. *Aggregates from natural sources for concrete.*
GARRETT, R. G. 1969. The Determination of sampling and analytical errors in exploration geochemistry. *Economic Geology*, **64**, 568–569.

HEY, R. W. 1980. Equivalents of the Westland Green Gravels in Essex and East Anglia. *Proceedings of the Geologists' Association*, **91**, 279–290.
—— & BRENCHLEY, P. J. 1977. Volcanic pebbles from Pleistocene gravels in Norfolk and Essex. *Geological Magazine*, **114**, 219–225.
HOHN, M. E. 1988. *Geostatistics and Petroleum Geology*. Van Nostrand Rheinhold.
MCGREGOR, D. M. & GREEN, C. P. 1983. Post-depositional modification of Pleistocene terraces of the River Thames. *Boreas*, **12**, 23–33.
READING, H. G. 1986. *Sedimentary environments and facies*. Blackwell Scientific Press.

READING UNIVERSITY 1992. *Petrographic examination of sand and gravel samples*. Unpublished report, Postgraduate Research Institute for Sedimentology, University of Reading.
ROSE, J. & ALLEN, P. 1977. Middle Pleistocene stratigraphy in south-east Suffolk. *Journal of the Geological Society, London*, **133**, 83–102.
——, —— & HEY, R. W. 1976. Middle Pleistocene stratigraphy in southern East Anglia. *Nature*, **263**, 492–494.
SELLEY, R. C. 1976. *An introduction to Sedimentology*. Academic Press.

Computer modelling of dewatering a major open pit mine: case study from Nevada, USA

R. I. CAMERON & H. MIDDLEMIS

Water Management Consultants Ltd, 2/3 Wyle Cop, Shrewsbury SY1 1UT, UK

Abstract: Operations began in 1985 at Amax Gold Inc.'s Sleeper Mine in Nevada, and the major open pit has been one of the largest North American gold producers. The orebody in Tertiary volcanic rocks is overlain mainly by the Desert Valley basin fill gravels. Natural groundwater levels at the mine are around 10–15 m below surface, and dewatering has been a major activity since initial overburden stripping. With a final projected mine depth of 215 m, a properly designed and cost-effective dewatering scheme, fully integrated with the mine plan, was required. Field investigation and numerical computer modelling proceeded in parallel, the feedback between the two activities being essential to the optimization of the dewatering scheme.

The basis of effective dewatering has been the interception of groundwater in the overburden before it can reach the pit. This paper describes the groundwater model, based on the USGS MODFLOW code, which was specifically designed to develop a wellfield design for optimum dewatering. The model enabled the determination of the most cost-effective wellfield configuration, pumping rate and distance from the pit. Subsequent field observations showed that real drawdowns were within 5% of those predicted.

The interceptor model has also been used to investigate the dewatering rate and cost implications of expanding the pit in the direction of the wellfield; and the implications of power failure at the wellfield for determination of standby generating capacity.

Sleeper Mine is a major open pit gold mine operated by Amax Gold Inc. and situated in the Basin and Range physiographic province in Nevada, USA. The mine has been in operation since 1985, and has been one of the largest North American gold producers.

The mine is situated on the east side of Desert Valley, Nevada, at a surface elevation of around 1280 m above sea level. Figure 1 shows the general location of Sleeper in the western United States. Desert Valley is about 20 km wide, and bounded to the east (above Sleeper) by the Slumbering Hills, rising to 1980 m and to the west by the Jackson Mountains, rising to 2740 m. Average annual rainfall varies from less than 120 mm on the valley floor, to 250–400 mm on the surrounding hills.

Figure 2 is a cross section of the eastern half of Desert Valley, showing the general geological situation of the mine. The gold-bearing orebody is situated in the Tertiary volcanic bedrock (rhyolites and andesitic basalts). Overburden at the mine mainly comprises the 'Older Alluvium' (basin fill gravels) and some 'Younger Alluvium' (intermediate sands with some lacustrine silts and silty clays). The older gravels and the lacustrine deposits, the latter being the legacy of the pro-glacial Lake Lahontan, fill the Desert

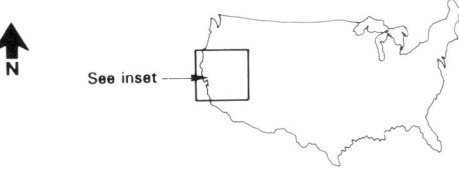

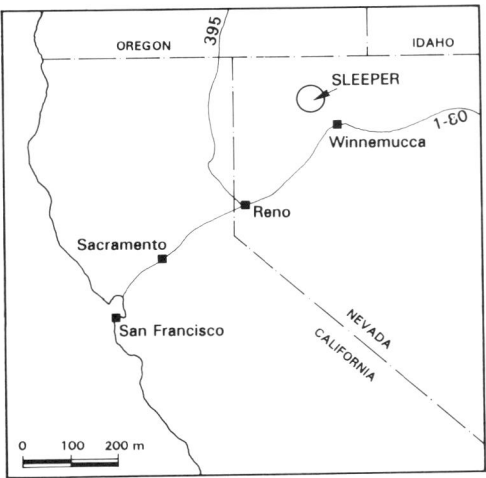

Fig. 1. Location map.

From Whateley, M. K. G. & Harvey, P. K. (eds), 1994, *Mineral Resource Evaluation II: Methods and Case Histories*, Geological Society Special Publication No. 79, 207–217.

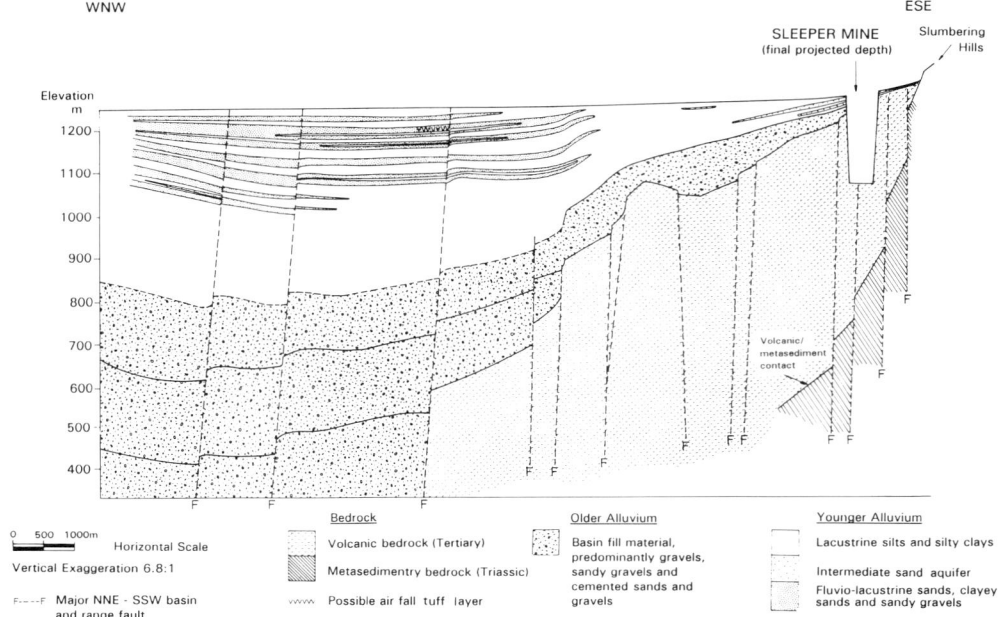

Fig. 2. General geological cross-section of Desert Valley.

Valley basin to thicknesses estimated at 1800–2500 m.

Natural groundwater levels are around 10–15 m below ground level in the vicinity of Sleeper, and dewatering became a major activity from the early days of overburden stripping and has continued since then. Initially this was carried out on an engineering basis, employing sump pumps and peripheral wells in the volcanic bedrock, but with a planned open pit depth of around 215 m it was clear that a properly designed and optimized dewatering scheme was required which was integrated with the mine plan, and which enabled dry workings to be maintained at minimum pumping rate and minimum cost.

The dewatering scheme was developed in a parallel programme of field investigation and numerical computer modelling. The field programme encompassed structural geology surveys, geophysics, hydrochemistry and the development of a comprehensive groundwater monitoring network, in addition to an ongoing programme of pilot well and production well drilling to provide the basis for dewatering and to keep ahead of the mine plan at all stages. Details of the field work are beyond the scope of this paper, but aspects have previously been reported (Beale & Tyler 1990). Results from the field were continuously fed back to the groundwater model, and the modelling results were in turn used to guide the field programme. In this way, dewatering was optimized in the context of a well-understood hydrogeological environment and a calibrated, robust model.

The basis of the models used was the United States Geological Survey (USGS) three-dimensional, finite-difference groundwater flow code MODFLOW (McDonald & Harbaugh 1988). This is an industry standard of its type, well documented and quality-assured. Moreover, the USGS themselves are developing a regional model of Desert Valley using MODFLOW, so comparison between results is facilitated. The specific conditions of Sleeper and the requirements of simulating dewatering and recovery necessitated some modifications to the basic model code, which were written and tested by Water Management Consultants.

Two basic models have been developed. The first, and the main subject of this paper, is a model designed specifically to develop an optimum dewatering scheme, concentrating on the hydrogeological regime in the vicinity of the mine itself. This is referred to as the interceptor model, because the basis of effective dewatering has been the use of wells in the overburden to intercept groundwater before it can percolate

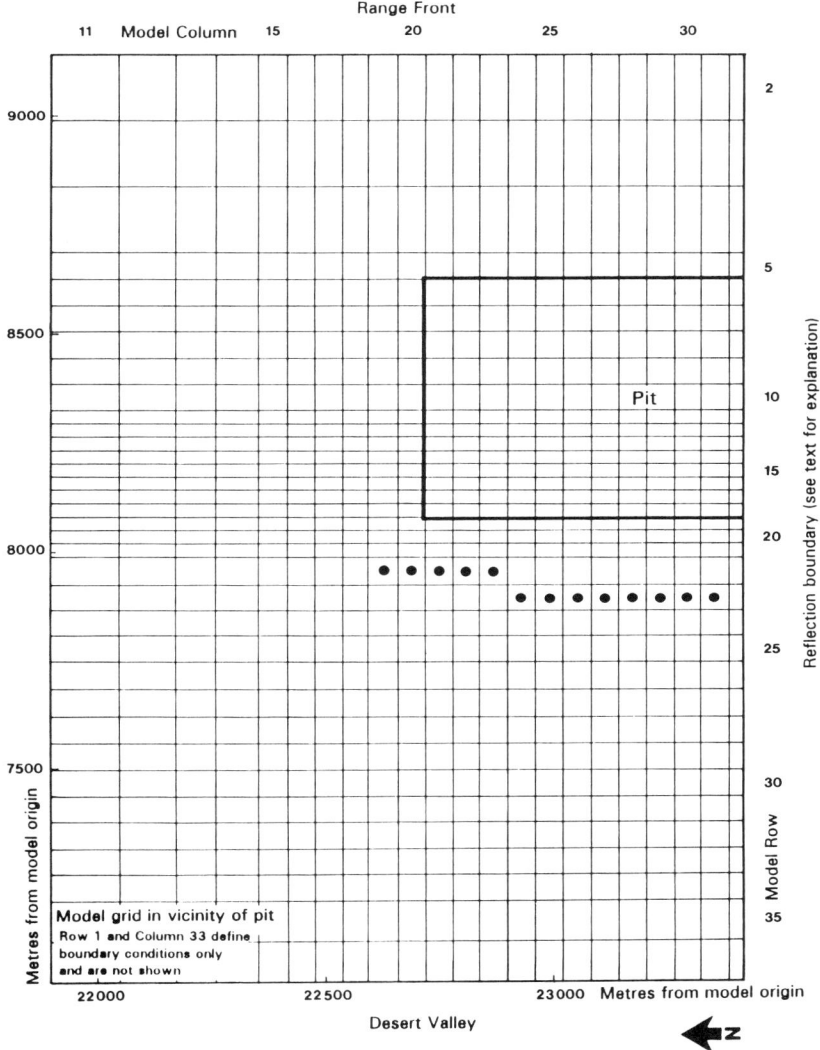

● Modelled well locations for optimum dewatering design

Fig. 3. Interceptor model grid in vicinity of mine.

into the bedrock and reach the pit. Two further developments of the interceptor model are also described, the first investigating the additional dewatering rates and costs of increasing the size of the pit; and the second investigating the implications of a loss of power supply resulting in the dewatering pumps switching out.

The second model is a regional model of Desert Valley, which seeks to place the Sleeper Mine dewatering in the overall context of the valley hydrogeological system and the other demands which are placed upon it. This model is yet to be reported on fully, and is only introduced in this paper.

Interceptor model

The original objectives of the interceptor well modelling were:

- to investigate the potential cost savings to be achieved by using interceptor wells in the long term mine dewatering scheme;
- to determine the optimum and most economic wellfield configuration for the mine dewatering scheme, including distances from the pit and well spacings;
- to allow initial predictions to be made of pumping rates required from the gravels

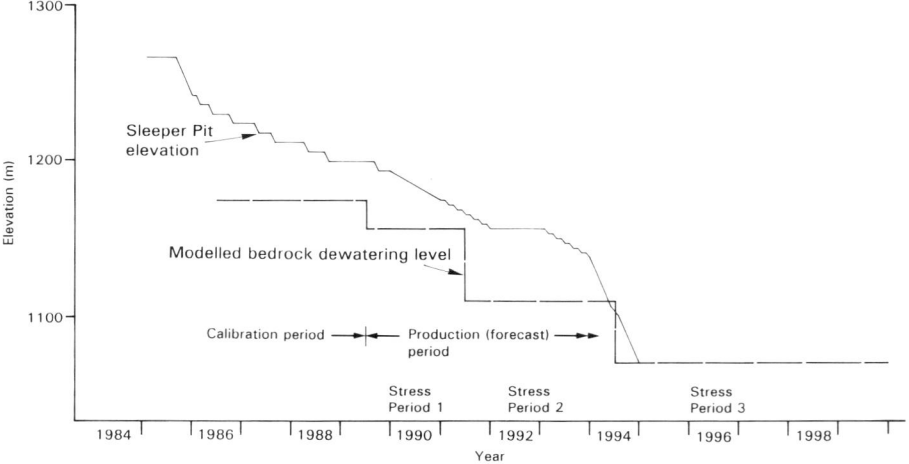

Fig. 4. Forecast and modelled pit floor elevations.

and from the bedrock over the life of the mine.

It was originally envisaged that the interceptor well modelling would take the form of a conceptual investigation, using different simplified model configurations and a range of wellfield designs. The optimum configuration derived from this model evaluation would then be passed forward to the regional model for further development. However, during initial development of the interceptor model, the practical requirements of the mine for results which could be used to assist in siting the early wells meant that the model had to be modified away from the 'conceptual', developing considerably in the direction of the regional model. This came about as a natural process resulting from the continuing feedback and the acquisition of new data from the mine hydrology field programme. As a result, the conclusions of the interceptor well modelling programme are consistent with known field conditions. Consequently they have been of considerable practical value in terms of the interceptor well drilling program. Moreover, the contribution of the interceptor well model to the regional model calibration process will allow a high degree of confidence to be placed on the function of the regional model as a management tool for the remaining life of the mine.

The model consists of two layers, an upper high permeability layer representing the basal gravels, and a lower low permeability layer representing the Tertiary volcanic bedrock (Fig. 2). The lacustrine silts and intermediate sands have been disregarded, since they were already virtually completely dewatered around the north and west of the pit and no longer formed a significant part of the active system. The west edge of the pit represented the eastern limit of the basal gravels. To reduce model size and running time, the area covered by the model was assumed to have bilateral symmetry, about an axis running east–west through the Sleeper pit. Consequently, the pit in the model is bisected and lies across the southern boundary of the model.

Figure 3 shows the rectilinear finite difference model grid in the vicinity of the mine. This is the final version of what was a lengthy evolutionary process. The model grid ultimately comprised 33 columns and 47 rows, with cells ranging in size from 6100 m × 1500 m approximately on the outer part of the model domain where high resolution was not required, down to 61 m × 30 m on the western edge of the pit itself where there was a need for considerable detail in defining the groundwater flow regime at the eastern edge of the basal gravels adjacent to the pit wall. The model domain covered a total area of 23.5 km × 9.1 km, the long dimension being effectively doubled when taking into account the reflection boundary. The modelled pit dimensions were 550 m east–west and 1460 m north–south, including the reflected image.

The model must incorporate specified boundary conditions to enable the solver to produce a unique solution to the equations describing groundwater flow in the defined model area.

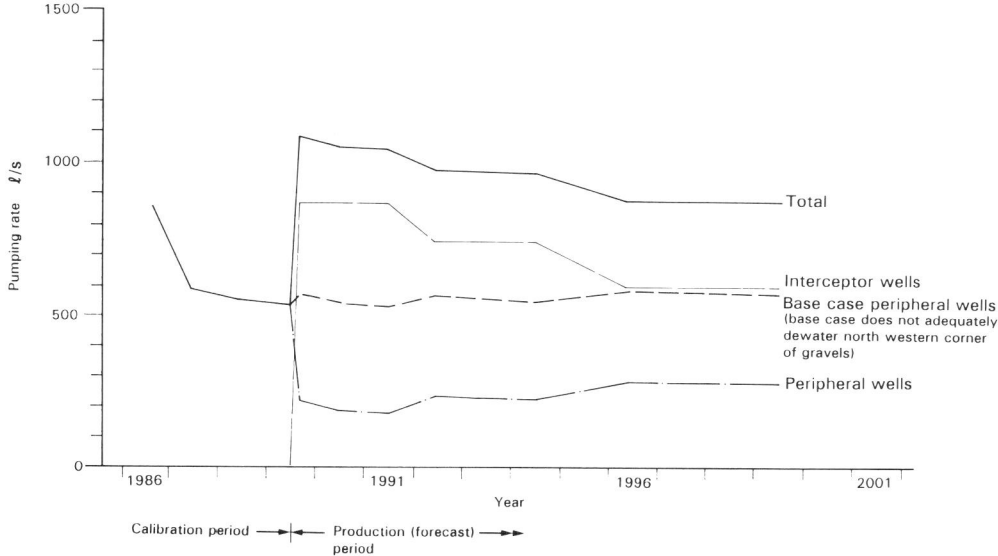

Fig. 5. Modelled flow hydrograph.

The axis of symmetry bisecting the pit east–west is a no-flow (barrier) boundary which enables the northern (real) half of the model to reflect an identical southern (virtual) half. Streamlines in the vicinity of this boundary run parallel to it. The eastern (Range Front) boundary is also no-flow since it is known that the gravels are absent east of the pit and only limited recharge occurs there. The northern and western boundaries were initially both specified as free-floating heads, producing (in the model) inflows on those boundaries dependent on the gradient and permeability conditions. In practice, these flows are negligible at the model margins, where gradients formed in response to stresses (dewatering abstractions) at the pit are insignificant. In the final model version the northern boundary was specified as fixed head to assist in numerical stability; again, however, the drawdowns at the boundary were so slight that fixed head inflows were negligible.

In order to produce a series of groundwater flow solutions over time, each model run comprises a number of stress periods. Within each stress period, abstraction conditions (including wells and fixed head cells) remain constant. Different stress periods must be used if abstraction conditions change, for example the production runs comprised three stress periods for the interval 1989–1999 to accommodate successive levels of the pit and their respective dewatering conditions (Fig. 4).

Aquifer parameters for the modelling were based on results from the hydrology field programme, and were intended to be standardized as far as possible. This made it much easier for the interceptor model to fulfil its principal function as a conceptual investigation tool, as opposed to attempting to incorporate all the aquifer geometry and parameter detail that was known. Had this latter approach been adopted, calibration would have been much more difficult and the robustness of the model (which made it so valuable as a guide to the field dewatering programme) would have been lost. Aquifer parameters adopted were:

bedrock transmissivity: $160 \, m^2$ per day N–S
$16 \, m^2$ per day E–W
bedrock storage coefficient: 0.0001
bedrock specific yield: 0.01
gravels hydraulic conductivity: 20 m per day
gravels specific yield: 0.1

The anisotropy in the bedrock transmissivity of 10:1 is a function of the structural fabric of the aquifer.

Calibration of the model was achieved for the period up to 1989 by simulating peripheral well drawdowns by means of fixed head cells around the pit (taking into account known well pumping levels and efficiencies) and matching the outflows thus generated in the model with the known pumping rates. As part of this process,

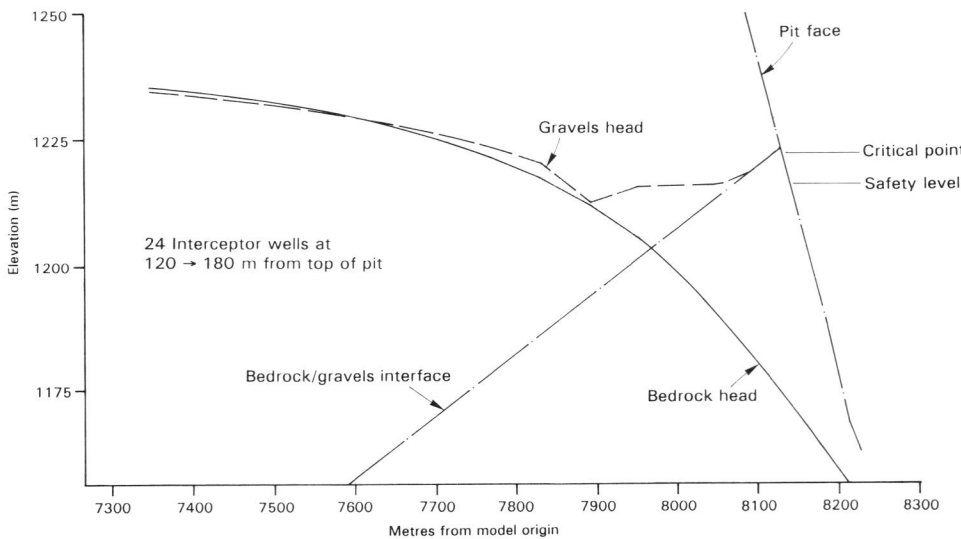

Fig. 6. Modelled heads in 1991; west–east section looking north.

the contribution of groundwater from the gravels into the bedrock in the vicinity of the pit was controlled by means of the vertical conductance term in the model between the two layers, using the observed head differential (from piezometers situated in different horizons) as a basis for calibration. The modelled flows of about 540 l s^{-1} were acceptably close to the actual total peripheral well pumping rate at that time (Fig. 5).

Wellfield optimization

To provide a basis for cost comparison and optimization between the use of peripheral wells in the bedrock only, and a combination of peripheral wells and a new interceptor wellfield in the gravels, it was necessary to use the model to extend simulated dewatering from bedrock peripheral wells beyond 1989, using the forecast pit floor levels shown in Fig. 4. The simulated pumping rates for this 'base case' are shown on the dashed line in Fig 5, although it is important to note that the base case does not achieve complete dewatering of the gravels in the northwestern corner of the pit. Although the pit deepens, actual flow rates do not increase significantly because the bedrock is becoming dewatered over a wide area and groundwaer gradients are becoming less steep towards the pumping wells. However, the cost of that pumping does increase because of the greater heads involved. Net present value (NPV) costs were developed for the simulated base case, incorporating capital and operating costs of the dewatering scheme for the 10 year mine life.

In practice, however, bedrock peripheral wells alone would never be a sound solution because of the need to dewater the gravels adjacent to the pit wall to a safe level that would always keep the face dry. This safety level is defined with reference to a critical point, which is the elevation of the gravels/bedrock interface at the pit wall. The actual margin of safety used in the interceptor model was 6 m vertically below a critical point elevation of 1222 m. This, therefore, represented the target elevation for heads in the gravels for the production runs of the model, to be achieved by 1991 when saturated gravels would, without interceptor wells, be exposed in the high wall. The critical point and safety level are shown diagrammatically on Fig. 6.

Having established a satisfactory base case, successfully calibrated to actual pumping rates and observed water levels, a matrix of production runs was undertaken using standardized wellfield configurations. For cost benefit analysis, well pumping rates were determined by fixing heads in the well cells, at a variety of distances from the pit, at a point between the 1989 calibrated heads and the base of the gravels, leaving 50% of the gravels saturated. A saturated thickness must be left in the model cell to accommodate true aquifer drawdown at a well point (the model cell head is only an average over the whole cell, and a correction must be applied based on pumping rate, aquifer permeability and cell size to obtain total aquifer drawdown at a well); and well losses due to non-Darcy (turbulent) flow in the vicinity of the well.

The cost benefit calculations took account of well siting and drilling, pumps and wellhead connections, and power costs. They did not include the costs of survey work, general exploration drilling, pipelines or staff. For peripheral wells, it was assumed for cost benefit analysis purposes that all wells are constructed to a depth of 270 m, with eight exploration wells for each production well of $50 \, \text{l s}^{-1}$ long term capacity and 60% well efficiency. For interceptor wells, capital costs for the cost benefit analysis assumed that two exploration wells are needed for each production well of $30 \, \text{l s}^{-1}$ long term capacity and 70% well efficiency.

It was evident from the cost analysis that wells located at a distance of 370 m from the pit wall represented roughly the breakeven point at which the cost of an interceptor well scheme becomes as expensive as the 'base case' (bedrock peripheral wells only). The closer to the pit the interceptor wells are, the bigger the potential savings over the base case situation, because of the reduced number of wells and reduced pumping rate required. However, there is also a breakeven point near the pit at which dewatering becomes ineffective due to insufficient saturated thickness and therefore low specific capacity in the gravels. Clearly, a balance situation must exist with an optimum wellfield configuration which meets the dewatering requirements for the gravels, but at minimum cost, and thereby providing maximum savings.

Having established the cost effectiveness threshold using standardized pumping based on fixed heads in the model, it was no longer necessary to continue to use such idealized pumping characteristics of the wellfield optimization evaluation. Subsequent new production runs therefore utilized well cells based on defined pumping rates (and therefore variable heads) rather than fixed heads and varying pumping rates.

Pumping rates in the interceptor model were generally reduced after 1991 since the required dewatering had been achieved in the required time. From 1991 onwards, merely sustaining the water levels (as opposed to dewatering further) allowed overall pumping rates to fall (Fig. 5).

It is not intended to detail the results of all the optimization runs, but general conclusions are:

1. wells at a distance closer than 120 m to the pit wall will not succeed in satisfactorily dewatering the gravels;
2. total interceptor well pumping rates of less than about $850 \, \text{l s}^{-1}$ will not produce sufficient dewatering of the gravels;
3. a wellfield in an arc, with outer wells (near pit corners) at 120 m distance and inner wells (near pit mid point) at 180 m, pumping between 880 and $950 \, \text{l s}^{-1}$ in total, appears to be the optimum idealized condition—this modelled pattern is shown in Fig. 3, and its results in Fig. 6.
4. total interceptor well pumping rates of over $1040 \, \text{l s}^{-1}$ result in too much interference in the wellfield, with individual wells becoming ineffective due to reduction in saturated gravels thickness;
5. the total capital and operating cost of the interceptor system realizes NPV savings of at least US$3 million compared to base case costs over the 10 years mine life; subsequent field work indicates that savings exceed this amount considerably.

Implementation and wellfield operation

Once operational, results from the initial interceptor wells suggested that the early effects of dewatering conformed generally to the predictions made by the model. In particular, the interaction between the interceptor wells and bedrock production wells had already reduced heads in the eastern extremities of the gravels to below the required target elevation set for the end of 1990. By April 1990, water levels in some areas of gravels within the pit had fallen by over 9 m since the beginning of the year.

At 120 m from the top of the pit wall in the model, the gravels/bedrock interface occurs at an elevation of 1202 m. At 180 m from the top of the pit wall in the model, the interface occurs at an elevation of 1195 m. Wells were generally sited in accordance both with the distances from the pit wall suggested by the model, and the associated real gravels/bedrock interface elevations. The first interceptor wells were targeted to intersect bedrock between 1192 and 1198 m, to hit the base of an obvious cliff line in the buried bedrock surface, which is semi-continuous immediately to the west of the mine and possibly represents the line of a major NNE–SSW 'range front direction' fault. The actual target elevations for interceptor well drilling will vary somewhat due to irregularities in the bedrock surface, irregularities in the elevation of the critical point, and the presence of buried canyon aquifers immediately to the west of the pit wall. Generally, the target elevation for drilling needed to be somewhat lower towards the north end of the pit since the elevation of the bedrock surface, and hence the elevation of the critical point, generally falls to the north. The presence of shallow fracture zones within the bedrock beneath the gravels has also been investigated as a possible means to site interceptor wells closer to the pit, while at the same time ensuring a sufficient saturated thickness is available to sustain the required pumping rates for the life of the mine.

The initial interceptor wells have generally been drilled at 180 m spacing, with some variation depending on buried topography of the bedrock. This allows, if required, infilling between the present wells to 90 m or 60 m spacing. The detailed wellfield configuration and pumping regime depends on detailed localized hydrogeological conditions, as well as logistical considerations on site.

Interceptor wells drilled to date have generally penetrated layers of coarser gravels (especially at the base of the gravels), with permeabilities up to 60–80 m per day, some three to four times higher than the global gravels permeability of 20 m per day used in the model. These localized high permeability zones mean in effect that some interceptor wells have much higher specific capacities. Interceptor wells have been completed with specific capacities of over $0.019 \, l \, s^{-1} \, m^{-1}$ of saturated gravel, more than double the specific capacity of $0.009 \, l \, s^{-1} \, m^{-1}$ initially assumed. Pumping rates from early interceptor wells were around $50-60 \, l \, s^{-1}$, against $30-45 \, l \, s^{-1}$, which was used in the model runs.

At the higher specific capacities encountered, actual pumping water levels will be higher than those predicted by the model. The specific capacity of the wells will, however, require close monitoring and control over the life of the mine if problems of low pumping water levels and declining yields in individual wells are to be avoided. If high specific capacities can be maintained, however, it may eventually mean that fewer interceptor wells are required, with higher individual pumping rates. In any event, the high specific capacities which are currently being achieved in the interceptor wells will allow for considerable flexibility in the pumping regime.

In November 1989, the total actual pumping rate at Sleeper was $600 \, l \, s^{-1}$, of which 50% was pumped from the basal gravels and 50% was pumped from the bedrock. Following implementation of the interceptor well scheme, 72% of the pumping rate was being derived from the basal gravels and 28% was being derived from the bedrock, close to that predicted (Fig. 5). On average, the drawdown per unit pumping rate of interceptor wells predicted by the model was within 5% of observed values. Therefore, although the model lacks the absolute physical realism of the exact geometry of the gravels/bedrock interface, confidence in the use of the model is enhanced by field verification.

The water produced from the dewatering system is of good quality and is discharged to a temporary wetland system created in the centre of the valley, which has become a haven for wildlife. State and Federal authorities are involved in the wetland project, which is serving as a beneficial use in environmental terms, as well as forming a high recharge area to help minimize semi-regional drawdown effects.

Pit expansion

Following the successful implementation of a dewatering scheme guided by the interceptor model, the outline feasibility of dewatering operations in the event of the pit west wall being moved up to 150 m to the west was investigated. This study provided for the identification of locations for additional interceptor wells and the evaluation of the increased pumping rates and costs required to dewater the expanded pit. Clearly, the maximum extension of the westwall 'pushback' would completely engulf most of the existing designed wellfield.

Assuming that dewatering to achieve the pushback constraints would be required by late 1992, simulation of the additional dewatering was specified to take place in the third of the ten years of simulation, after two years of dewatering according to the previously specified scheme. Figure 7 shows how the interceptor wellfield would have to be expanded to maintain dry working conditions in a pit expanded by different amounts. The only pushback case which can make some use of the existing interceptor wells is the 30 m case. Greater pushbacks require deeper dewatering, which means that there will not be sufficient saturated thickness of gravels under the existing interceptor wells to permit the pumping rates required. With increasing pushback distance, an increase is required in both the pumping rate and number of interceptor wells to meet the dewatering goals. For 150 m pushback, the total pumping rate would increase to over $2000 \, l \, s^{-1}$, nearly double the rate with the existing design, and with a corresponding cost implication.

Security of power supply

An additional requirement of the interceptor

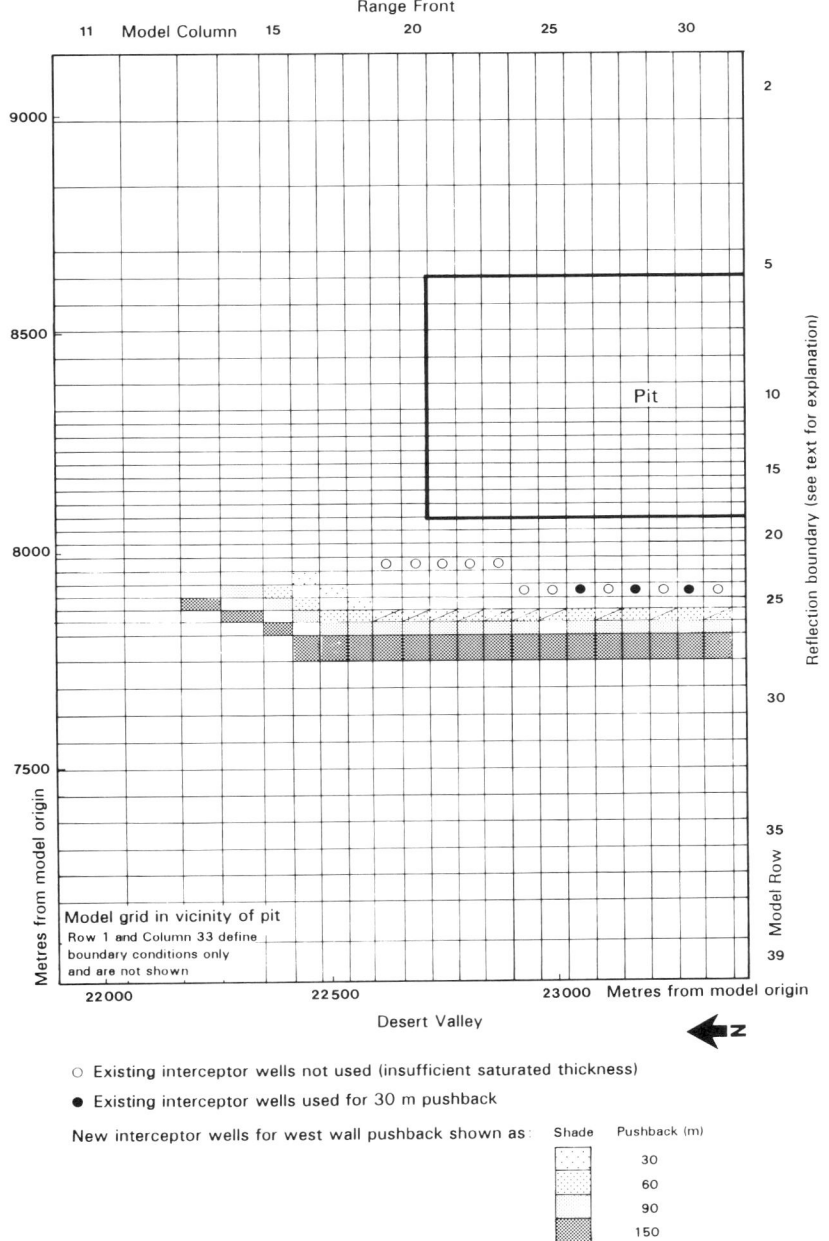

Fig. 7. Wellfield design for west wall pushback.

model was to investigate the impact on groundwater levels of a loss of power supply at the mine, to enable a decision to be made as to whether to maintain expensive standby generating capacity on site, or bring in emergency generators should the need arise. Sleeper is at the very end of the power supply grid, so any power disruption is likely to affect operations at Sleeper immediately. Information regarding the estimated duration of the power outage should be available to mine management within a few hours of the event. Consequently, the rewatering

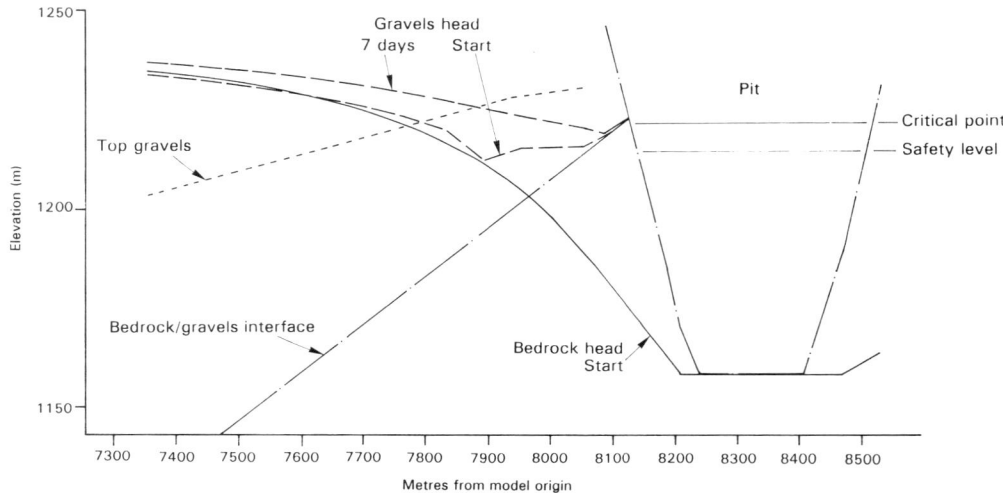

Fig. 8. Predicted recovery of aquifer heads at seven days after power outage.

of the gravels and bedrock aquifers in and around the Sleeper pit was simulated following one, three, and seven day periods of power outage.

In order to evaluate the rate of rewatering effectively, some modifications were made to the model, changing the aquifer type for the gravels layer in certain rows from unconfined to semi-confined/unconfined, to represent the effect of the overlying silts on the hydrogeological regime. The piezometric heads in the dewatered gravels begin to exceed the elevation of the top of the gravels about 300 m west of the pit, at the feather edge of the Lahontan clays (Fig. 8). West of this point, the gravels are specified as semi-confined. The rewatering of the gravels near the pit from aquifer storage in the semi-confined gravels close to the pit produces higher heads than if the gravels were represented as being unconfined everywhere. Higher heads mean that the gravels exposed in the pit wall are more likely to rewater during the power outage, and the stability of the high wall may be at risk as a result.

Figure 8 shows predicted recovery in the gravels aquifer after a pump shutdown of seven days. Water levels have risen above the safety level and are approaching the critical point. There is therefore a risk of the gravels becoming saturated in the pit wall in places. This risk is greater if zones of permeability greater than 20 m per day are widespread.

It is understood that the return frequency of long duration power outage events is very low. This being so, there was no need to carry out power outage simulations at different times through the life of the mine, as restarting the dewatering scheme should achieve the safety level constraints by about three months after a power outage of seven days duration.

In the areas of very fractured and permeable bedrock (generally beneath the western side of the pit), pumping rates are being set to dewater the bedrock to maintain heads at some 15–30 m below the level of the pit floor. This allows for drainage of the lower permeability bedrock beneath the eastern side of the pit. It was determined that the recovery of heads in the bedrock beneath the pit over seven days of power loss is relatively insignificant at about 15 m, and is not sensitive to the actual stage in the mine life that the power outage occurs.

Regional model

A regional groundwater flow model has also been developed to quantify the impacts of the dewatering operation and mine closure activities on the natural hydrologic system of Desert Valley. Results from the model identified that the regional impacts were limited to the close vicinity of the mine (Fig. 9).

Conclusions

Dewatering at the Sleeper Mine in Nevada became a major activity from the early days of overburden stripping of basin fill gravels and lacustrine deposits which are saturated from 10 to 15 m below surface. With the open pit gold

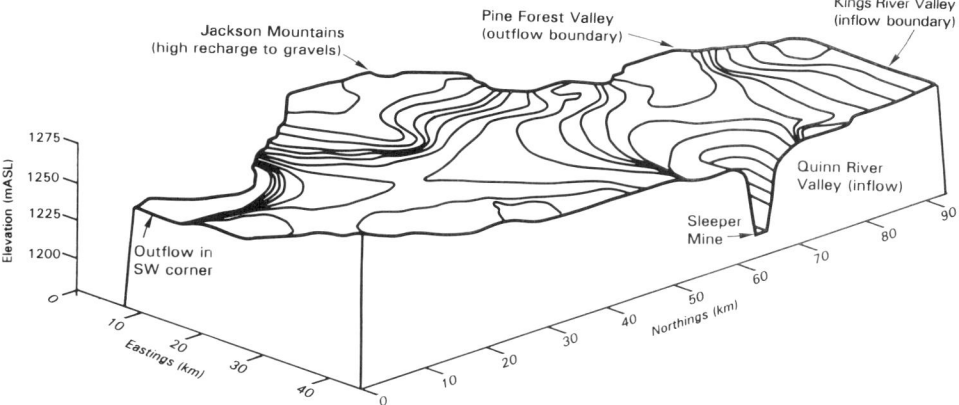

Fig. 9. Regional model; predicted aquifer heads after 15 years dewatering.

mine eventually to be excavated to over 200 m depth, into Tertiary volcanics hosting the ore, it was clear that a properly designed and optimized dewatering scheme was required to enable dry working to be maintained at minimum pumping rates and costs. A three-dimensional numerical groundwater flow model was developed in parallel with a field programme encompassing structural geology surveys, geophysics, hydrochemistry, pilot and production well drilling and groundwater monitoring. Results from the field work were continuously fed back into the groundwater model and the modelling results were in turn used to guide the field programme, resulting in a well understood hydrogeological environment and a calibrated, robust, cost-linked model.

A detailed multi-layered model was developed using the USGS MODFLOW code, and calibrated to the initial peripheral (deep) well dewatering system that was being implemented. Aquifer parameters incorporating anisotropy and vertical conductance terms used in the modelling were based on results from the hydrology field programme. The model was developed into a cost-linked management tool from its initial function as a conceptual tool to identify the most appropriate dewatering system.

Various wellfield layouts were simulated by the interceptor model and cost-benefit analysis incorporating capital and operating costs over the 10 year mine life were carried out to determine the optimum wellfield layout and pumping rate to achieve dewatering targets commensurate with the mine plan. NPV cost savings of some US$3 million were identified for the adopted two-stage dewatering system of shallow interceptor wells and deep peripheral wells, with a total pumping rate of some $900–950 \, l \, s^{-1}$. The water produced from the dewatering system is discharged to a temporary wetland system created in the centre of the valley, which serves as a beneficial use in environmental terms as well as a high recharge area to help minimize semi-regional drawdown effects. A regional model has also been developed which confirms that the drawdown effect does not extend beyond the centre of the valley, and does not affect irrigation activities in the valley.

The interceptor model was also used, with minor modifications, to determine the outline feasibility of dewatering operations to facilitate a proposed pit expansion. It was then used to identify targets for additional wells for implementation of pit expansion plans. The model was also used to determine the impact on groundwater levels of a complete loss of power at the mine. Based on these results, decisions were made regarding the timing of commissioning emergency power generation for the dewatering system to ensure the gravels exposed in the pit wall remain dry, thereby minimizing the risk of the stability of the high wall.

The authors wish to acknowledge the considerable contributions of the geologists and other staff of Amax Gold Inc. at Sleeper Mine, without which this paper would not have been possible.

References

BEALE, G. & TYLER, W. E. 1990. Dewatering—A Practical Approach. *GOLDTech 4 Symposium*, Reno, Society for Mining, Metallurgy and Exploration, Inc.

MCDONALD, M. G. & HARBAUGH, A. W. 1988. *A Modular Three-Dimensional Finite-Difference Ground-Water Flow Model*. United States Geological Survey, Techniques of Water-Resources Investigations, Book 6 Chapter A1.

Opencast coal mining: a unique opportunity for Clee Hill Quarry

L. A. CRUMP & R. DONNELLY

ARC Ltd (Central), Shepshed, Leicestershire

Abstract: Clee Hill Quarry is an active hardstone (dolerite) quarry located 5 miles to the east of Ludlow, Shropshire. The quarry is in Carboniferous Coal Measures which have been intruded by a thick (c. 60 m) conformable fine-grained olivine-dolerite sill. The whole sequence is now folded into a broad synclinal structure and is extensively faulted. The area is blanketed by glacial deposits, 2–12 m thick. Historically, quarrying operations have been closely allied to the geological structure being confined to the margins of the syncline where the dolerite is devoid of overlying Coal Measures. In 1973 a programme was undertaken to assess the viability of using Coal Measures material for the production of lightweight aggregate. This programme was initiated because virtually all remaining, albeit substantial, planned reserves of dolerite were overlain by considerable thicknesses of Coal Measures and glacial deposits. While initial results from a technical viewpoint were encouraging, the project was abandoned for commercial reasons. The Coal Measures sequence overlying the dolerite sill comprises in excess of 30 m of mudstones, siltstones and sandstones as well as four recognized coal seams. Piecemeal mining of this coal had taken place on the Clee Hills for centuries, and in the area for future quarrying abundant evidence was available at surface of former bell-pit workings. In the early 1980s a decision was taken to evaluate the coal deposits at Clee Hill Quarry. The area for detailed assessment was defined using existing borehole data and evidence from old workings identified from site survey and aerial photographs. Exploration in two separate phases comprised the drilling of 44 open holes with spot coring and borehole geophysical logging. Coal quality was determined by analysing all relevant borehole core samples. Although coal reserve estimates assumed a high degree of past shallow mining activity, the economic viability of working the coal seams by opencast methods was established. In 1986 a planning application was submitted to Shropshire County Council to work and remove all coal overlying the dolerite, over a 3 year period. The proposal would allow exploitation of a valuable mineral asset, release substantial reserves of hardstone (dolerite) and allow restoration at an early date of large areas of former mining and quarrying dereliction. Planning permission was obtained in August 1988. Negotiations with the Opencast Executive of British Coal culminated in the granting of a licence to work both British Coal (vested) and ARC (alienated) coal. Coal mining was commenced in October 1988. Predictions of seam thickness and faulting were consistently accurate and vindicated the site investigation programme. In two aspects more detailed/accurate information would have been useful. Firstly, with regard to coal quality, borehole core samples did not accurately reflect *in situ* moisture content. Secondly, the extent of old workings in the upper coal seams was significantly overestimated. Neither aspect significantly affected the viability of the project. The mining phase including the major restoration works was completed in July 1992.

Clee Hill Quarry, located 5 miles to the east of Ludlow, Shropshire (Fig. 1), is an active hardstone (dolerite) quarry owned and operated by ARC Limited (Central Region) a wholly owned subsidiary of Hanson plc.

The quarry is now a major source of construction materials for the whole of the West Midlands region. The stone also meets specialist requirements in areas as far afield as South Wales.

This paper is a case history tracking the progress of resource evaluation at Clee Hill Quarry over the past thirty years and relating it to the development of the quarry.

Geology

The Clee Hills form a discrete outlier of Carboniferous rocks which occupy the core of a broad NE–SW trending (Caledonide) synclinal structure. The simplified geological succession is illustrated in Fig. 2.

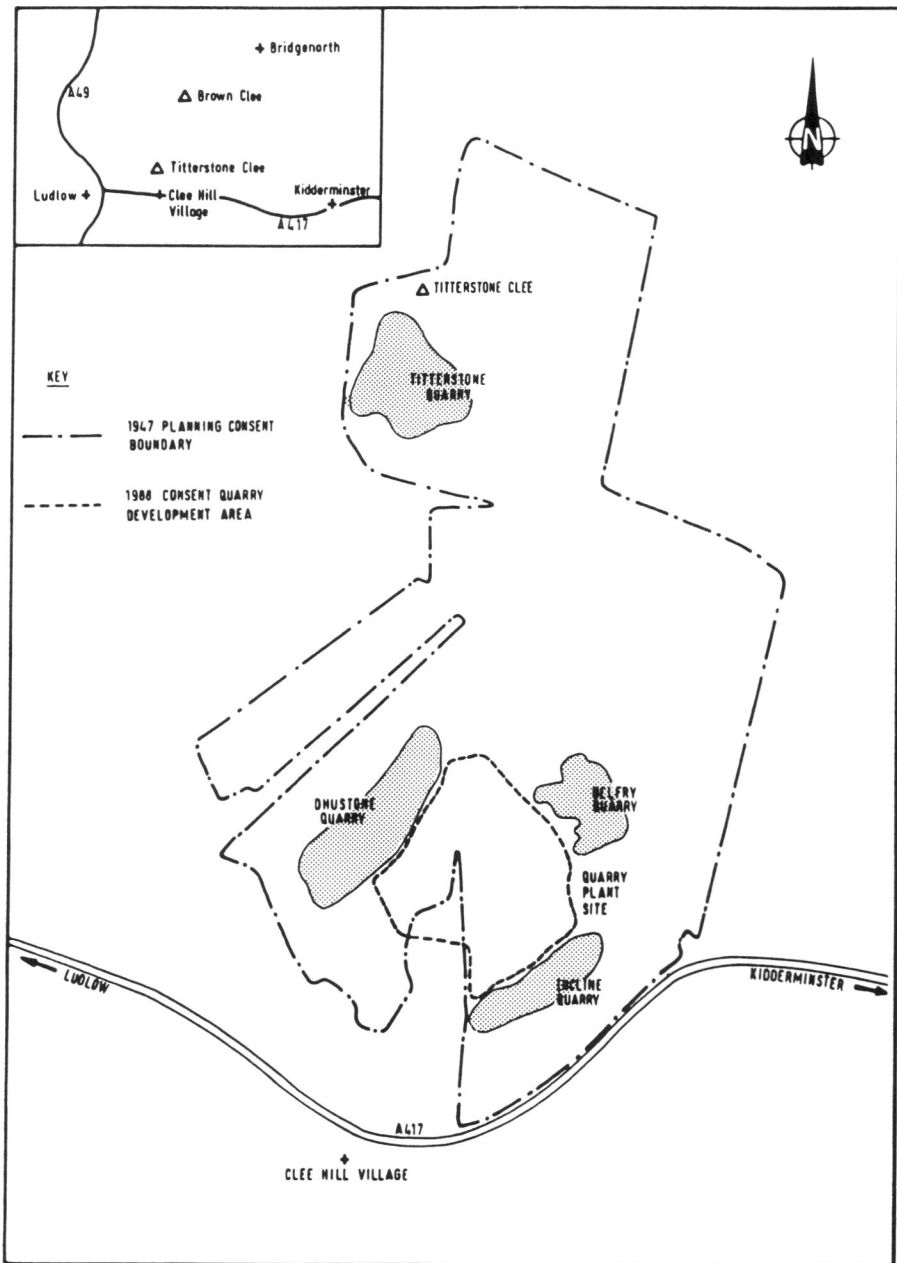

Fig. 1. Clee Hill site location plan.

The Upper Carboniferous Coal Measures comprise an alternating sequence of predominantly shales/mudstones, with subordinate sandstones and thin coal seams which have been intruded by a thick (c. 60 m) conformable olivine-dolerite sill.

The Coal Measures rest unconformably on the sandstone, conglomerates and marls of the Lower Old Red Sandstone Clee Group. These in turn overlie the red marls and sandstones of the Ditton Series and Downton Series. Silurian Wenlock Limestones and Shales and the Precambrian Stretton Series outcrop to the northwest.

GLACIAL DEPOSITS (Head and Till?)	QUATERNARY
COAL MEASURES INTRUSIVE DOLERITE COAL MEASURES	UPPER CARBONIFEROUS (WESTPHALIAN)
CLEE GROUP DITTON SERIES DOWNTON SERIES	DEVONIAN (LOWER O.R.S.)

Fig. 2. Clee Hill geological succession

Where the dolerite sill is unprotected by overlying Coal Measures, it has been subjected to deep (tropical?) weathering during the Quaternary (warm interglacial periods). The uppermost 2–3 m are commonly weathered to a thoroughly decomposed orange-red residue with occasional blocks of less weathered dolerite (roche). For considerable depths (30–40 m) below the surface, the dolerite may also be weathered along discontinuities, principally subvertical joints and high-angle faults. This deep weathering may be pervasive or only affect exposed joint surfaces with little detrimental effect in terms of aggregate quality.

When fresh, the dolerite is dark blue-grey or blue-black in colour, displays concoidal fracture and exhibits a fairly well developed system of columnar jointing. Petrographically the rock is fine-grained consisting of olivine phenocrysts set in a groundmass of plagioclase feldspars (labradorite–bytownite) and augite with minor magnetite and rutile. Analcite may occasionally be found interstitially in the freshest dolerite. Olivine shows complete to partial serpentinization whilst plagioclase may be albitized and replaced by carbonate.

Radiometric dating suggests an age of 295 Ma (Westphalian D) for the West Midlands dolerite sills (Kirton 1984). The Coal Measures are equivalent to Westphalian A–C. The sills are therefore considered to have been intruded into ductile unconsolidated Westphalian C Coal Measures prior to or during folding. Lateral offshoots of the sill into the Coal Measures, the presence of chilled margins and narrow bands of coarse-grained material within the sill, and palaeomagnetic evidence confirm an intrusive origin.

The youngest sediments, glacial deposits, cover the whole area. The traditional interpretation (Greig *et al.* 1968; Hains & Horton 1969) of the 'glacial materials' on the Clee Hills is that they comprise predominantly of solifluction deposits (head). However, although considerable thicknesses of deeply weathered head have been observed overlying the dolerite in the Clee Hill Quarries, the Coal Measures sequence is often overlain by material resembling a glacial till, which by implication would belong to an older (Anglian?) glaciation.

Historical background

The Clee Hills contain a wealth of potentially exploitable rocks and minerals, including limestone, dolerite (locally known as dhustone) iron ore and coal. These have been extensively and variously worked during the past eight hundred years.

The earliest recorded industrial activity in the area is coal mining (Jenkins 1983). In 1235 Wigmore Abbey received five shillings for the sale of coal at Caynham on Titterstone Clee. Piecemeal mining continued on the Clees for centuries and by the early eighteenth century a large number of shallow 'Bell Pit' workings were in existence. During the nineteenth century coal became much more important commercially as the only suitable source of local power for the increasing development and rapid mechanization of stone quarrying. Coal mining became progressively rationalized and in 1858 a railway was constructed from Ludlow to Clee Hill for development of the coal trade. By 1860 a number of deep underground pits had developed around the hills.

Stone gradually replaced coal as the most important product and in 1910 output from the Clee Hill quarries exceeded 400 000 tonnes per annum with over two thousand employed in the industry. The last working colliery in the area, Barn Pit, was closed at about the time of the General Strike in 1926.

Historically, quarrying operations have been closely allied to the geological structure, being confined to the margins of the syncline where the dolerite escarpments are devoid of overlying Coal Measures. This was recognized in the planning consent granted in 1947, which included all the major quarries in the Clee Hill complex (Belfry, Titterstone, Dhustone and Incline; Fig. 1) and covered an overall area of 251 ha (620 acres). Until relatively recently, quarrying continued to be controlled by this pattern of dolerite outcrop.

Quarrying and resource evaluation: 1950–1972

During the 1950s and 60s, the reserve situation progressively deteriorated as quarrying was forced into areas with either significant Coal Measures overburden or deep weathering. Quarry production faces were frequently characterized by a mixture of fresh and weathered dolerite. Contamination of the rock pile following blasting was common and consequently led to a high percentage (20–25%) of waste.

Titterstone Quarry closed in the late 1950s and in the mid-1960s, Incline Quarry ceased working due to increasing thickness of Coal Measures overburden-sandstones and shales. Quarrying at Dhustone was also severely restricted to the south by Coal Measures shale and mudstones. Main production moved to Belfry Quarry which was in fact 'Hobson's Choice', since although overlying Coal Measures were absent, workings were hampered by extremely variable rock quality due to weathering, faulting and a policy of not stripping overburden as a separate advance operation.

From the mid-60s through to the early 70s, quarry development at Clee Hill was broadly based on the findings of a number of separate geophysical surveys centred on the Belfry Quarry area. Early core drilling at Clee Hill was unsuccessful (and expensive) and for geological control considerable reliance was placed on percussion (open hole) drilling and sampling, and on the experience of Clee Hill management and drillers in interpreting the results.

In 1963 combined ground magnetic, resistivity and seismic refraction surveys were undertaken over an area of approximately 12.5 ha north of Belfry Quarry (Brown et al. 1963).

(i) The magnetic survey comprised closely spaced readings, taken along pegged lines, using an Elsec Proton magnetometer. Whilst there was a strong magnetic susceptibility contrast between overburden and unweathered dolerite in laboratory trials, many field readings actually fell within a common range. Results indicated the whole survey area was underlain by dolerite, but the technique was of little use qualitatively.

(ii) In the ground resistivity survey, measurements of ground resistance were made using both constant separation and expanding electrode layouts. The theory that unconsolidated damp overburden would have a lower electrical resistivity than the crystalline dolerite was sound. However, in practice, the resistances were highly variable and the results inconclusive.

(iii) Seismic survey proved the most successful geophysical technique. The equipment used a facsimile seismograph, with a 3 kg sledgehammer and steel plate as the source of energy and single fixed geophone as detector. As a control, a 25 m outcrop line was first 'shot' in Dhustone Quarry and this showed dolerite velocity increasing with depth. Quantitative interpretation of the survey lines was made and profiles of rock-head deduced. Reported depths correlated reasonably well with those given from check boreholes.

In summary, the surveys were only partially successful because of the general presence of an upper layer of weathered dolerite and the varying thickness of overlying glacial deposits, which contain a high proportion of dolerite boulders. Both factors tend to mask the deeply incised weathered or faulted zones. A general interpretation of the results of the various surveys concluded that the area immediately surrounding the Belfry Quarry was probably faulted and areas of sound, comparatively clean dolerite would be limited in extent.

Quarrying and resource evaluation: 1972–1993

Dolerite

There was a reluctance over a long period of time to assess the overall situation by core drilling (with cost being the most important factor). However, in 1972 a major new initiative was taken to evaluate the Company's hardrock resources. The programme was instigated by the fact that virtually all remaining, albeit very substantial, planned reserves of dolerite were overlain by considerable thickness of Coal Measures and/or glacial deposits which were at that stage unquantified. The object of the survey was to define long term reserves of good quality clean dolerite, with economic overburden/mineral stripping ratios.

Initially it was decided to investigate the area between Incline and Dhustone quarries where outcrop contacts of dolerite with the overlying Coal Measures record dips of 20–30 towards the southeast and northwest respectively, indicating a synclinal area between the two quarries. Assuming a constant dip, Coal Measures in excess of 90 m could overlie the dolerite in the centre of the syncline. However, it was reasoned that beneath the Coal Measure, fresh unweathered dolerite would be present.

During the period July 1972 to June 1973, geological work comprised a core drilling survey

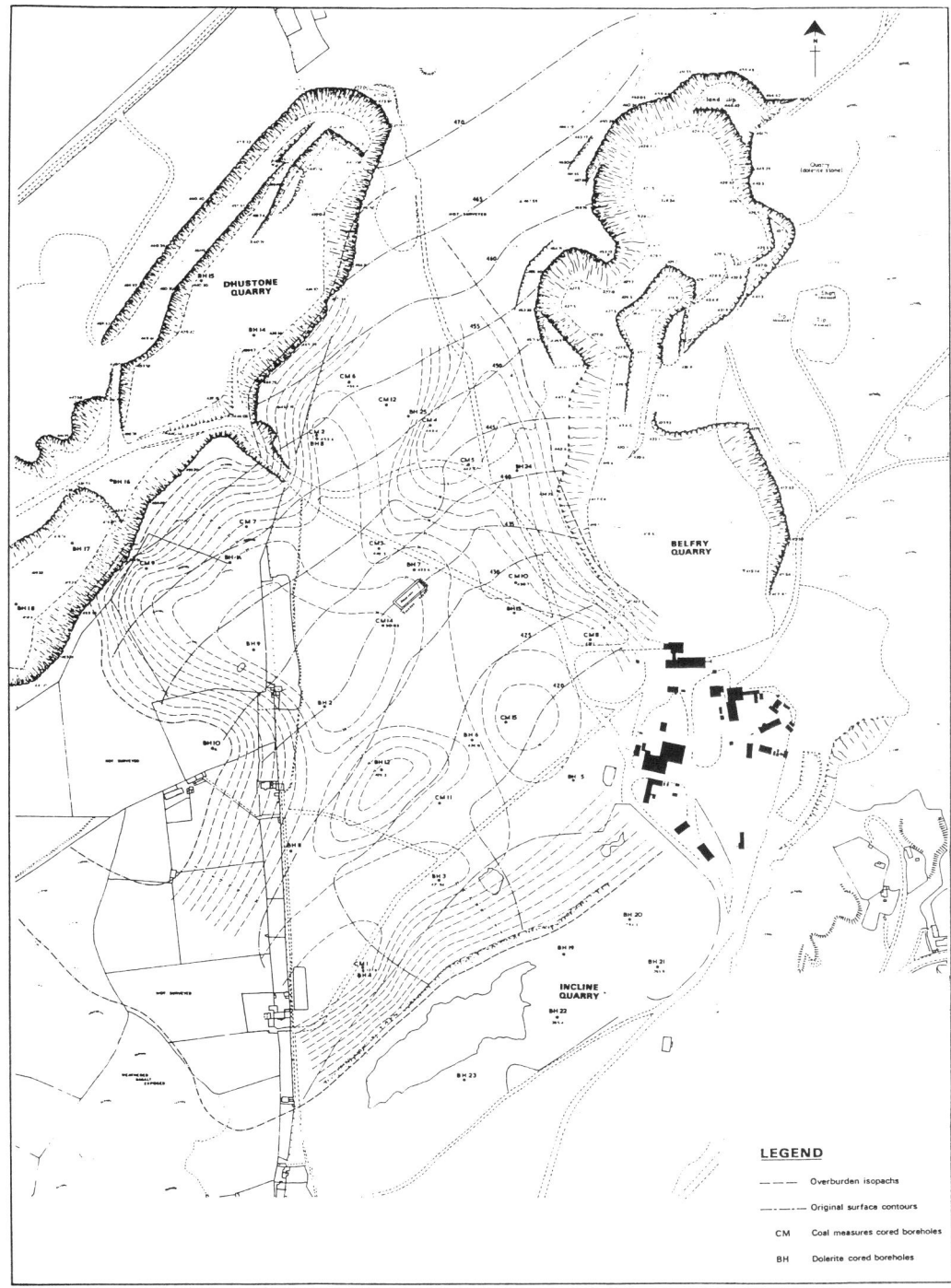

Fig. 3. Clee Hill overburden (glacial deposits and Coal Measures) isopachs.

Table 1. *Summary of borehole information for the area between Dhustone and Incline Quarries*

Borehole	Drift/dumped material (M)	Coal Measures (M)	Total overburden thickness (M)	Dolerite thickness (M)	Remarks
1A	6.45	19.00	25.45	57.91	Dolerite thickness not proved
2	3.28	26.96	30.24	64.67	
3	6.7	19.66	26.37	50.29	Dolerite thickness not proved
4	0.82	28.9	29.72	59.74	
5	1.27	20.12	21.38	47.24	Dolerite thickness not proved
6	3.35	19.2	22.56	56.77	
7	3.66	21.34	25	56.24	
8	2.04	20.51	22.56	53.34	Dolerite thickness not proved
9	8.32	20.99	29.31	57.91	
10	6.92	6.92	nil	54.86	Dolerite thickness not proved
11	3.35	19.42	22.77	54.96	
12	3.26	9.99	13.24	65.58	
13	3.73	17.13	19.95	61.28	
Average thickness	3.96	18.6	22.56	57.00	

over approximately 30 ha within and between Dhustone and Incline quarries, and a seismic refraction survey of 10 ha in the area east/northeast of Belfry Quarry.

Using an ARC drill rig, 31 boreholes were drilled with a total metreage of *c.* 1200 m. Drilling generally involved open holing through the overlying Coal Measures and coring of the underlying dolerite. Of the 13 boreholes drilled between Dhustone and Incline, eight boreholes accurately determined the full thickness of both Coal Measures and dolerite.

Table 1 summarizes borehole results and shows the relative thickness of glacial deposits, Coal Measures and dolerite between Dhustone and Incline Quarries. For borehole locations see Fig. 3.

The drilling survey proved that the area between Dhustone and Incline quarries is underlain by a well-defined faulted asymmetric syncline. The dip on the upper dolerite contact with Coal Measures rapidly flattens out from the margins to the central part of the syncline. The synclinal fold appears to close both to the southwest and in the area to the NE between Dhustone and Belfry Quarries.

A maximum thickness of 30.24 m total overburden was proved. The underlying dolerite sill was very consistent with an average thickness of 57.0 m. Contours drawn on the base of the dolerite indicated widespread faulting. Substantial dolerite displacements were mirrored by changes in the thickness of the Coal Measures overburden.

The theory that good quality, clean, massive unweathered dolerite existed beneath Coal Measures was proved correct. In the area between Dhustone and Incline Quarries, eight boreholes proved massive unweathered dolerite with very limited zones of broken material. Two boreholes (on the margin of the syncline and devoid of overlying Coal Measures) proved alternating broken and massive material.

Material representative of boreholes showing different grades of weathering and frequency of jointing/fracturing was sampled and subjected to a full range of physical and mechanical tests. The results showed very little variation, indicating good quality rock suitable for general structural and roadstone purposes and having identical properties to those of current production material. Broken dolerite did not produce inferior quality chippings.

In summary, the borehole survey was successful in proving over 20 million tonnes of predominantly massive, clean dolerite in an area between Dhustone and Incline quarries. This area was defined for future quarry development. The resource was, however, overlain by approximately 6.5×10^6 m^3 of Coal Measures and glacial materials.

During this period of exploration, production

Table 2. Coal Measures (lightweight aggregate) boreholes: summary data

Borehole	Total cored length (metres)	Sandstone %	Siltstone %	Clay/mudstone %	Carbonaceous clay/mudstone	Coal %
CM.1*	13.28	40.4	30.1	29.5	Nil	Nil
CM.2	16.65	Nil	10.5	70.1	10.8	8.5
CM.3	15.42	3.4	16.5	40.9	33.7	5.5
CM.4	(No recovery of Coal Measure–total O/B thickness 5.9 metres)					
CM.5	3.94	Nil	58.38(?)	Nil	41.62	Nil
CM.6	15.01	Nil	25.5	42.8	13.3	18.4
CM.7*	15.09	5.39	26.84	32.8	30.09	5.04
CM.8*	11.81	Nil	Nil	44.4	37.5	18.05
CM.9*	7.71	Nil	Nil	76.38	23.82	Nil
CM.10	13.12	Nil	Nil	51.25	18.75	30.0
CM.11	12.29	4.68	44.7	48.65	1.95	Nil
CM.12	11.48	27.12	11.11	22.49	22.66	16.7
CM.14*	18.37	6.43	29.64	34.07	16.21	13.75
CM.15	9.51	Nil	52.86	42.93	4.31	Nil
BH.2A	18.45	7.25	Nil	44.4	29.03	19.38
BH.7A	18.96	Nil	21.45	55.16	16.02	8.23
BH.11A	19.22	18.22	2.47	62.6	12.77	3.66
BH.12A	10.74	Nil	54.5	12.21	27.94	5.34
Average percentages		6.64	22.15	41.80	21.28	9.53
Average percentages discounting CM.1 (located in extreme southwest part of area – not to be worked for light aggregate)		4.53	22.62	42.57	20.03	8.97

* Coal Measures not cored to base.

continued to be concentrated from faces established at the northern end of Belfry Quarry. Results of advance (pre-production) percussion drilling were inconclusive and it was originally intended to extend the core drilling survey to this area. However, to save time and expense a further seismic refraction survey was undertaken adjacent to the current production area. The survey concluded that the area east and southeast of Belfry was underlain by dolerite relatively close to surface, but much of it would be highly weathered and broken and likely to contain a high dirt content. Further east and northeast another Coal Measures basin would appear to be present (Roberts 1964).

Coal Measures (lightweight aggregate)

Preliminary test work on the suitability of Clee Hill Coal Measures as a potential source of raw material for lightweight aggregate production was undertaken as early as 1970. Initial results were very encouraging. These results combined with the now recognized overburden problem led to an extension of the 1973 borehole survey, and at four locations the Coal Measures were core drilled to determine their precise composition, in particular the amounts of mudstone present.

Predictably, drilling proved the Coal Measures to be highly variable, both laterally and vertically, due to the characteristic rhythmic style of sedimentation. Five main rock types were recognized: sandstone, siltstone, brown-grey-blue mudstone, black carbonaceous mudstone and coal. It was concluded that a high percentage of the Coal Measures sequence was mudstone, and sandstone did not form a major part of the succession.

Two complete sets of core samples were submitted to a commercial lightweight aggregate producer and a number of tests were carried out on mudstones from different depths. They indicated the material was suitable for the manufacture of good quality lightweight aggregate and a bulk sample (approximately 4 t) of blue-grey mudstone was dug from a shallow pit and subjected to pilot plant trials. A quantity of coal from South Wales (Abernant Middlings) was used for fuel. The initial results from borehole samples were substantiated and the Clee Hill blue-grey mudstone readily sintered to a good quality lightweight aggregate-type product (Thrasher 1974).

Following the successful plant trials, it was decided in early 1974 to extend the investigation to gain detailed information over a much wider area. During the period April–August 1974 an additional 15 boreholes (c. 272 m, in part cored), were drilled to base of Coal Measures in the area between Dhustone and Incline quarries. Summary data obtained from these boreholes are contained in Table 2

The composition, distribution and quantity of Coal Measures rock types became much clearer. Mudstone proved the dominant component forming the major part of the sequence. Lateral variation was particularly evident with respect to sandstones and coal. Sandstone was well developed towards the southwest limit of the area, but poorly developed elsewhere. Coal was absent over the southern and southwestern areas, but well developed over the central and northwest part of the area where the sandstone is poorly represented.

The pilot plant trials showed different mudstones produced different quality lightweight aggregates. The most notable contrast was between black carbonaceous mudstone which produced excellent quality lightweight aggregate and the yellow silty mudstone which produced a heavier material, albeit within a general specification.

Quality appeared to be primarily a function of the silica content. The high silica sandstones would not sinter and were therefore useless. The siltstones and silty mudstones were found to sinter readily but produced a heavier aggregate than the mudstones. The carbonaceous mudstones produced excellent lightweight aggregate. The coal would not sinter itself, but was useful as a source of fuel.

On completion of the final phase of drilling a further set of laboratory-scale tests was carried out on three borehole samples:

(i) normal mudstone, sintered with Abernant Middlings of known calorific value, to establish fuel demand and aggregate quality;
(ii) normal mudstone sintered with carbonaceous mudstone and coal or shaley coal;
(iii) siltstones/silty mudstones sintered with carbonaceous mudstone and coal or shaley coal.

It was concluded that 60% of Coal Measures sequence would produce good quality lightweight aggregate and a further 20% (siltstones and silty mudstones) would produce material of a somewhat inferior but still acceptable quality. A further 10% coal could fulfil part of the fuel requirement and a remaining 10% would be waste sandstone.

During the mid-1970s despite all the above site investigation, laboratory test work, plant trials and evaluation, the lightweight aggregate project floundered and was finally abandoned for reasons of a non-technical (commercial) nature.

It was reluctantly accepted that future development of Clee Hill Quarry, would take place between Dhustone and Incline Quarries with removal of substantial volumes of Coal Measures overburden an integral part of future quarrying.

A programme of phased development was initiated and overburden stripping was undertaken on an irregular basis at Dhustone southwest. A benched system was re-established with the objective of trying to minimize high stripping ratios whilst maintaining operating efficiency and slope stability.

Pre-development drilling (using the quarry rig) was undertaken to predict and determine overburden thicknesses and volumes, essential for budget planning. To illustrate the scale of overburden removal, a 30 m advance development over a 200 m face length involved stripping c. 100 000 m^3 overburden.

Coal measures (coal project)

In late 1980/early 1981 ARC received a number of separate enquiries regarding the Company's interest in possibly exploiting the underlying coal deposits. This new initiative was ultimately to prove to be a unique opportunity for development at Clee Hill Quarry.

A major part of the area between Dhustone and Incline had been bell pitted for coal, evidenced by a mosaic of flooded hollows in collapsed/infilled bell pits and mounds of clay spoil. It was previously thought that there were no viable coal reserves remaining. However, in conjunction with H. J. Rorke Ltd, a subsidiary of ARC's parent company and an opencast coal contractor, ARC met the British Coal Opencast Executive (formerly The National Coal Board) to discuss a proposed project. Despite earlier reservations the initial reaction from British Coal (BC) was positive and following preliminary and separate desk-top exercises both parties arrived at similar estimates of the coal reserves within the area between Dhustone and Incline Quarries (260 000 t BC/278 000 t ARC).

A joint working party was set up to oversee the progress of the project. Interpretation of the structure and correlation of seams was not possible from existing records and no quality assessments were available. BC stressed conclusive evidence of coal quantity and quality would be required before agreeing marketing terms and they also wished to have confirmation of ARC's rights of ownership. Whilst they would not contribute financially in the evaluation, they were prepared to offer technical advice and laboratory facilities.

During the period August–September 1981, Phase I of a coal exploration drilling programme was undertaken (Crump 1981). The area for evaluation was defined using existing borehole data from the lightweight aggregate drilling survey and evidence of old workings as shown on aerial photographs. This area totalled some 14 ha, although it was open ended to the northeast. Regarding coal ownership, BC (vested) coal and ARC (alienated) coal extended to 2.8 ha and 11.3 ha respectively.

In consultations with BC it was decided that a limited drilling programme, combined with the previous borehole information would give the right level of assessment initially required, and provide details on the following aspects:

(i) accurate correlation of the coal seams;
(ii) quality of the coal;
(iii) extent to which seams had previously been worked;
(iv) the economic potential of the other Coal Measures strata—particularly the mudstones.

The method of drilling and borehole layout was decided in consultation with BC geological staff. The drilling programme was contracted out and comprised 13 open holes located at 60 m intervals along two profiles at right angles. In addition, two cored borehole intersections of 'productive' Coal Measures strata were completed. The total metreages of open hole and core drilling were 405.0 m and 20.34 m respectively. The results of the drilling programme are summarized in Table 3.

For accurate correlation and strata identification geophysical logging was undertaken on the open holes. A standard coal composite log was run incorporating gamma ray, short spaced density and long spaced density. Geolog data was interpreted in conjunction with BC.

Table 3. *Coal Measures (coal project) boreholes summary data*

BH no.	Total coal thickness (M) (J, K, L & M seams only)	Total overburden thickness(M) (to base of bottom seam)	OB:coal ratio
1	0.86	14.69	17:1
2	1.07	14.24	13:1
3	2.63	19.59	7:1
4	3.45	21.83	6:1
5	2.44	15.08	6:1
6	2.16	7.53	3:5:1
7	0.50	5.65	11:1
8	—	—	—
9	1.08	17.92	17:1
10	—	—	—
11	2.89	15.54	5:1
12	2.34	13.02	6:1
13	1.53	6.50	4:1
14	—	—	—
15	1.35	15.15	11:1
16	—	—	—
17	4.26	18.48	4:1
18		Not drilled	
19	2.0	16.98	8.5:1
20	2.0	11.88	6:1
21	1.31	3.84	3:1
22	3.8	11.4	3:1
23	2.14	8.78	4:1
24	1.67	8.15	5:1
25	2.38	6.32	2.5:1
26	2.16	6.84	3:1
27	—	—	—
28	3.2	8.88	3:1
29	—	—	—
30	2.11	14.01	6.5:1
31	0.26	5.66	22:1
32	1.86+	18.78	10:1
33	2.00+	15.05	7.5:1
34	2.02	10.86	5:1
35	3.38	9.12	2.5:1
36	4.06	14.32	3.5:1
37	c.2.40	c.10.1	c.4:1
38	c.2.7	9.3	c.3.5:1

The drill cores were logged on site jointly by ARC and BC geologists and samples of coal material were taken to BC laboratories at Cannock where the cores were dried, crushed, split and analysed (sufficient material was retained for in-company test work). The remaining cores were sent separately to BC for an assessment of the mudstone fraction.

Results of the open hole drilling, geologs and drill cores allowed recognition of four main coal seams (illustrated in Fig. 4), in stratigraphic

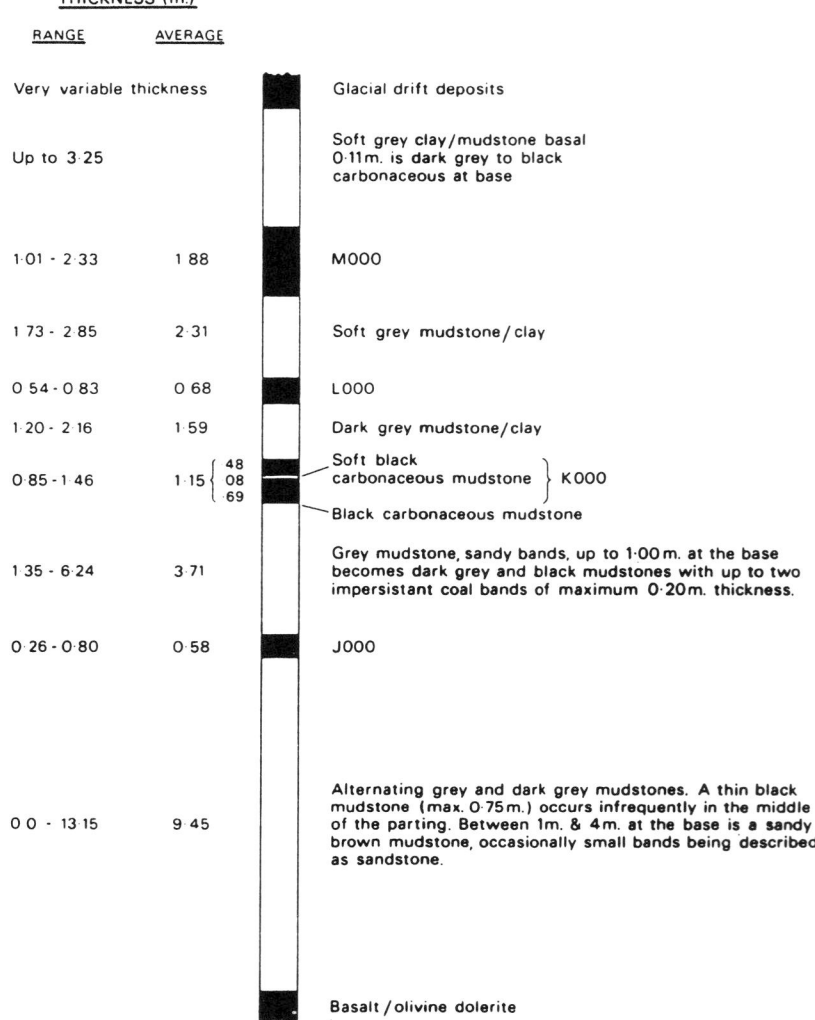

Fig. 4. Clee Hill Coal Measures (coal project) generalized geological interpretation.

order; the Fourfeet Seam (referred to in this project as J000), Smith Seam (K000), Three-quarter Seam (L000), and the Great Seam (M000). Other unnamed coal seams were intersected in the boreholes, but these are thin and discontinuous.

The asymmetrical synclinal structure in which the coal seams occur is illustrated in Fig. 5. The vertical displacement of coal seams and rapid attenuation of Coal Measures particularly at the margins of the area containing coal illustrates the significance of faulting.

With limited data, the overall site stripping ratio was difficult to assess, but estimated to be in the region of 6:1 (overburden thickness:coal thickness). Initial coal quality data indicated general compliance with limits applied for commercial viability (power station requirements).

Based on the results obtained to date, the mining prospects at Clee Hill, subject of course to planning permission, were very favourable. It was therefore agreed that a second phase of drilling should be carried out at 120 m centres

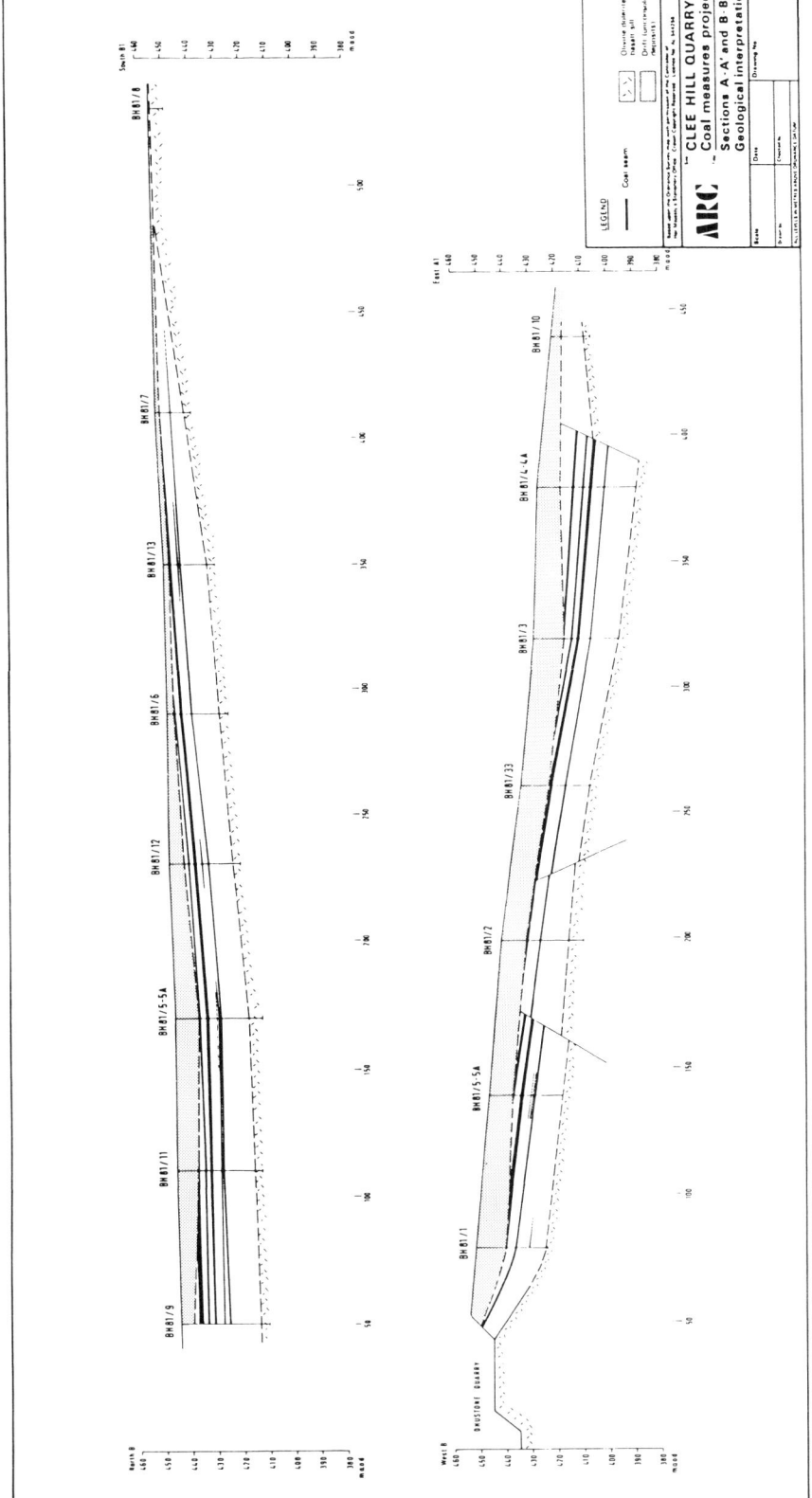

Fig. 5. Clee Hill Coal Measures (coal project) geological interpretation.

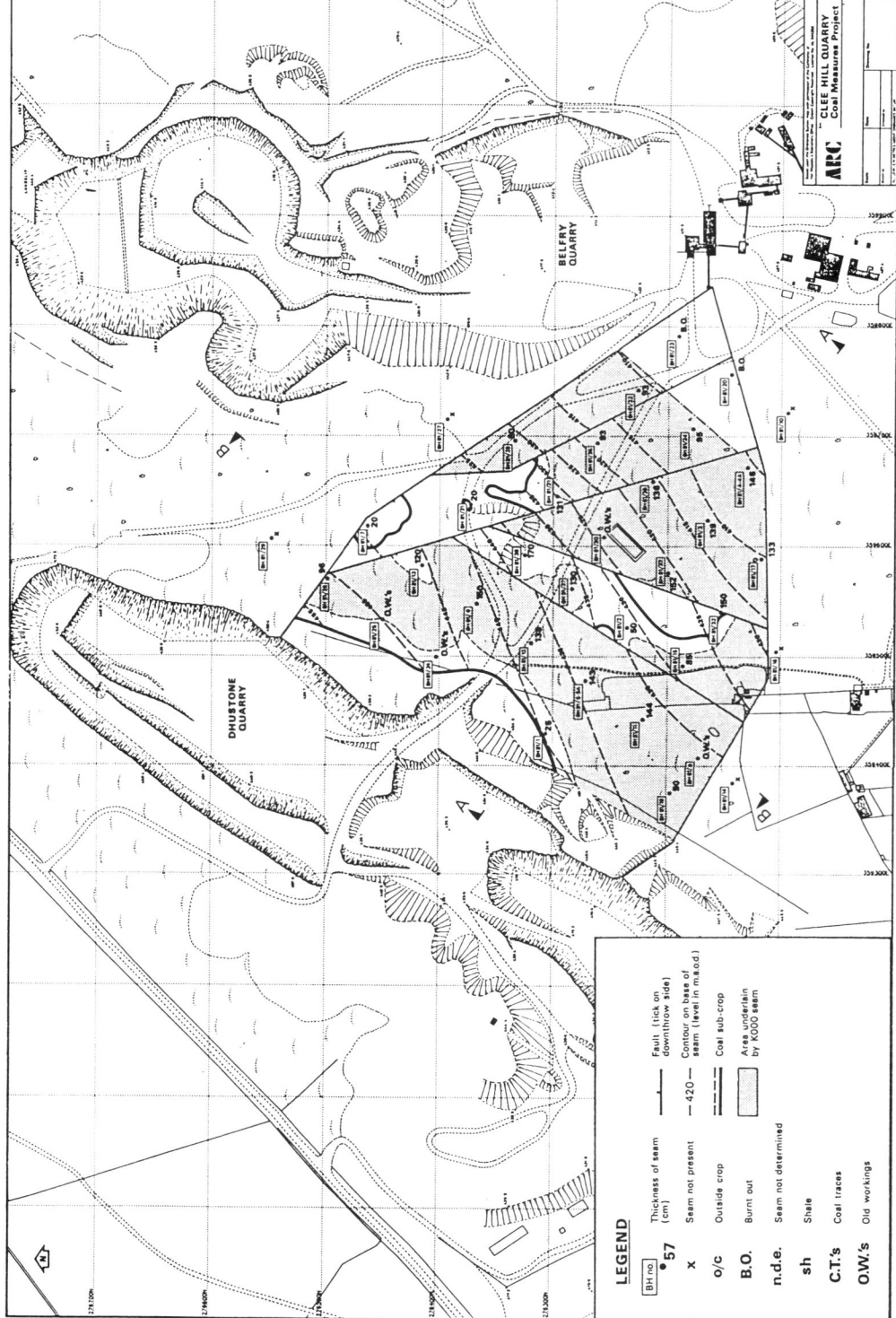

Fig. 6. Clee Hill Coal Measures (coal project) geological interpretation as illustrated by K000 seam.

Table 4. *Coal quality: mean analytical value*

	No. of samples	M	A	S	Chl	VM	CV
M000	11	10.7	7.9	0.34	0.03	30.7	24 680
L000	6	6.6	15.0	0.37	0.03	24.2	25 480
K000	19	6.1	9.6	0.4	0.03	23.1	28 200
J000	4	4.9	13.7	0.3	0.03	15.2	27 420
Whole site	40	7.1	11.5	0.36	0.03	23.3	26 445

M, moisture content; A, ash content; S, sulphur; Chl, chlorine; VM, volatile matter; CV, calorific value.

over the whole of the potential coal bearing area. This grid was closed up at the margins of the coal areas, adjacent to known fault lines and in the vicinity of bell pit workings.

Following the need to change drilling contractors to achieve satisfactory coal recovery, the programme was finally undertaken in January and February 1981 and consisted of an additional 24 open holes (451.6 m) and 5 cored boreholes (59.76 m). All the open holes which encountered coal were subsequently logged geophysically using the coal composite log. These and measurements from cored intersections were used to determine seam thicknesses.

An interpretation of the exploration data was made in conjunction with BC Opencast Executive geological staff (Figs 5 and 6). The coal area is affected by at least six faults, three of these forming boundaries to the coal area. The dip of the strata in each fault block is generally gentle at around 5–10°.

Additional coal quality determinations were carried out on drill core samples (Table 4). Analyses show the effect of thermal alteration in increasing rank of coal from the M000 seam to the J000 seam, although rank in any individual seam varied quite widely due to the proximity of the seam to the sill. Sulphur and chlorine were consistently low and the coal was 'environmentally acceptable'.

The schedule of estimated quantities for coal and total overburden was produced using conventional graphical methods with average seam thicknesses for each faulted block calculated from the borehole intersections (Table 5).

Within the defined coal extraction area the overburden thickness: coal thickness stripping ratio was extremely favourable at 9:1. However, advantage was taken of the fact that a considerably higher ratio may be adopted for economic coal exploitation. Between the base of the J000 seam and the upper dolerite contact is a sequence of barren Coal Measures which average approximately 10 m. Removal of an additional volume of Coal Measures down to the dolerite contact was identified in the area proposed for quarrying immediately SE of Dhustone Quarry. The balance of the overburden was made up with a volume of glacial material, waste weathered dolerite and barren Coal Measures outside the coal extraction area (Bainton 1982).

Table 5. *Clee Hill schedule of estimated quanties.*

Seam	Coal *in situ* (m^3)	Recoverable coal (t)*
Great (M000)	11 702	14 695
Three Quarter (L000)	29 052	31 968
Smith (K000)	78 061	95 060
Four Feet (J000)	48 571	54 307
	167 386	196 030

Total extraction area = 138 500 m^2 (−34.2 acres)
Coal extraction area = 107 200 m^2 (−26.2 acres)
*Tonnage Factor 1.33 tonnes/m^3

Allowances
15% Old workings in K000
60% Old workings in L000
80% Old workings in M000
5% Faulting in each seam
10 cm thickness reduction for incomplete extraction
ARC (alienated coal) = 154 680 t
NCB (vested coal) = 41 350 t
Total overburden/interburden volume in
coal extaction area to basal seam
(includes batters) = 1.70 mm^3
Total overburden volume in coal
extraction area beneath basal seam
(includes batters) = 0.60 mm^3
Total overburden volume outside coal
extraction area = 0.17 mm^3
                   2.47 mm^3

In October 1986, a planning application was submitted to Shropshire County Council to both rationalize future quarrying at Clee Hill and extract and remove from site all coal overlying the dolerite within the application area, over a three year period.

The overall benefits of the scheme were:

- to restrict future quarrying at Clee Hill within a relatively small and well-defined area (c. 33 ha) over which mineral extraction would be undertaken over the next 60 years (Fig. 1);
- recovery of coal during a carefully controlled special overburden stripping operation;
- utilization of a valuable national energy resource that would otherwise have been lost;
- restoration of large areas of former quarry and mine workings at an early date, not possible without the overburden released in this programme;
- continued production of high-grade hardstone aggregate secured by reducing potentially onerous overburden stripping costs.

Prior to the application and during the planning determination period, negotiations were held with British Coal Opencast Executive and a licence application submitted to work both coal nationalized under the 1938 Coal Act and the Coal Industry Nationalisation Act 1946 (vested coal), and coal in ARC ownership (alienated coal).

A licence was subsequently obtained from BC and Planning Consent was granted in August 1988. Opencast coal mining commenced in October 1988.

During the period October 1988–July 1992, 322 000 tonnes of coal (saleable) was mined as part of the removal of c. 2.5×10^6 m^3 of overburden.

There were two main reasons for the disparity in estimated and actual recoverable coal reserves. Firstly, the extent of previous coal workings was substantially overestimated. Shallow bell pitting appears to have been largely confined to the working of M000 and K000 seams along subcrops below the drift. The percentage extraction from the bell pitting was much lower than anticipated. Secondly, at the time of the planning application it was not anticipated any coal would be washed, crushed or screened. The installation of a washing plant substantially improved clean coal recovery, particularly for coal associated with old workings, along subcrops and faults.

Potential reserves of fireclay underlying the four coal seams (J000, K000, L000, M000) were estimated at c. 470 000 t. The fireclays beneath seams J000 and K000 were particularly marketable. Unfortunately, commercial agreements could not be finalized within the required timescale.

Conclusions

Opencast Coal extraction at Clee Hill is now complete. Considerable volumes of Coal Measures, mudstones and siltstones which formerly comprised a substantial proportion of the overburden have been used to restore large areas of quarrying dereliction (Humphries 1982). In addition, extensive reserves of high quality dolerite aggregate have been exposed, thus ensuring the viability of Clee Hill Quarry. As other strategic reserves in the West Midlands become exhausted, it is likely that Clee Hill will continue to supply local and national markets well into the next century.

References

BAINTON, C. S. 1982. *The mining of coal by open-pit methods with special reference to the Clee Hill project (ARC), Shropshire.* MSc dissertation Royal School of Mines.

GREIG, D. C., WRIGHT, J. E., HAINS, B. A. & MITCHELL, G. H. 1968. Geology of the Country around Church Stretton, Craven Arms, Wenlock Edge and Brown Clee. (Sheet 166) Memoirs of the Geological Survey.

HAINS, B. A. & HORTON, A. 1969. *British Regional Geology—Central England.* British Geological Survey.

BROWN, H. C., HAMILTON, N. & KHAN, M. A. 1963. *Report on Magnetic Resistivity and Seismic Surveys at Clee Hill.* Unpublished Report for ARC.

CRUMP, L. A. 1981. *Clee Hill—Coal Measures Project (1981), Summary report.* ARC Internal Report.

HUMPHRIES, R. N. 1982. *Soil Report, Clee Hill, Shropshire.* Unpublished Report for ARC.

JENKINS, A. E. 1983. *Titterstone Clee Hills Everyday Life Industrial History and Dialect.* A. E. Jenkins.

KIRTON, S. R. 1984. *Carboniferous volcanicity in England with special reference to the Westphalian of the E and W Midlands. Journal of the Geological Society, London,* **141**, 161–170.

ROBERTS, D. I. 1964. *Clee Hill Quarries, Shropshire.* ARC Internal Report.

THRASHER, F. E. 1974. *Geological Survey at Clee Hill 1972–73.* ARC Internal Report.

—— & BEARDSMORE, S. 1975. *Diamond Drilling Surveys at Clee Hill Quarry.* ARC Internal Report.

Database management at the Lisheen deposit, Co. Tipperary, Ireland

J. BARRY,[1] J. GUARD[1] & G. WALTON[2]

[1] *Crowe, Schaffalitzky & Associates, Newstead, Clonskeagh, Dublin 14, Ireland*
[2] *Chevron Mineral Corporation of Ireland, Newstead, Clonskeagh, Dublin 14, Ireland*

Abstract: The Lisheen Zn-Pb-Ag deposit was discovered by the Chevron Mineral Corporation of Ireland/Ivernia West Plc Joint Venture in 1990. The mineral resource estimate is 20.5 Mt grading 13.3% Zn, 2.3% Pb and 35 g/t Ag. Massive sphalerite–galena–pyrite mineralization occurs at the base of the dolomitized, lower Carboniferous Waulsortian limestones and is bounded to the south by a shallow-dipping normal fault. The deposit is subdivided into three areas: the Main, North and Derryville Zones.

Lisheen diamond drill hole data are stored in a relational database management system called TECHBASE with graphical output to AutoCad. Coding and in particular 'nested' coding facilitates data filtering and efficient editing and interrogation of data. The coding system has the flexibility to evolve as the multi-disciplined and multi-sourced data at Lisheen become more complex. Routine tasks which require frequently-used sequences of commands such as the generation of stick sections and data reports are automated by the use of macros (runlogs). Faster processing and editing of data considerably increases the generating speed of sections, maps and data reports which results in faster, and more accurate calculation of mineral resource estimations. 'Technicn' is a facility withing TECHBASE which permits parameter substitution within macros through a simple memo-driven user interface.

The purpose in writing this paper is to discuss the implementation and development of computerization at the Lisheen Zn-Pb-Ag mineral deposit in County Tipperary, Ireland and its benefits in managing a diamond drillhole database from resource delineation to evaluation. As a framework for the central topic a brief outline of the geology and structure is presented, a more detailed account of which can be reviewed in Hitzman *et al.* (1992).

Diamond drillhole from the Lisheen Project data is loaded into TECHBASE which is a database management system developed by MINEsoft Ltd, based in Colorado, USA. TECHBASE facilitates efficient validation, editing, and interrogation of the 'Lisheen diamond-drilling database' which acts as a core supporting peripheral application modules used to generate, for example, drillhole stick-sections, mineral intensity contour maps and geostatistical parameters (Fig. 1).

Geology

Lisheen is located about 10 km northwest of Thurles in County Tipperary, Ireland and lies at the southwestern end of the Rathdowney Trend, a 40 km belt of dolomitized Waulsortian lime-

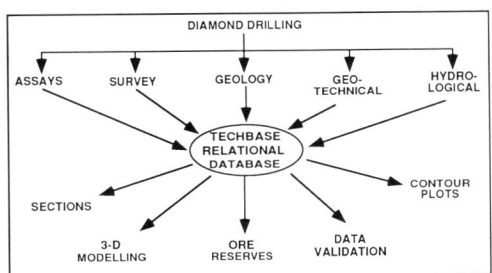

Fig. 1. Data processing flow chart.

stones extending northeastwards to Abbeyleix (Fig. 2). The Lisheen mineral resource, estimated at 20.44 Mt grading 13% Zn, 2% Pb and 35 g/t Ag (Ivernia West Plc 1992) was discovered by a Chevron Mineral Corporation of Ireland/Ivernia West Plc Joint Venture (IPL Joint Venture) and their geological consultants Crowe, Schaffalitzky and Associates (CSA) in April 1990 (Hitzman *et al.* 1992). About 8 km further northeast of Lisheen along the Rathdowney Trend lies the 6.5 Mt Galmoy Zn-Pb-Ag mineral deposit, grading 12% Zn and 1.26% Pb (Doyle *et al.* 1992). There are two mineral prospects at

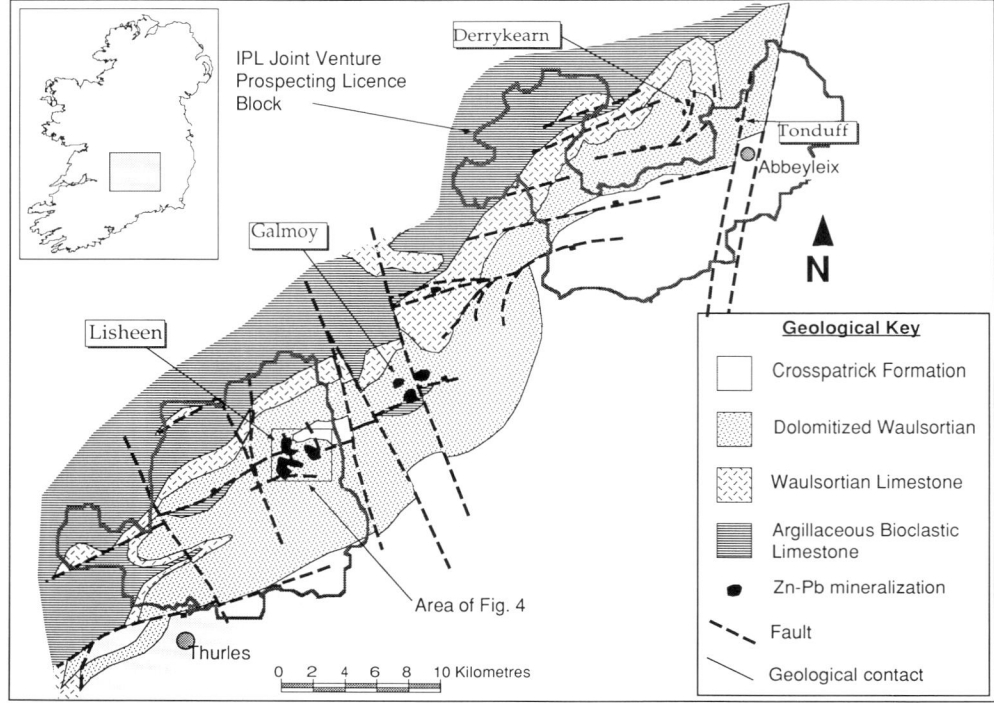

Fig. 2. Location and geological setting of the Lisheen base metal deposit.

Derrykearn and Tonduff about 25 km further northeast along the trend near the town of Abbeyleix (Fig. 2). CSA are retained by the IPL Joint Venture as geological consultants on the mineral exploration, resource delineation and database management of the Lisheen deposit.

The Argillaceous Bioclastic Limestone (ABL), Waulsortian Limestone and Crosspatrick Limestone are the main lithological formations recognized at Lisheen. The ABL is considered to be approximately 400 m thick in the Lisheen area and the overlying Waulsortian Limestone between 170 and 190 m. The Waulsortian Limestone including the lower part of the Crosspatrick Limestone have been regionally dolomitized and hydrothermal ferroan dolomite occurs as narrow veinlets cutting regional dolomite (Hitzman et al. 1992). A breccia called the Black Matrix Breccia (BMB) forms a more laterally extensive sheath to mineralization. The BMB can be clast supported or matrix supported depending on the degree of development and consists of dark grey angular to subrounded clasts of dolostone in a black matrix consisting of minor sulphide and both ferroan and non-ferroan dolomite (Hitzman et al. 1992).

For a more detailed description of the geology and stratigraphy of the Lisheen mineral deposit see Hitzman et al. (1992).

Structure and mineralization

At Lisheen Zn, Pb and Fe sulphides are mainly concentrated in one, or locally two, stratiform lenses at, or close to, the base of lower Carboniferous dolomitized Waulsortian limestone which dips gently to the southeast at about 5°. Mineralization occurs at depths of 180–200 m as massive replacement of the host rock or as dense ramifying stockworks of veinlets. Thickest sulphide intersections of up to 35 m occur in the proximal hanging wall of the Killoran fault which is a normal fault striking approximately 090° and dipping 30° to 55° north with a throw of approximately 200 m (Fig. 3). Mineralization increases in grade and thickness close to sub-parallel E–W faults in the hanging wall of the Killoran fault. Sub-vertical, north-trending strike-slip faults with throws of up to 15 m transect the deposit. The Lisheen deposit is divided into three main mineralized zones: the Main, North and Derryville Zones (Fig. 4). The Derryville Zone is about 1 km ENE of the Main Zone and about 200 m east of a

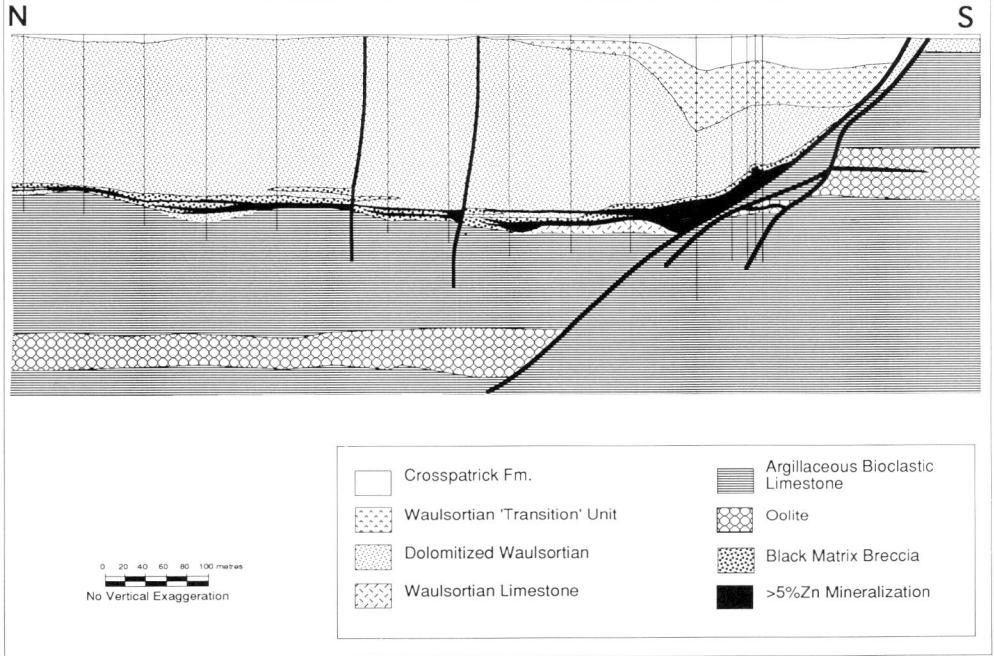

Fig. 3. North–south cross-section (facing east) through the Lisheen base metal deposit.

Fig. 4. Borehole plan of the Lisheen base metal deposit indicating the main ore zones.

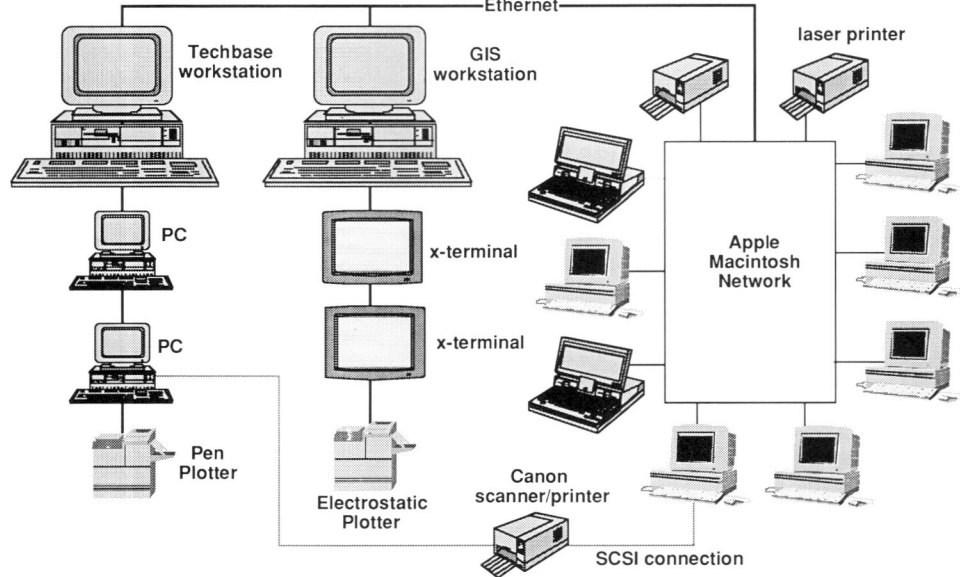

Fig. 5. Lisheen open computer system.

relay zone which accommodates a 'left step' in the Killoran fault. Two important observations concerning mineralization at Lisheen are that it is spatially associated with east–west-striking, moderately dipping normal faults which appear to form part of an ENE-trending 'left stepping' en echelon fault system (Fig. 1) and secondly that mineralization occurs just at or within metres of the contact between the ABL and the regionally dolomitized Waulsortian limestone where there is an abrupt change in the material and chemical properties of the rock.

Computerization

Data

To date more than 390 diamond drillholes have been drilled totalling 60 km of drill core. Diamond drilling is on a north–south grid at 30 m, 60 m and 120 m centres. In three areas of the Main Zone boreholes were drilled on a 7.5 m, 15 m and 30 m spaced 'L' pattern to test for sulphide grade and thickness continuity. In the Main Zone inclined boreholes have been drilled to determine the attitude of joint sets. Downhole survey readings are taken at 50 m intervals with a Sperry Sun magnetic single shot instrument.

Core samples, commonly 50 cm in length, are assayed for Zn, Pb, Cu, Ag, Fe, As, Cd, Ba and Hg. Specific gravity measurements are made for all assayed intervals because the wide variation in pyrite content throughout the deposit results in a specific gravity range of 2.8–5.3.

Background to the current system

At an early stage in the evaluation of the Lisheen mineral resource, the IPL Joint Venture purchased an Apple Macintosh version of the TECHBASE database management system. Initially TECHBASE was loaded on CSA's in-house Apple Macintosh network but problems with the software made it necessary to convert to a DOS-based system loaded onto a 386 IBM desktop computer.

TECHBASE places high demands on hard-disk access times and requires fast processing speeds, and within 6 months the drillhole database had rapidly increased in size to the point where the 386 computer was inadequate. TECHBASE was upgraded to a UNIX-based operating system housed in a 35 MIPS Solbourne 500 workstation which provided greater memory and processing speed together with multi-user/multi-tasking capabilities.

Currently at CSA two Solbourne Unix workstations provide a combined memory capacity of 3 Gbytes. The workstation is linked to IBM compatible and Apple Macintosh desktop computers by Ethernet providing the capability for it to function as a network fileserver. The Lisheen database is also integrated with regional geological, geochemical and geophysical data using a geographical information system (GIS) (Fig. 5).

Output devices at CSA include an A0-colour pen plotter, an A1-colour electrostatic printer, an A4-colour inkjet printer/scanner and laser printers. Besides TECHBASE and GIS systems, other ancillary software packages include AutoCad, Correldraw and Canvas drafting packages.

Lisheen database

Two different data input methods are used to load data into the Lisheen database. The first input route involves loading data from ASCII text files generated from software packages such as Lotus, Excel and Word. A second input mechanism is by a customized user interface through which data is typed directly into the database.

In TECHBASE, diamond-drillhole data are stored in columns called 'Fields' within files called 'Table'. Every table has a 'key' through which it is related or linked to the other tables in the database. This 'relational' aspect of the database means that it is necessary to enter an information item only once, with data spread across many Tables being simultaneously accessed for various module applications such as stick-section generation. One or more fields in a table can be selected as the 'key' to the table provided that every record in a table relates to a unique key value or combination of values. Tables are updated by loading new records 'targeted' at the unique key address. As an example some tables such as COLLAR have collar coordinate and elevation records which relate to a unique key value which in this case is the drillhole number. Other tables such as ASSAYS and LITHOLOGY have many records relating to the same drillhole number and in such cases two fields are required to define a unique address for each record in the table. In this case the key consists of the two key fields which store the drillhole number and the 'from' of the interval to which the data now relates.

If we consider an example from the ASSAY table where for a particular sample Zn, Pb, Cu, Ag, Cd, and Fe assay values are loaded into the respective element fields in the database together with the respective borehole number, sample number and sample interval (from,to). If additional assay values need to be loaded it is only necessary to specify a borehole number and a 'from' so that during the loading of new data it can be matched unambiguously with the correct data row.

Every field in the database has certain attributes assigned to it by the user which can be modified as required. Fields may be text, numeric or calculated and are assigned a maximum number of characters. Numeric fields can be assigned maximum and minimum values and degrees of resolution. Data which does not conform to a predetermined format cannot be loaded into the target field thus ensuring a 'first line of defence' in data validation.

Three or four digit code numbers are assigned to data items such as the type of lithology, type of dolomitic alteration and the degree of breccia development. This coding facilitates the filtering of data for efficient management, editing and validation of the Lisheen database.

During the early exploration stage of the Lisheen project a relatively simple coding system reflected a relatively simple geological database. A more comprehensive and detailed coding system was necessary due to the rapid increase in size of the drillhole data during the deposit evaluation stage and greater complexity in this data which reflected a better understanding of the geology, metallogenesis and structure. As examples of the increased complexity in the geology at Lisheen, nine facies have been identified within the Waulsortian Limestone and eight within the ABL. Black Matrix Breccia is coded according to whether it is clast or matrix supported and whether the predominant clast type is limestone, hydrothermal dolostone, regional dolostone or polymict.

The degree of detail in cross-sections needs to be flexible so that detailed continuity sections can be plotted at large scales of 1:500 and more synoptic structural sections can be plotted at smaller scales of 1:2500.

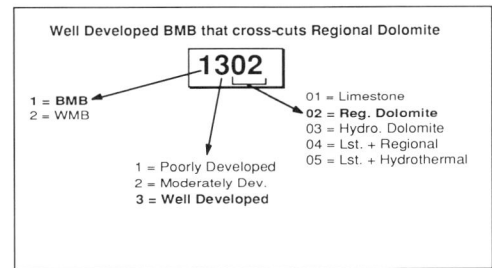

Fig. 6. Example of nested coding for Black Matrix Breccia data.

Nesting of coding facilitates greater flexibility in the extraction of data, for example a four-digit code for all types of BMB was used with the first digit identifying the information item as BMB, the second digit denoting the degree of development whether it is clast or matrix-supported and the last two digits indicating the lithology or lithologies of the predominant clasts (Fig. 6). Information can be extracted for reports or cross-sections at various levels of detail using data filters. If as an example it is only required to determine the total thickness of BMB in a borehole a filter such as:

Field	Argument	Value
BMB CODE	>	999
BMB CODE	<	2000

will exclude all non BMB data. If, however, one wanted to 'flag' where the location in the database where matrix supported BMB contains predominantly regionally dolomitized clasts, then a filter such as:

Field	Argument	Value
BMB CODE	=	1302

would be applied (Fig. 6).

Peripheral modules of TECHBASE are utilized for data output and analysis. As an example, the 'SECTION' module creates stick-sections, with data plotted down the hole as coloured bars, histograms, hatched bars or text. Multiple strands of information can be plotted alongside each drillhole.

The drillhole stick-section generated by TECHBASE is in the form of a metafile. This can either be plotted directly or exported into a CAD package such as AutoCad, where digitizing of interpreted cross-sections and graphical enhancement of the drawing takes place to create a finished cross-section (Fig. 3). The drawing exchange files (DXF) generated by TECHBASE can be operated in 'true' map coordinates which can be imported directly into AutoCad template drawing files which contain ordinance data and borehole locations plotted at the required scale.

Data can be composited in TECHBASE on code, interval or length composite options to calculate, for example, lithological thicknesses for isopach maps or assay composites for mineral resource estimations. To calculate the total thickness of Waulsortian limestone in a borehole it is necessary to aggregate the thicknesses of the Waulsortian facies intersected in that borehole. All Waulsortian facies have three digit code numbers beginning with the number '1' so that a filter can be applied to the lithologies which excludes all data outside of the Waulsortian code numbers. A 'Length' composite will aggregate all Waulsortian facies thicknesses and calculate a point value for a particular borehole which can then be used to generate an isopach of the Waulsortian limestone.

To composite assays it is necessary to flag the assays above a certain cutoff grade together with any subgrade assays required to yield a minimum mining width. The 'Code' composite is then used to aggregate and average the flagged assays for a particular borehole. If the composite interval is already known repetitive assay compositing using the 'Interval' composite option is automated and easy to use by utilizing a 'Technicn' programme where the only parameter substitution is for borehole number, 'from' and 'to'.

Geological composites are grade controlled and there is a sharp cutoff between economic and background Zn and Pb values in both the hanging wall and footwall. TECHBASE ore reserve facilities include 3D surface modelling, kriging and inverse and minimum curvature estimations. Data can be estimated into polygons, cells, or 3D blocks.

One of the most powerful and useful aspects of TECHBASE is its macro generating programme 'TECHNICN'. Runlogs (macros) can be created to run various task routines, such as stick-section generation and mineralized intersection composites. TECHNICN can be set up where menu parameters can be edited quickly allowing a user-friendly interface for non-TECHBASE users.

Conclusions

- The Lisheen computer system is flexible and effective in editing, interrogating and validating multi-sourced and multi-disciplined data.
- TECNICN macros with TECHBASE can be quickly modified using a customized user interface to produce maps and stick sections at the required scale.
- Faster extraction and processing of data increases the speed and accuracy of mineral resource calculations and processing of survey data.
- A coding system needs the flexibility to evolve as data becomes more complex and a nested coding system is required to extract data at various levels of detail.

References

DOYLE, E., BOWDEN, A. A., JONES., G. V. & STANLEY, G. A. (1992). The geology of the Galmoy zinc-lead deposits, Co. Kilkenny. *In:* BOWDEN, A. A.,

EARLS, G., O'CONNOR, P. G. & PYNE, J. F. (eds) *The Irish Minerals Industry 1980–1990*. The Irish Association for Economic Geology, Dublin.

HITZMAN, M. W., O'CONNOR, P., SHEARLY, E., SCHAFFALITZKY, C., BEATY, D. W., ALLAN, J. R. & THOMPSON, T. 1992. Discovery and geology of the Lisheen Zn-Pb-Ag prospect, Rathdowney Trend, Ireland. *In:* BOWDEN, A. A., EARLS, G., O'CONNOR, P. G. & PYNE, J. F. (eds) *The Irish Minerals Industry 1980–1990*. The Irish Association for Economic Geology, Dublin.

IVERNIA WEST PLC 1992. Annual Report. Limerick, Ireland.

Laboratory evaluation of kaolin: a case study from Zambia

C. J. MITCHELL

Mineralogy and Petrology Group, British Geological Survey, Keyworth, Nottingham NG12 5GG, UK

Abstract: Kaolin is principally used as a white pigment in the manufacture of paper and whiteware ceramics and in paints, rubbers and plastics. The desirable properties of kaolin in these end uses include chemical purity, high kaolinite content, fine particle size, euhedral kaolinite platelets, high brightness values and appropriate rheology. This paper outlines the laboratory evaluation of a kaolin from Chilulwe, near Serenje, Central Province of Zambia. The kaolin occurs in a hydrothermally-altered feldspar pegmatite within a granite-gneiss basement. Initial laboratory characterization of the kaolin showed it to consist mainly of microcline feldspar (80%) with 17% kaolinite and trace quantities of muscovite, beryl and tourmaline. A kaolinite concentrate, produced by wet screening and hydrocloning, contained 79–87% kaolinite, with a clay ($<2\,\mu m$) content of 58%, a brightness of 70–76% (86–87% on firing) and a viscosity concentration of 68%. Transmission electron microscopy showed the kaolinite to consist of rolled and hexagonal crystals. The results of this study showed that the Chilulwe kaolin, and by-product K-feldspar, have potential as a raw material for manufacture of ceramic products.

Kaolin is a commercial term used to describe a white clay composed essentially of the clay mineral kaolinite ($Al_2Si_4O_{10}(OH)_8$). The term is typically used to refer to both the raw clay and the refined commercial product. It is pricipally used in paper and ceramic products and to a lesser extent in paints, rubbers, plastics, and agricultural and pharmaceutical products. In paper, kaolin is used as a filler to reduce the cost of the wood pulp and as a coating agent to improve printing qualities such as smoothness, gloss and printability. In ceramic products, kaolin is used to confer both whiteness and good casting properties to the body. The properties of kaolin that determine its suitability for such applications include chemical purity, high kaolinite content, fine particle size, euhedral kaolinite platelets, high brightness values and appropriate rheology. The evaluation of kaolin for commercial use involves the detailed investigation of these properties in representative field samples.

Samples of the Chilulwe kaolin were evaluated in 1990 and 1991 (Mitchell *et al.* 1992) and this was followed up by a visit to the deposit in May 1992. The work described in this paper forms part of the BGS/ODA research and development project 'Minerals for Development' the aims of which include the encouragement of small-scale mining of indigenous mineral resources in less-developed countries. The two samples of Chilulwe kaolin were processed by hydrocycloning to produce kaolinite concentrates and these were then evaluated for their use as an industrial raw material. The evaluation described in this study is appropriate for a reconnaissance appraisal of a kaolin deposit such as might be carried out by a geological survey or mines department to give an indicator of commercial potential. It is not intended to match the highly specialized use-related testing carried out by suppliers in order to meet the rigorous specifications required by consumers in the paper and ceramic industries.

The Chilulwe kaolin deposit

The Chilulwe kaolin is located 40 km SE of Serenje in the Central Province of Zambia (Fig. 1) in rolling bush with occasional dambos (broad grassy depressions). The kaolin occurs in a hydrothermally-altered feldspar pegmatite within the basement, which is a sequence of biotite-rich granulites and migmatites, garnet–mica schists and granulitic quartzites. The pegmatite consists of a massive quartz core (2 m in diameter) surrounded by a zone of microcline feldspar with muscovite intergrowths. Its contact with the country rock is obscured by ramifying veins of fine muscovite and quartz.

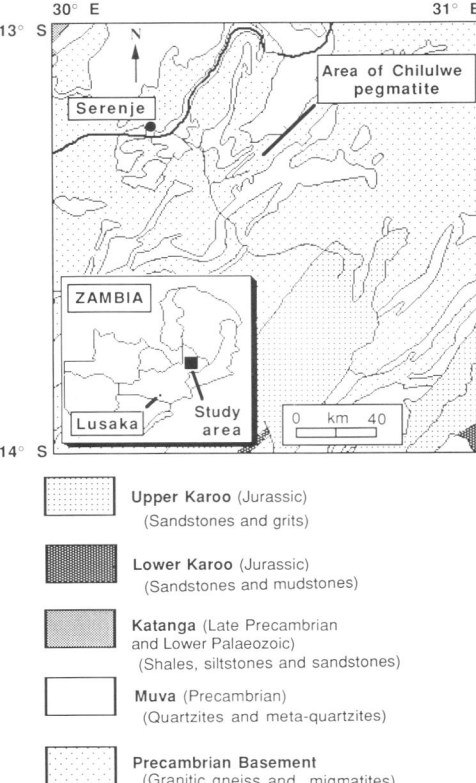

Fig. 1. Location and geology of the Chilulwe kaolin deposit, Serenje area, Central Province, Zambia (after Thieme & Johnson 1981).

The pegmatite is one of a series oriented roughly parallel to the regional north easterly strike, but which cut across it in places.

Pitting by the Zambian Mining Exploration Department (Minex) has confirmed the Chilulwe pegmatite to be at least 1 km long, 15 m deep and 10–15 m wide (thinning locally to a minimum of 5 m). The deposit is overlain by approximately two metres of lateritic clay (Figs 2 & 3). Samples for evaluation were collected from the altered feldspar zone of the pegmatite and this would be amenable to selective extraction as part of a small-scale mining operation. The feldspar has been kaolinized to a similar degree throughout and can be easily extracted.

Evaluation methods

Raw material characterization

Kaolin evaluation involves determination of mineralogical, physical and chemical properties relevant to potential applications of the clay. The mineralogy of the kaolin was determined by X-ray diffraction (XRD) using a Phillips PW 1700 X-ray diffractometer with Co-Kα radiation at 45 kV and 40 mA. A randomly-oriented powder mount was scanned over the range of 3–50°2θ. An oriented mount of $<2\,\mu$m material, obtained by sedimentation, was also analysed by XRD. Kaolinite content was determined by thermogravimetry (TG) using a Stanton Redcroft TG 770 thermobalance with a sample size of 10 mg and a heating rate of 50°C per minute. On heating, kaolinite loses 14% of its mass due to dehydroxylation between 500–600°C.

Fig. 2. Chilulwe kaolin deposit, Copperbelt province, Zambia.

Mineral processing

Kaolinite tends to concentrate in the $<2\,\mu$m, or clay, fraction and therefore the concentration of kaolin is essentially a size fractionation process. A combination of wet screening and hydrocycloning is commonly used in laboratory investigations of kaolin. Wet processing is preferred as it is a more efficient method of concentrating kaolin and is commonly used commercially. Initially the kaolin was attrition scrubbed (two hours at 250 RPM using a paddle

Fig. 3. Sampling of Chilulwe kaolin (note lateritic staining on kaolin).

mixer) and then wet screened down to 63 μm (240 mesh). The <63 μm kaolin was processed as a suspension (5–10% w/v in water) using a 30 mm glass hydrocyclone (Liquid–Solid Separations Ltd) of conventional design (Bain & Morgan 1983). The kaolin suspension was pumped into the hydrocyclone at about 4 litres per minute at a pressure of 2.5 kg cm^{-2} using a small rotary pump. The hydrocyclone split the kaolin at about 10 μm (the 'cut point') into a coarse-grained product (the 'underflow') and a fine-grained product (the 'overflow'). The overflow was repassed to maximize the kaolinite content and subsequent evaluation was carried out on the final overflow product.

Product evaluation

The processed kaolin was evaluated using a range of techniques. Mineralogy was determined by XRD and TG analysis. Particle-size distributions of samples dispersed in a solution of sodium hexametaphosphate (Calgon) were determined with a Micromeritics X-ray Sedigraph. Major element chemistry was determined by X-ray fluorescence using a Phillips PW 1480 X-ray fluorescence spectrophotometer.

The brightness (or 'whiteness', a qualitative measurement of colour) was measured by EEL reflectance spectrophotometer using filtered light (4700, 4900, 5500 and 5800Å) which was reflected from the surface of a powdered sample and compared to that from a barium sulphate (DIN 5033) calibration standard. Brightness was also measured of a sample of the kaolin that had been fired to 1050°C for 2 hours.

The viscosity (the resistance of a liquid to flow) of slurry made from kaolin was determined using a Brookfield RVF 100 low-shear viscometer. Several properties were measured, using tests adapted from English China Clays Ltd (ECC, undated) for paper-coating grade kaolin including flowability (solids content at which a slurry begins to flow), deflocculant demand (amount of deflocculant, Calgon, required to obtain minimum viscosity) and viscosity concentration (solids content of a fully deflocculated kaolin slurry with a viscosity of five poise at 22°C). This last parameter, which is the most significant, is equivalent to the casting concentration for ceramic kaolins. Viscosity measurement of ceramic-grade kaolin is carried out using a similar set of tests using sodium silicate as the deflocculant.

The kaolinite crystal morphology was examined using an AEI transmission electron microscope (TEM). The samples were 'metal shadowed' (Pt deposited at 5°) in order to determine particle thickness and aspect ratio (diameter/thickness).

The techniques used are appropriate for evaluating kaolin for both paper and ceramic

Table 1. *Mineralogy and processing characteristics of Chilulwe kaolin.*

Product	Mineralogy		Particle-size < 10 μm	< 2 μm	Kaolinite recovery
Raw material	K-feldspar	80%	~14%	~6%	100%
	Kaolinite	18%			
	Mica	2%			
Wet screening products					
> 63 μm	K-feldspar	95%	na	na	22%
	Kaolinite	5%			
	Mica	<1%			
< 63 μm	Kaolinite	63–68%	54%	22%	78–96%
	K-feldspar	~30%			
	Mica	~5%			
Hydrocyclone products					
Underflow	Kaolinite	60–62%	19–33%	3–10%	58–60%
	K-feldspar	~35%			
	Mica	~5%			
Overflow	Kaolinite	79–87%	96–97%	55–58%	20–30%
	K-feldspar	~10–15%			
	Mica	~5%			

Mineralogy of the kaolin products was determined by X-ray diffraction, thermal analysis and binocular microscopy. All mineral contents are weight percentages. The relationship between the particle-size distribution of the hydrocyclone feed material (< 63 μm wet screening product) and the hydrocyclone products is distorted by the breakdown of clay aggregates due to the high shear stresses generated during processing. Recovery is the amount of kaolinite present in the product as a proportion of the kaolinite content of the raw material. na, not analysed.

applications, with the specific addition of particle-shape determination for paper-coating and fired brightness for the ceramic evaluation. This paper does not cover all aspects of kaolin evaluation; other tests that could be carried out include modulus of rupture (tensile strength of ceramic test pieces), surface area measurement, and ceramic forming (drying shrinkage) and fired properties (changes in shrinkage, porosity, specific gravity and bulk density with temperature gradient firing). Also the sieve residues, from wet screening, can be examined by binocular microscope and SEM to provide useful information on mineral processing efficiency and petrogenesis of the kaolinite-bearing rock.

Properties of commercial kaolin and feldspar

Paper-grade kaolin

Paper-coating grade kaolin requires a very high kaolinite content (>93%), a high brightness (>78% 4570Å), a very fine particle-size (>78% <2 μm), flat euhedral (or plate shaped) kaolinite crystals and strictly defined rheological properties. This will ensure that the paper produced will be smooth, glossy and have good printability. Plate-shaped crystals are required as they will naturally orient parallel to their basal spacing to form a smooth, even coating. Roll-form kaolinite will tangle in suspension (adversely affecting viscosity) and lead to an irregular, uneven coating. Paper-filler grade kaolin also requires a high kaolinite content (>89%), high brightness (>70% 4570Å) and fine particle-size (>30% <2 μm). Deleterious impurities include quartz and feldspar (which are abrasive and result in excessive wear in paper-coating equipment), iron oxides (which reduce brightness) and other clays (such as smectite which will drastically increase viscosity).

Ceramic-grade kaolin and feldspar

Ceramic-grade kaolin requires a high kaolinite content (>87%), a fine particle-size (>39% <2 μm), a high fired brightness (>86% 4570Å) and strictly defined rheological properties. Ceramic forming properties and firing properties are important. Deleterious impurities include alkalis (which affect the vitrification temperature) and iron and titanium oxides (which reduce the fired brightness).

Ceramic-grade feldspar requires a high silica and alumina content, a high K_2O content and a low Fe_2O_3 content.

Table 2. *Properties of Chilulwe kaolin, Zambia and commercial kaolin products.*

Property	Chilulwe kaolin	Paper coating kaolin	Paper filling kaolin	Ceramic kaolin
Kaolinite content	79–87	93–100	89–97	87–97
Particles <10 μm	96–97	98.5–99.9	59–97	82–98
Particles <2 μm	55–58	78–97	30–78	39–70
Brightness				
Unfired	70–76	78–93	70–90	na
Fired (1050°C)	86–87	na	na	86–91
Viscosity concentration	68	67–70	na	58–64 (casting concentration)
Aspect ratio	6.1	5–22	na	na
Chemistry				
SiO_2	46.24	45–49		47–50
TiO_2	0.03	0.03–1.5		0.02–0.06
Al_2O_3	36.99	36–38		34–38
$Fe_2O_3{}^t$	0.23	0.5–1.15		0.4–1.0
MnO	0.004	na		na
MgO	0.03	0.13–0.25		0.2–0.3
CaO	0.17	0.04–0.07		0.02–0.10
Na_2O	0.07	0.08–0.14		0.1–0.15
K_2O	1.55	1.1–2.84		0.8–4.0
P_2O_5	0.08	na		na
LOI	13.19	11.19–14.3		10–13
Total	98.58	100.00		100.00

Kaolinite content, particle-size, viscosity and chemistry are all weight percentages. Brightness values are percentage reflectance using a $BaSO_4$ calibration standard. Paper coating and filling kaolin are represented by one set of chemical analyses. na, not analysed.

Results

Raw material

The Chilulwe kaolin samples consist predominantly of microcline feldspar (80%) with 17% kaolinite and small amounts of muscovite mica, beryl and tourmaline. The XRD trace of the <2 μm material showed the kaolinite to be poorly-ordered, and quartz and gibbsite were also detected.

Processing

Wet screening of the kaolin removed most of the feldspar, which was mainly >63 μm in diameter and concentrated the kaolinite into the <63 μm fraction (increasing the kaolinite grade to 63%). Hydrocycloning of the <63 μm kaolin further increased the kaolinite content although recovery (i.e. the amount of kaolinite in the product as a proportion of that present in the raw material) of kaolinite to the overflow product was low, only 20 to 30%. The remainder of the kaolinite was present in the underflow product and sieve residue material (Table 1).

Kaolin and feldspar products

The Chilulwe kaolin product (hydrocyclone overflow) has a high kaolinite grade (Table 2). The kaolin product contains only small amounts of alkalis, TiO_2 and Fe_2O_3. It has a fine particle-size, mainly <10 μm in diameter and a high clay, <2 μm, content. The brightness of the kaolin was not particularly high but after firing it increased to >80% (4570Å). The flowability of the kaolin was measured as 65% solids content, the deflocculant demand was 0.23 g Calgon per 100 g kaolin and the viscosity concentration was 68% solids content. TEM examination (Fig. 4) revealed subhedral, rounded plate and roll-form ('halloysitic') kaolinite crystals. The kaolin particles ranged in diameter from 0.1 μm to 0.44 μm (average of 0.26 μm) and 0.02 μm to 0.11 μm in thickness (average of 0.05 μm), with an average aspect ratio of 6.1.

The wet screening residue contained 95% K-feldspar, with a brightness of 86% (4570Å) and a high silica and alumina content, a high potash content and a low iron content (Table 3).

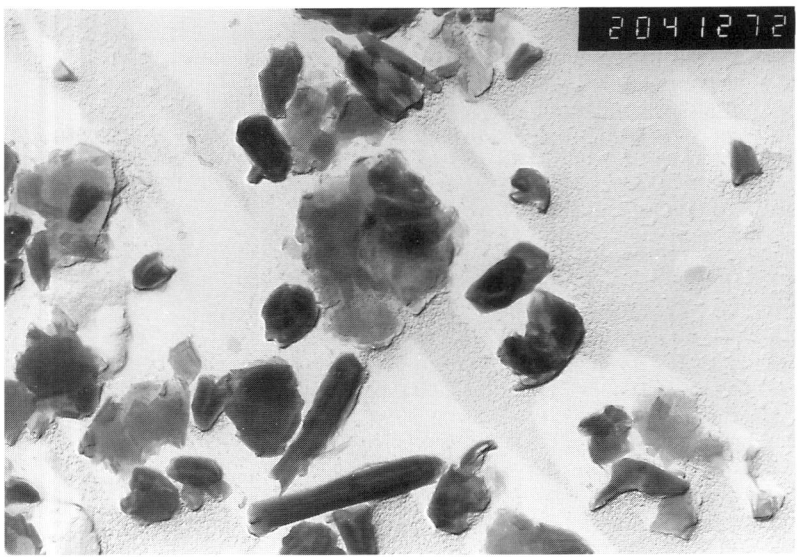

Fig. 4. TEM photomicrograph of kaolinite crystals, shadowed for aspect ratio determination (magnification × 27 300).

Table 3. Properties of Chilulwe feldspar, Zambia and commercial feldspar product.

Chemistry	Chilulwe feldspar	Ceramic feldspar
SiO_2	64.32	66–67
TiO_2	0.01	na
Al_2O_3	19.14	18.2–18.6
$Fe_2O_3 t$	0.04	0.03–0.1
MnO	0.003	na
MgO	0.09	Trace
CaO	0.03	0.1–0.15
Na_2O	0.71	3–3.3
K_2O	15.27	1.3–11.5
P_2O_5	0.24	na
LOI	0.28	0.01–0.3
Total	100.13	100.00

All data are weight percentages. na, not analysed.

Kaolin use evaluation

Paper-grade suitability

The Chilulwe kaolin has a lower kaolinite content, a lower brightness and is coarser than that required for paper-coating grade kaolin. Also it contains roll-form kaolinite crystals which are not suitable for paper-coating. The kaolin has all the properties required for use as a paper-filler, except that it does not contain enough kaolinite. This kaolin is therefore not suitable for use in paper.

Ceramic-grade suitability

The Chilulwe kaolin contains sufficient kaolinite, has a fine enough particle-size, and adequate fired brightness and rheological properties for a ceramic-grade kaolin. It also has low Na_2O, K_2O, Fe_2O_3 and TiO_2 contents. This kaolin has potential as a ceramic-grade kaolin; however, further testing, including modulus of rupture and fired properties, would be required to confirm this.

Feldspar suitability

The residue from wet screening of the kaolin is pure K-feldspar that has potential for use as ceramic-grade material, as it has a high brightness and suitable chemistry.

Further processing

The results indicate that the Chilulwe kaolin is unsuitable for use in paper applications but it may meet the requirements for ceramic-grade kaolin. However, it is anticipated that with further upgrading the properties of this kaolin would be improved. Removal of a larger proportion of the material coarser than $2\,\mu m$ from the kaolin, during hydrocycloning would have the effect of increasing its kaolinite content, clay content and brightness values.

Viability

The local demand for kaolin and feldspar would be led by the Zambian ceramics industry (Moore Pottery in Lusaka and Zambia Ceramics in Kitwe). The Chilulwe pegmatite would, as estimated by Minex, yield upwards of 12 000 tonnes of refined kaolin and 250 000 tonnes of feldspar. Currently such deposits are mined by the consumers to replenish stocks and little, if any, processing is carried out. Several kaolinized pegmatites are known to exist in the region and their combined output would more than satisfy demand. The viability of such deposits would depend upon economic considerations such as the establishment of a central processing facility and improved access to the deposits.

Conclusions

The laboratory evaluation of a kaolin for use in paper and ceramic applications involves the determination of its mineralogical, chemical and physical properties. Results of the laboratory processing and testing show that the Chilulwe kaolin is unsuitable for use as a paper-coating and filler grade kaolin. It has a relatively low kaolinite content, contains too much coarse material and has a relatively low brightness. However, the material does have potential for use as a ceramic-grade kaolin. Additional refining of the kaolin, using high-performance hydrocyclones, would improve its properties and increase its commercial potential for use in ceramic applications. The by-products from processing include a K-feldspar potentially suitable for use as a ceramic raw material, There is demand for kaolin and feldspar from the local ceramics industry.

Thanks to: A. Banda & J. Malindi, Zambian Geological Survey Department, S. Simasiku, Zambian Mining Exploration Department, E. Roberts, EM Unit, Leicester University, P. H. Miles, A. G. Scothern, M. N. Ingham & A. S. Robertson, Analytical Geochemistry Group, BGS, D. J. Morgan, D. J. Harrison, A. J. Bloodworth & S. D. J. Inglethorpe, Mineralogy and Petrology Group, BGS. This paper is published by permission of the Director, British Geological Survey.

References

BAIN, J. A. & MORGAN, D. J. 1983. Laboratory separation of clays by hydrocycloning. *Clay Minerals*, **18**, 33–47.

BRISTOW, C. M. 1987. World kaolin—genesis, exploitation and application. *Industrial Minerals*, **238**, 45–59.

BUNDY, W. M. & ISHLEY, J. N. 1991. Kaolin in paper filling and coating. *Applied Clay Science*, **5**, 397–420

ECC. Test methods P106—Viscosity Concentration. *In: Products for the Paper Industry*. ECC International.

ECC. Test methods P107—Deflocculation Demand. *In: Products for the Paper Industry*. ECC International.

HARBEN, P. W. 1992. *The Industrial Minerals Handy-Book*. Metal Bulletin.

HIGHLEY, D. E. 1984. *China clay*. Mineral Dossier No. 26, Mineral Resources Consultative Committee, HMSO, 22–29.

MITCHELL, C. J., BRIGGS, D. A. & BLOODWORTH, A. J. 1992. Mineralogy and technical appraisal of kaolinite-bearing rocks from Zambia. *Zambian Journal of Applied Earth Sciences*, **6**, 32–45.

THIEME, J. G. & JOHNSON, R. L. 1981. *Geological map of the Republic of Zambia—1:1,000,000*. Geological Survey Department of Zambia.

Cia Minera Los Pelambres: a project history

J. O'LEARY

Montagu Mining Finance Ltd, 10 Lower Thames Street, London EC3R 6AE, UK

Abstract: The first documented discovery of copper mineralization at Los Pelambres was by William Braden, the American geologist, who noted and explored the area in 1920. He drove a series of short adits into the canyon wall, but these did not penetrate beyond the leached cap of the deposit.

After the work of Braden, there was no further activity until 1955 when two Chilean companies, Minera Protectora and Minera Los Pelambres, staked claims in the area. The Protectora and Los Pelambres claims were surveyed in 1960 and 1970 respectively.

The United Nations first became involved in 1964 with a surface examination of the property. This was followed by a further examination by the Chilean Institute of Geological Investigation financed by Corfo (a Chilean government organization) in 1967 and 1968.

Beginning in 1969, a partnership arrangement between the United Nations and Enami (the Chilean government organization dedicated to small-scale mining) undertook serious exploration of the property including diamond drilling programmes. This work developed a combined probable and possible ore reserve of 428 Mt of material with a grade of 0.78% copper and 0.033% molybdenum. The drilling was confined largely to the valley bottom. This work was finished in 1971. In 1978 an international licitation for bids on the claims prompted examination by Anaconda geologists. Anaconda (then a subsidiary of Atlantic Richfield) was successful in the bidding and in 1983 they produced a study which examined the alternative means of exploiting the deposit. The conclusion of this study was that it was technically feasible to mine the deposit as an open pit operation at a scale of 60 000 tpd using conventional froth flotation techniques to recover the copper minerals. The study took three and a half years to complete at a cost of $59 M. The study, however, concluded that although the project could produce a satisfactory rate of return, the project economics were marginal in the light of the copper prices projected at that time. In this study the 'geological reserves' were stated to be in excess of 3 Gt at a grade of 0.62% copper. Shortly thereafter Atlantic Richfield made a strategic decision to divest itself of all its mining assets and in 1986 Anaconda Chile was acquired by the Luksic Group.

Montagu Mining Finance (MMF) were appointed, to advise the Luksic group, in late 1987. It was MMF's view that the opportunity might exist to finance the project using a debt for equity swap under Chile's Chapter XIX rules. At that date the largest debt swap, in the mining industry, that had been completed in Chile was for $35 M and it was MMF's view that the Chilean central bank would not countenance a swap much larger than this. MMF decided therefore to examine developing the project with a capital budget limited to what might be achievable through a debt swap, and in early 1988 consultants were appointed to examine the feasibility of mining the deposit on a small scale.

As part of their brief, the consultants examined the distribution of copper throughout the deposit and concluded that there exist two high grade zones, the East Zone and the West Zone, which together contain 40 Mt at an average grade of 1.52% copper. It is these higher grade zones that have been the target for initial mining and at present the mine is being operated as a 5000 tpd underground operation. Initial operation and addition drilling on the East Zone has confirmed the feasibility study estimates. However, a further high grade zone, The Central Zone, has now been confirmed by drilling and trial mining and a study has indicated the potential for an increase in production to 15 000 tpd by mining the West and Central Zones by open pit methods.

Although copper mineralization was discovered in the area in the early 1900s, the deposit at Los Pelambres was not studied in any detail until the mid-1960s when the UN conducted exploration drilling. In 1983 Anaconda Chile (then a subsidiary of the American oil company Atlantic Richfield) produced a study (Anaconda Minerals Co. 1983) which examined the alternatives for exploiting the deposit. The conclusion of this study was that it was technically feasible to mine the deposit as an open pit operation at a scale of 60 000 tpd using conventional froth flotation techniques to recover the copper minerals. The study took three and a half years to complete at a cost of US$59 M. Shortly thereafter Atlantic Richfield made a strategic decision to divest itself of all its mining assets and in 1986 Anaconda Chile was acquired by the Luksic group.

The Midland Bank group's introduction to the project was in 1987 when Montagu Mining Finance (MMF) were appointed to help Antofagasta Holdings plc (a London listed company controlled by the Luksic group) to develop ways to exploit the deposit. Following a feasibility study, conducted by RTZ Consultants (1988), MMF were appointed to raise the capital required for the project development. MMF invited the Midland Bank to underwrite a financing using a debt/equity swap under Chapter XIX of Chile's foreign investment rules. In the event Midland decided to provide all of the financing and MMF were appointed to manage the ongoing investment. Documentation was completed in July 1989 and in October 1989 a debt/equity swap under Chile's Chapter XIX was approved by the Central Bank. Cia Minera Los Pelambres Ltd, a single purpose company, was created for the development of the Los Pelambres mine.

The initial investment into the project was made in November 1989 when US$ 62.8 M of Chile sovereign debt was swapped for US$ 52.8 M in local currency. This amount together with interest received and foreign exchange gains was used to fund the capital expenditure on the project.

After a two year construction period, production from the mine commenced in January 1992 and the first copper concentrate was shipped in April 1992.

As currently configured the mine will operate at 5000 tonnes per day (tpd) for over 20 years and produce concentrates in excess of 60 000 tonnes per year with about 23 000 tonnes of contained copper metal. Additionally, gold, silver and molybdenum will be produced as by products. Studies have indicated that there is considerable scope for expansion which, within the present infrastructure alone, could be up to three times the present operation i.e. 15 000 tpd.

Chile: country profile

Political profile

Chile emerged from the regime of general Pinochet (who was retained and continues as the head of armed forces) in December 1989 and began the road to democracy under President Patricio Aylwin, who was elected for a period of four years. This period is transitional. President Frei was elected in December 1993, for four years. Chile has, in the period since 1989, established a new democracy with a strong Executive, Legislature, Senate, Chamber of Deputies and an independent judiciary. This new unitary republic would follow the Chilean republican tradition keeping the well established legal system along French and Spanish structures. The present legislation is well developed to accommodate the needs of the mining industry as Chile has had a long mining tradition.

The transition to democracy has been surprisingly smooth. The present government comprises a coalition of centre-left parties which, while maintaining the market-oriented economic policies of the previous regime which have been so successful in Chile, has been more active in distributing their benefits to poorer parts of the population. Taxes have been increased modestly and spending on social and infrastructure projects raised. Workers' rights have also been improved. The latest step in the process of political liberalization came in June 1992 with the first municipal elections in over twenty years.

Economic profile

Chile's GDP has grown by 55% since the 1982/83 recession—an average annual growth rate of over 5.5% compared with population growth of 1.7% pa. Nevertheless, with a GDP of US$30 G, the Chilean economy remains one of the smaller South American economies. Copper remains Chile's largest single export, having consistently provided 40–50% of exports by value through the 1980s and into the 1990s. Chile is estimated to have one quarter of the world's copper reserves and is the world's largest producer. The two largest mines, Chuquicamata and El Teniente, together provide 70% of Chile's output and dominate the state copper sector, organized under Codelco. Output from the state sector has been falling recently, however, due to declining ore grades and technical problems. Major investment will be required to reverse this trend. By contrast, output from the private copper sector has increased strongly after foreign investment regulations were eased in 1983. La Escondida, which came on stream in the 1990s, is the largest of the private mining projects which together with a number of other projects will take the private sector production to almost 50% of the total in the next few years.

Apart from mining, the other sectors with a significant state presence are oil and petrochemicals and some transport and utilities. In general, however, Chile's public sector has been drastically reduced by an aggressive privatization programme since 1985, in parallel with a successful debt-reduction policy. Public finances are in a good state due to comprehensive tax reforms and strict expenditure control, with overall budget surpluses since 1988. The finances of the Central Bank are less robust due to substantial non-performing loans acquired from commercial banks as part of a banking rescue in the early 1980s. After several years of real exchange rate depreciation starting in 1982, the exchange rate strengthened in real terms last year due to the strength of capital inflows attracted by Chile's buoyant economy and relatively high interest rates. In order to alleviate upward pressure on the exchange rate the authorities have taken steps to liberalize capital outflows, thus continuing the process of liberalization and deregulation of the economy, begun in the mid-1970s. With average tariffs of only 11% and foreign trade comprising over a quarter of GDP, the Chilean economy is one of the most open in Latin America. In 1991 a free trade agreement was signed with Mexico with the objective of eliminating most tariffs by 1998. Similar agreements were signed during April 1993 with the United States, Argentina, Venezula and Colombia. The external debt burden, once one of the heaviest in Latin America, has been reduced to one of the lowest. Chilean private investors have begun to access international capital markets on a voluntary basis although most financing needs are satisfied domestically. Reflecting the growing strength of the private sector and Chile's prospects in the medium-term, the market capitalization of the stock market virtually doubled in US dollar terms in 1991.

After a policy induced slowdown in 1990, designed to curb serious overheating of the economy, growth resumed at 6.0% in 1991, and at 10.4% in 1992. Private investment was very strong, responding to falling interest rates and inflation. The latter was 13% in 1992, down from 27% in 1990 and the lowest since 1986. Private consumption responded to significant real wage growth earlier in the year and public spending continued to accelerate, reflecting the new priorities of the democratic government. The trade surplus in 1992 was US$ 0.7 billion with exports growing by 12% in that year despite a fall in the value of copper exports.

The strength of capital inflows has continued to pose a dilemma for the authorities during 1993. Resisting the upward pressure on the peso has generated excess liquidity which the authorities have been unable to absorb for fear of pushing interest rates too high. They have responded instead with a combination of measures to discourage capital inflows and encourage capital outflows. Chilean banks are now able to use their US$ deposits to finance trade with third countries and Chilean pension funds are now able to invest up to 1.5% of their assets overseas. 1991 saw a surge in Chilean investment in other Latin American markets. Nevertheless, a further 5% revaluation of the exchange rate was required in January 1992. This helped to attain their target of lower inflation. However, the continued buoyancy of non-copper exports suggests that for the time being the adverse effect of the stronger exchange rate has been outweighed by the effects of investment in new capacity, attracted by the combination of Chile's natural resources and positive policy environment.

Tax

In 1989 the transitional government of President Aylwin and Finance Minister Alejandro Foxley, temporarilly raised tax levels to assist the new democracy, the first category tax was raised from 10% to 15% and VAT from 16% to 18% until 1994. The new democracy has, however, welcomed foreign investment and is maintaining a free market economy. Rules, laws and statutes are used as guide-lines and exemptions are frequently negotiable to accommodate new business or project expansions.

Foreign investment

Decree Law No. 600, Chapter XIV and Chapter XIX are the statutes governing foreign investment and repatriation of funds out of Chile. Chile's debt rescheduling programme has been enormously successful and the booming economy fuelled by political reform has caused the value of Chilean debt to rise from about 60 cents at the time the Los Pelambres investment was made to over 90 cents in the dollar at present. At these levels most foreign banks have chosen to hold their Chilean paper in preference to transference to equity under Chapter XIX regulations

Foreign exchange is monitored by The Central Bank which maintains the exchange rate within certain margins. Two markets are used for currency transactions, the 'formal' and 'informal' markets; liquidation of foreign currency and remittance of profits are through the formal market. The foreign investment statute provides for the registration of foreign capital.

The stock market

The stock market has been experiencing a bull market over the last few years. Between 1985 and 1991 the market capitalization grew by almost 800% to US$ 25 G. Underpinning the bullish sentiment is the position of the Chilean pension funds in the market. Currently all employees must pay 14% of their earnings into retirement funds. Between 1981 and 1990 pension funds have increased at a compound rate of 47.4% real. These funds are allowed to invest up to 30% of their value in the equity market. The future of the Chilean stock market under these conditions looks well supported.

Despite the large gains seen on the Chilean Bourse, there has been a reluctance by industry to list on the market. Strict regulations imposed on companies accessing funds in this way have dissuaded many and others fear their stock may become illiquid as only 50 of the 230 stocks are normally actively traded. Of these stocks a select 10–15 companies make up the bulk of issues traded each day. Many of these listings are recent privatizations by the new democracy, encouraging wide employee share ownership. Many stocks are only part floated as owners have used the Bourse to improve their credit rating rather than truly come to the market.

Chilean equity is becoming more expensive but shares still outperformed all other investment instruments in 1992. Some listed companies riding on the economic boom showed a 30% real increase in profits during 1992. Chile's economic stability makes Santiago's Bolsa one of the strongest within the emerging markets in Latin America with a total of thirteen foreign investment funds now actively bringing in about US$400 M in the last two years.

Cia Minera Los Pelambres current operations

Project history

The first documented discovery of copper mineralization at Los Pelambres was by William Braden, the American geologist, who noted and explored the area in 1910. He drove a series of short adits into the canyon wall, but these did not penetrate beyond the leached cap of the deposit.

After the work of Braden, there was no further activity until 1955 when two Chilean companies, Minera Protectora and Minera Los Pelambres staked claims in the area. The Protectora and Los Pelambres claims were surveyed in 1960 and 1970 respectively.

The United Nations first became involved in 1964 with a surface examination of the property. This was followed by further examination by the Chilean Institute of Geological Investigation financed by Corfo (a Chilean government organization) in 1967 and 1968.

Begining in 1969, a partnership arrangement between the United Nations and Enami (the Chilean government organization dedicated to small- and medium-scale mining) undertook serious exploration of the property including diamond drilling programmes. This work developed a combined probable and possible ore reserve of 428 Mt with a grade of 0.78% copper and 0.033% molybdenum. The drilling was confined largely to the valley bottom. This work was finished in 1971. In 1978 an international licitation for bids on the claims prompted examination by Anaconda geologists. Anaconda (then a subsidiary of Atlantic Richfield, the American oil company) was successful in the bidding and in 1979 they purchased the claims from a group of private Chilean investors.

Shortly after the acquisition, Anaconda improved the road to the site and began a diamond drilling exploration programme which was completed in mid-1981. Anaconda also formed a Chilean subsidiary, Anaconda Chile SA, to explore and develop the property. Authorization was received from the Chilean government to invest up to US$1.5 G should the project prove to be feasible. In early 1980, a project team was assembled to specify, supervize and coordinate various engineering studies and prepare a feasibility report. The feasibility report was completed in June 1983 and to that date Anaconda had spent $59 M and demonstrated a geological reserve of over 3 Gt and grade of 0.63% Cu at a cut off grade of 0.4% Cu.

Atlantic Richfield made a strategic decision to divest itself of its mining assets and Anaconda Chile was bought by the Luksic Group in 1986 reportedly for about $6 M.

General features

Location and access. Los Pelambres is situated in Choapa Province of the IV Region about 200 km from the country's capital Santiago (Fig.

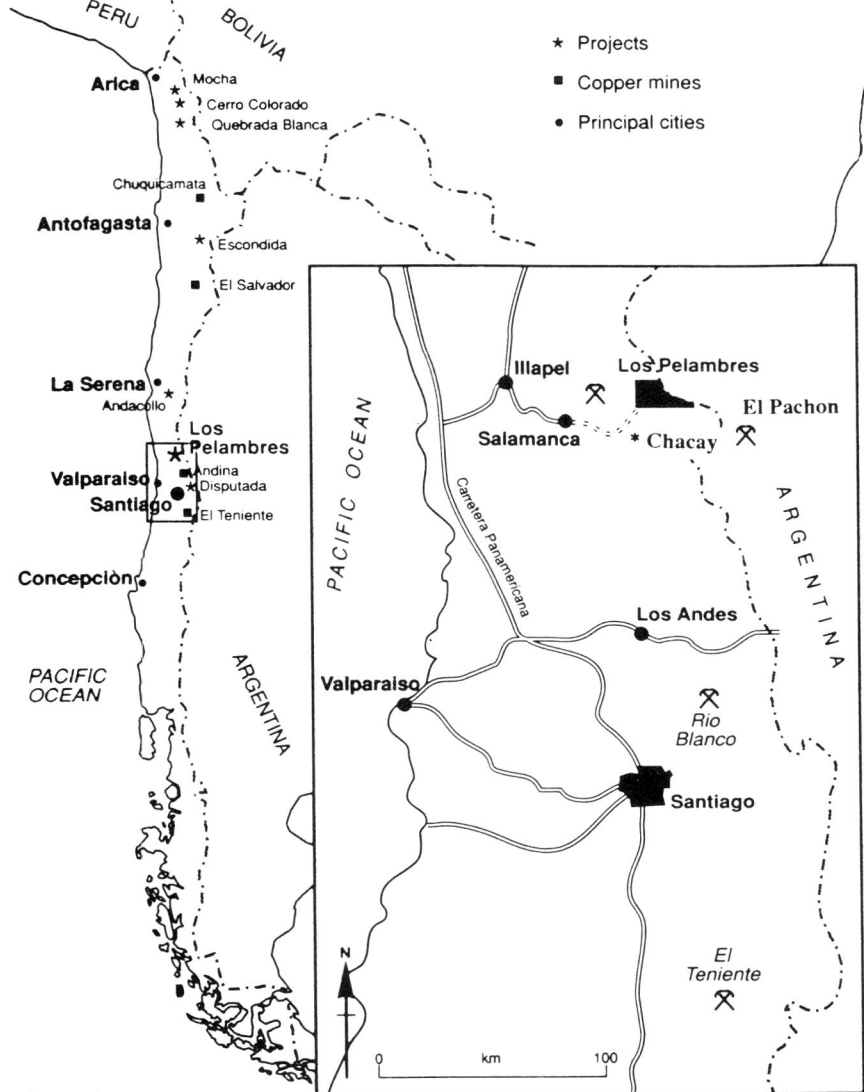

Fig. 1. Location and access.

1). The deposit is located in steep mountainous terrain at the head waters of the Los Pelambres river, a tributary of the Choapa River. It is close to the border with Argentina, separated by a high ridge which forms the continental divide.

The nearest town is Salamanca which has a population of about 10 000 and is 45 km to the west of the deposit. Salamanca is the seat of local government. Access is via a paved road to Salamanca and then on a good to moderate dirt road to the concentrator site at Chacay. Chacay is at an elevation of 1600 m and this is where the mine offices and the bulk of the accommodation are located. From Chacay to the Los Pelambres mine site is via a dirt road. The road climbs steeply to an elevation of 3000 m to the mine site, where canteen and accommodation for the mine staff are located.

Climate. A weather station has been operated at the mine area since 1980. The climate in the region is Mediterranean with most of the precipitation falling in the winter months of May, June, July and August. The remainder of the year is normally dry. At the mine site precipitation falls as snow with snowfalls in excess of 2 m having been recorded. There is an avalanche hazard in the mine area because of the steep terrain. The locations of the crusher and mill sites have been chosen to minimize this potential hazard and dykes have been built to protect the campsite. An avalanche control system has been installed.

Seismicity. The central part of Chile has frequent and sometimes strong seismic activity. Within the last 100 years there have been major earthquakes, measuring up to 8.0 on the Richter Scale as close as 100 km from the project area. With this in mind the buildings in general and the tailings dam in particular have been built to withstand such seismic activity and a programme has been developed with the University of Chile for seismic monitoring.

Geology

Regional. The Los Pelambres porphyry copper deposit is related to a late Tertiary multiple-phase quartz diorite stock which intruded Cretaceous and Tertiary stratified volcanic rocks (Faunes *et al.* 1992). A partial cover of non-consolidated Quaternary sediments also exists. The Cretaceous sequence (the Los Pelambres Formation) contains intermediate lavas, breccias and volcanoclastic rocks interbedded with marine sediments which strike in a northerly direction and have a sub-vertical dip.

Overlying the Los Pelambres Formation are the upper Cretaceous–lower Tertiary Vinita and Los Elquinos Formations which consist of andesites flow breccias and tuffaceous rocks of intermediate composition. These units strike north–south and dip to the West.

East of Los Pelambres, along the border with Argentina, a flat-lying sequence of intermediate volcanic rocks represents the upper Tertiary Farellones Formation. This sequence in general strikes N25W and dips slightly to the west.

The most prominent regional structure is the north–south Pocuro Fault Zone. This occurs west of Los Pelambres, and is not genetically related to the deposit. There are other north–south normal faults and NE and NW trending secondary structures.

All the units have been eroded by glacial action which has produced U-shaped valleys and moraine deposits.

Local geology. A north–south elongated quartz-diorite stock intrudes a sequence of dark grey to black andesitic lavas. Relicts of a quartz feldspar porphyry occur in the contact zone which are older than the quartz-diorite.

A body of breccia occurs in the central portion of the stock together with porphyry dykes. The latter are oriented NE and have peripheral hydrothermal breccias containing high grade copper and minor Mo–Au–Ag mineralization. Radiometric dating of the rocks indicates an age of 9.5–10.2 Ma. The intrusive complex covers an area of 7 km^2 and is the principal host for mineralization. Typically the quartz-diorite is a light to medium-grey, medium-grained, sub-equigranular rock with disseminated biotite and biotitized hornblende. Locally, a porphyritic and a fine-grained phase are present. To the south and east the mineralized zone is thought to be continuous with the El Pachon deposit, in the Republic of Argentina and about 7 km distant.

Primary minerals are chalcopyrite, pyrite, bornite, and molybdenite with chalcocite being dominant in the enriched zone.

Reserves

Anaconda study July 1983. The reserve estimates, at the time that Anaconda Chile produced their feasibility study, were based almost entirely on information derived from diamond drill holes, with minor input from exploration tunnels and raises. At that time 211 diamond drill holes had been completed for a total of 62 625 m of drilling. These were calculated both manually and via a kriged block model to give the 'geological reserve' (sic). The kriged model was used to generate a series of mineable reserves. These are summarized in Table 1.

Table 1. *Summary of mineable reserves*

	Cut off	Millions of tonnes	Grade % Cu
Geological	0.4%	3 311	0.60
Mineable stage 1	0.4%	604	0.79
Mineable total	0.4%	1 033	0.75

Anaconda selected the 0.4% cut off as being the most appropriate thus their total mineable reserve was 1.033 Gt at an average grade of 0.75% which was sufficient to support a 60 000 tpd operation for almost 50 years.

RTZ Consultants study October 1988. After acquisition by the Luksic group extensive underground development and sampling was undertaken. An additional 22 diamond drill

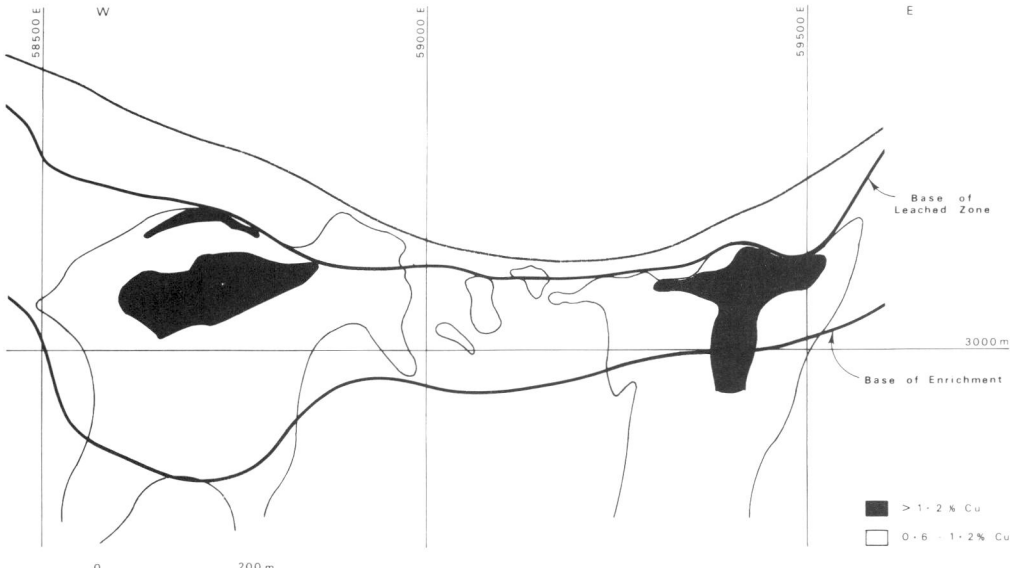

Fig. 2. Typical E–W cross section showing copper distribution.

holes totalling over 3000 m were drilled. In 1988 RTZ Consultants, as part of an engineering study to examine the feasibility of mining the deposit on a small scale, examined the distribution of values throughout the orebody and concluded that there exists two higher grade zones, the East Zone and the West Zone, which contained at a 1.2% cut-off (Fig. 2) 37 Mt at an average grade of 1.52% Cu.

Subsequent to this study further data have become available from both additional drill hole information and sampling of the development headings and a review of the reserve base by RTZ Consultants has estimated the mineable reserve, at a cut-off of 1.2%, to be 40.9 million tonnes at an average grade of 1.51% Cu.

Table 2. *Calculated reserves of the Central Zone*

Cut-off % Cu	Millions of tonnes	Grade% Cu
0.8	33.2	1.03
1.0	15.4	1.18
1.2	6.0	1.34

Central Zone study 1992. During the construction period a drilling programme was carried out on a newly identifed area of high grade mineralization between the East and West Zones, the so called Central Zone. The attraction of the area is that the mineralization crops out and occupies a knoll and is amenable to easy exploitation by open pit methods. The reserves of this central zone have been calculated as listed in Table 2.

Operations

Mining. Mining has commenced on the East Zone using sub-level caving. With this method production levels initially were developed at 11 m vertical intervals within the orebody. From these levels the orebody is drilled and blasted and the broken ore is drawn from below and loaded. The overlying material is induced to cave as the ore is removed and is drawn down as the mine advances. Subsequently the sub-level interval has been increased to 13 m. There have been some difficulties in maintaining the grade at the predicted levels with the grade being 4% below forecast during 1993. This has been attributed to dilution coupled with sampling problems. A programme is underway to study the problem.

The mine is fully equipped to support mining at 5000 tpd using trackless equipment. Broken ore is loaded from the working face to a series of ore passes and dropped down to the 3000 m level where the ore is loaded onto trucks and trucked 1300 m through the main access level and thereafter 2 km on surface to the crusher

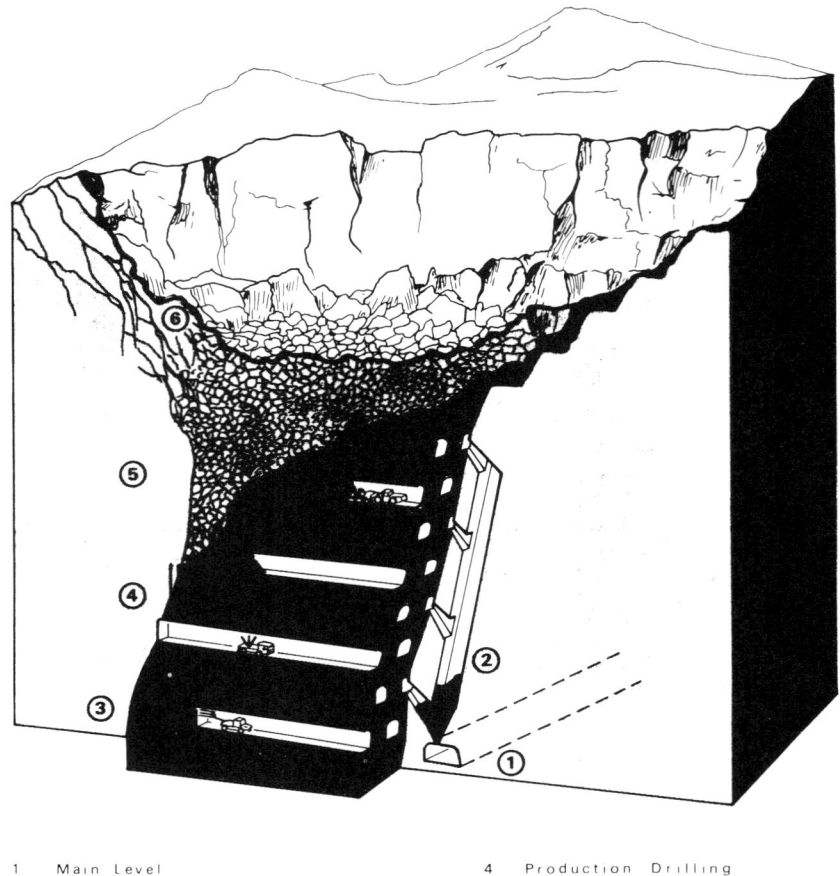

Fig. 3. Illustration of the sublevel caving mining method. 1, main level; 2, ore pass; 3, development of new sublevels; 4, production drilling; 5, mining (blasting and loading); 6, caved hanging wall.

stockpile (Fig. 3). As currently configured the East Zone can support mining for five years during which time mining from the larger West Zone is planned to commence. Currently, consideration is being given to increasing the vertical sub-level interval to 22 m which will reduce costs by about 10% in the present underground operations. Additionally the orebody in the Central and West Zone is under examination. It is anticipated that the reserve in these areas can support open pit mining at low stripping ratios with minimum pre-stripping and provide an additional source of ore over and above that contemplated in the RTZ Consultants study.

Process. Operations at the processing facility commenced in January 1992 and by the end of April 1992 had achieved at its nominal design capacity of 4500 tpd on a monthly basis. Following an interruption in June 1992 due to bad weather the plant achieved target production rate of 5000 tpd in August 1992.

The processing facilities have been built in two parts (Fig. 4). Crushing and grinding takes place at the Los Pelambres mine site in a location chosen to minimize the avalanche hazard. Ore from the mine is fed to a 60 in × 48 in jaw crusher which in turn feeds a 30 000 t crushed ore stockpile. A 20 ft × 11 ft SAG Mill and 12 ft × 16 ft Ball Mill in closed circuit reduce the ore to

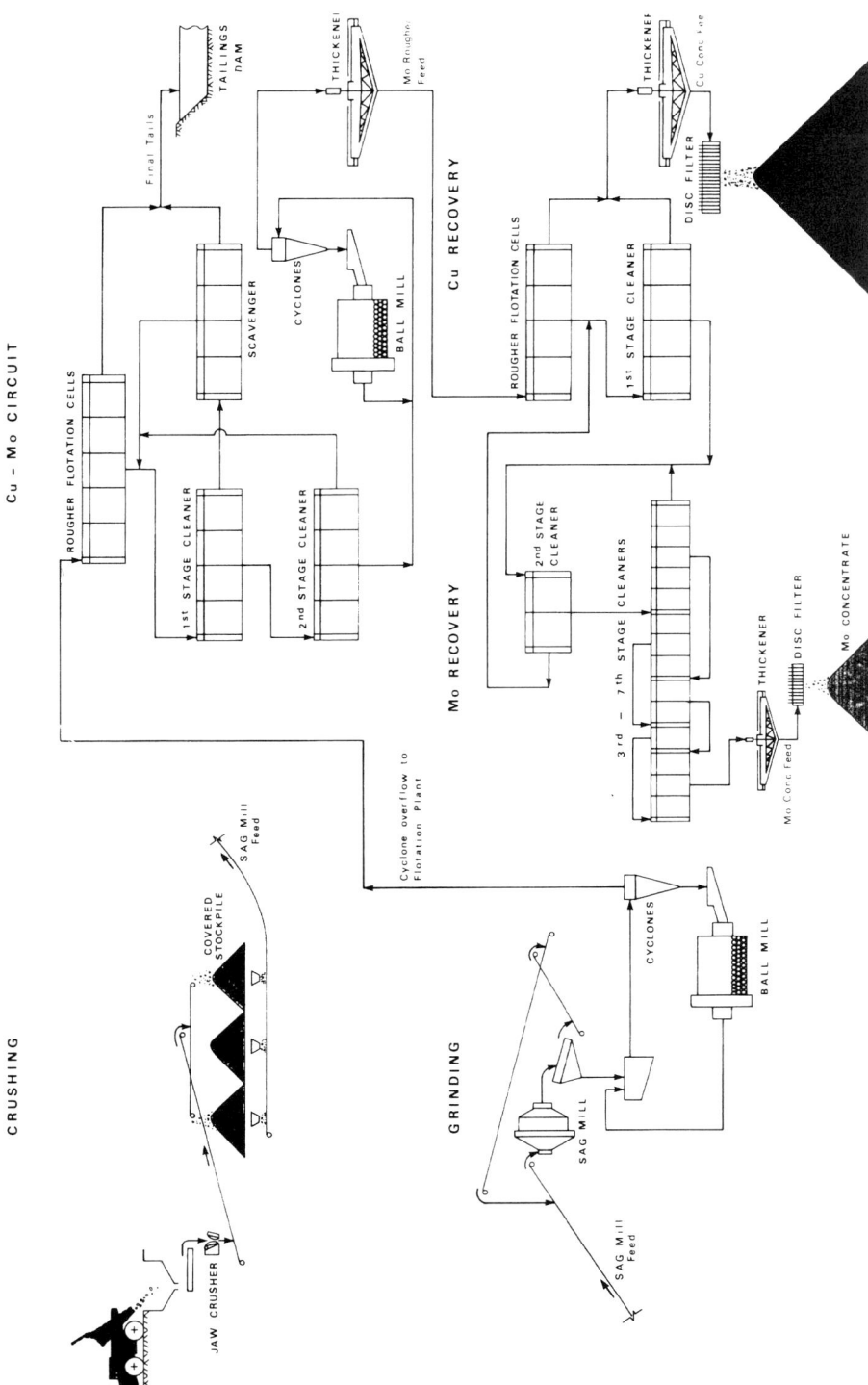

Fig. 4. Metallurgical process flowsheet.

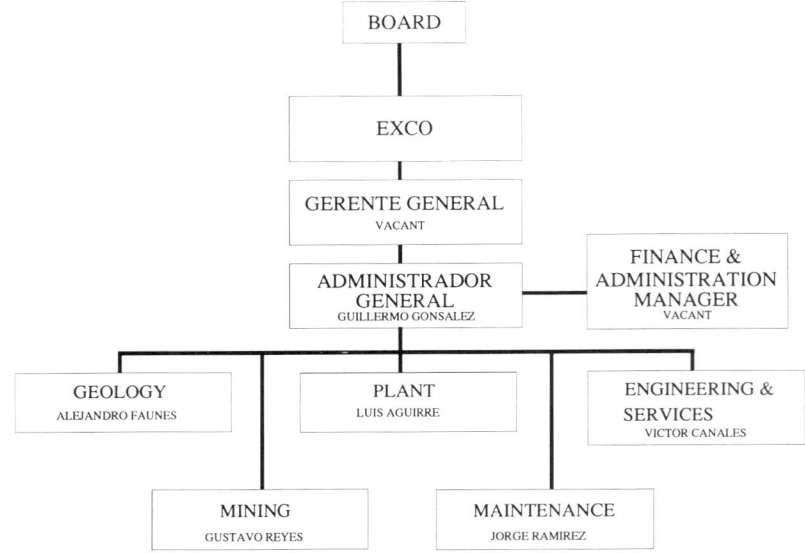

Fig. 5. CMLP organization.

80% minus 120 μm. This is thickened to 50% solids and delivered to the benefication plant at Chacay through a 14 km, 10 in steel pressurized pulp line. The crushing and grinding plant is at 3000 m elevation and the beneficiation plant at 1600 m. Four pressure reduction stations are an integral part of the pulp line system.

Beneficiation takes place at Chacay and is by conventional froth flotation. The beneficiation process has reached its target level of 92.6% recovery and 38% Cu grade. The original design called for the preparation of separate Cu–Mo concentrates, but due to the initial low Mo grade and low price the circuit is not operational. The original plant design contemplated an early expansion of the operation and the layout of the beneficiation facilities reflects this.

Concentrate transport. Concentrate is dried on site using a Larox pressure filter to about 8–9% moisture and stored under cover in a concentrate loading facility. About 3000 t can be stored at Chacay. Concentrate is transported by truck to the port of Ventanas about 350 km to the south where it is stored in Codelco's facilities under a contract with Codelco. Full loading, weighing and sampling facilities are available at this site. Under present concentrate sale agreements the concentrate is shipped in 5000 t or 10 000 t shipments in Panamax vessels to far eastern ports.

Infrastructure

Power. Power is delivered from Illapel to Chacay (Fig. 1), a distance of 75 km, on an 110 kV surface transmission line. Thereafter this line continues to within 5 km of the mine site where it is taken underground as a precaution against avalanche damage. The line has a 25 MW capacity as against the 7 MW required for the project as presently configured and is a major asset in relation to any expansion plans. Generator back-up is provided at both Chacay and Los Pelambres sites.

Roads. Access from the mine gate to Chacay is on a good quality dirt road. This road has a minimum running width of 6 m and has shown itself capable of withstanding winter storms.

Tailings disposal. The tailing disposal area is in the Chinche Valley between the 1300 m and 1500 m elevations. The site has been designed as a zero discharge facility with clear water overflow being pumped into a nearby channel for a downstream irrigation project.

The area in general is seismically active and a non-active fault was identified during early construction on the original dam wall site. International consultants have been used to advise on the location and design of the site. Seismic monitoring is carried out in conjunction with the University of Chile and will continue to be standard practice. The design has incorporated the latest technology and will allow the dam to survive one in one thousand year storms and earthquakes.

Other. Domestic and industrial water for the mine and plant sites are being extracted from the Rio La Cascada, Rio Pelambres and Rio Piquenes. Intakes and header tanks have been constructed to manage the supply. All sources are double supply sources. Economy in the use of water has been a special concern of the project and is reflected in the provision of thickeners to recover water for milling and concentration.

Fire water is provided at both locations in ring mains. Potable water is provided by liquid chlorination.

Snow removal and avalanche control has been a major concern to CMLP. The mill site is situated under a near vertical rock face on the north side of the valley. The slope is sufficiently steep so as to prevent an accumulation of snow from occurring. On the south side of the valley the coarse ore stockpile and a series of trenches and dykes are designed to prevent an avalanche from entering the site area. Additionally the Gasex avalanche control system has been installed. This is an explosive gas system that can be operated remotely and when operated is designed to prevent a dangerous build up of snow. An emergency shelter has been provided for personnel stranded at the mine site in bad weather. A meteorological station is manned full time and is capable of producing the latest satellite images and receiving up to the minute weather forecasts.

Management

The CMLP management team is independent of the partners. The organization of CMLP is outlined in Fig. 5. The Administrator General reports at present directly to Exco, an executive committee, that has the power to approve expenditure up to $2 M. Exco meets frequently and is composed of members from the three partners. Decisions over $2 M require the approval of the Board. The Administrator General has authority to commit to expenditure up to $100 000.

Concentrate marketing

Contracts have been agreed for the sale of all CMLP's production through 1994. Treatment charges are negotiated annually and the charges for 1993 reflect current world prices.

CMLP concentrates are of high quality and low in environmentally unacceptable elements such as arsenic. As such it attracts no penalties under the existing sales contracts. Typical full analysis is given in Table 3 below.

Table 3. *Cia Minera Los Pelambres typical copper concentrate analysis*

Copper (total)	35–40%
Molybdenum (total)	0.052%
Iron (total)	18.66%
Sulphur	33.9%
Zinc	0.045%
Lead	0.006%
Bismuth	<0.002%
Nickel	0.004%
Arsenic	0.010%
Antimony	<0.005%
Aluminium	1.25%
Silica	4.20%
Magnesium	0.10%
Gold	1.5–4.0 g/t
Silver	20–80 g/t
Mercury	<1.00 g/t
Tellurium	<0.001%
Rhenium	NP
Sodium	NP
Potassium	NP
Phosphorus	NP
Insoluble	11.47%

NP, not present.

Environmental

It is not mandatory at present to produce an Environmental Impact Statement before proceeding with a project such as Los Pelambres in Chile. The company, however, regarded it as responsible and prudent to examine the environmental consequences of the project, particularly in the tailings dam area and as a result an environmental 'audit' was commissioned. The study indicated that the installation of the Los Pelambres tailings facility would not constitute an ecological risk to the surrounding environment.

Additionally a technical audit was commissioned which identified some environmental risks. These relate mainly to the thickeners, pulp-line and tailings facility. These results were discussed with the consultant and their recommendations implemented.

Future operations

The feasibility study conducted by Anaconda (Atlantic Richfield) and presented in mid-1983

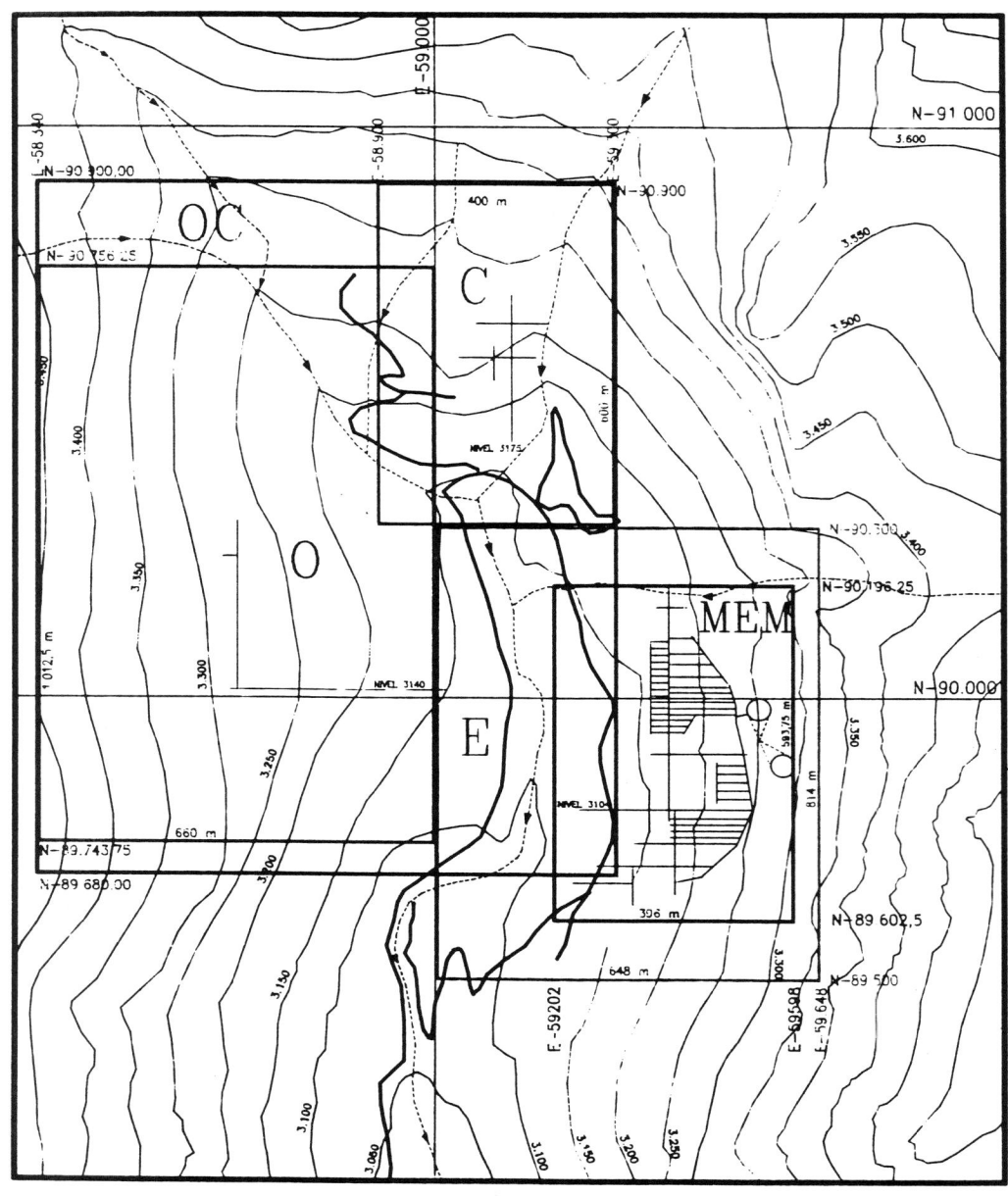

Scale 1:10 000

O	West Zone
C	Central Zone
E	East Zone
MEM	Current Short Term Mining Reserve

Fig. 6. Reserve matrix 15 000 tpd project.

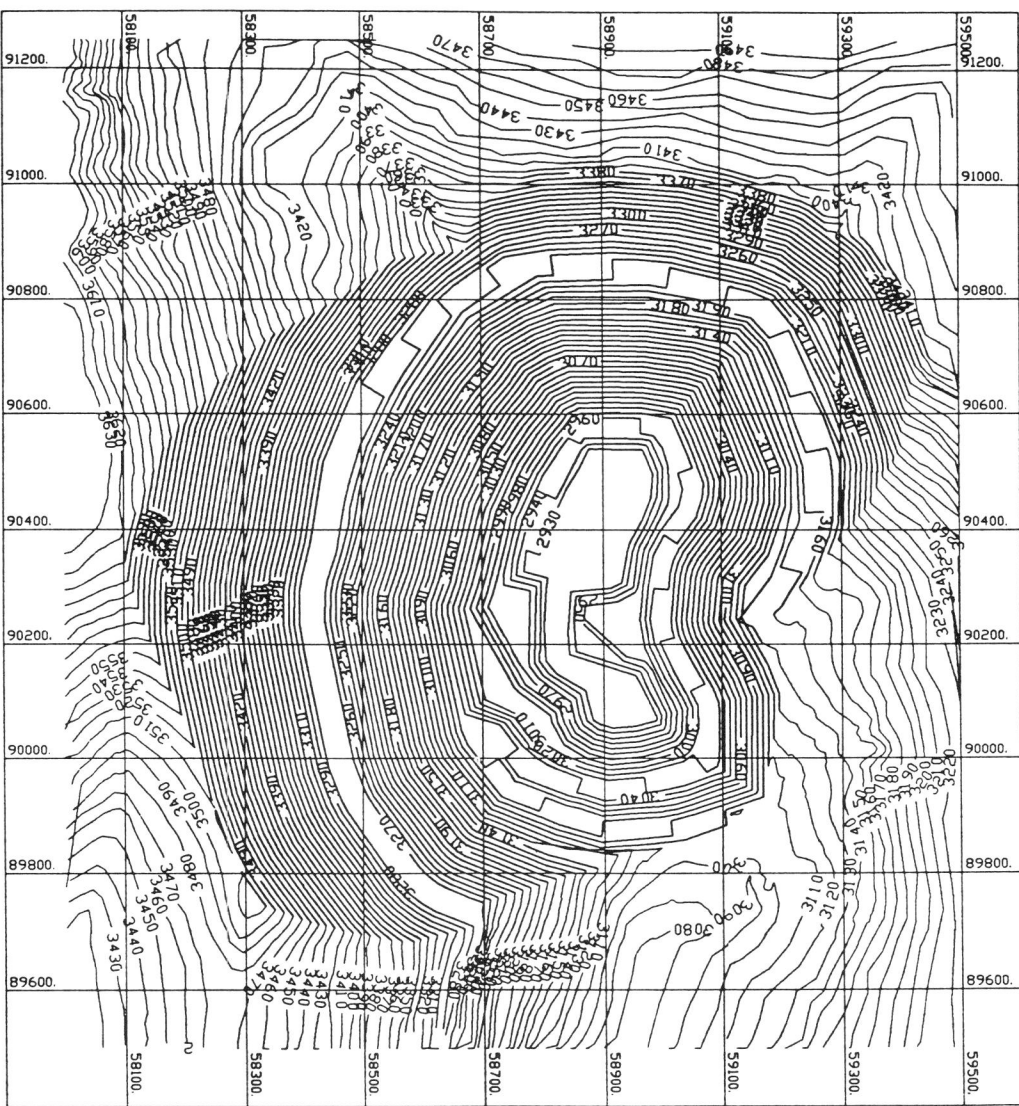

Fig. 7. Final pit 15 000 tpd project.

called for mining over 1000 Mt tonnes of ore over almost 50 years at a rate of 60 000 tpd. As configured presently mining will be at a rate of 5000 tpd on a reserve of 40 Mt (i.e. less than one twentieth of the overall reserve).

Fluor Daniel (Chile) were commissioned to prepare a feasibility study to examine the viability of a limited expansion. This work, which started in October 1992 and was completed in July 1993, included an audit of work previously completed by the company. The study examined an expansion to 15 000 tpd which Fluor consider could be supported within the present general infrastructure of the operation. Secondly they examined an 8500 tpd case. This is a minimum expansion which would be confined only to minor capital works in the plant. Thirdly they examined an initial expansion to 8500 tpd followed immediately with an expansion to 15 000 tpd.

Reserves

Copper mineralization covers an area 3.5 kms north to south, by 2 kms east to west. Atlantic Richfield calculated the geological reserves of the deposit as a whole to be 3300 Mt at an average grade of 0.63 Cu at a 0.4% Cu cut off.

The expansion project considers a reserve matrix containing those reserves, exploitable by open pit methods and in the Central and West Zones which in total contain 610 Mt at an average grade of 0.79% Cu at a 0.4% Cu cut off. A summary of the reserves within the expansion project and including the East Zone (Fig. 6) are shown in Table 4.

Table 4. *Summary of reserves with the expansion project, including the east zone.*

Cut off % Cu	Millions of tonnes	% Cu
0.0	960	0.59
0.2	771	0.71
0.4	673	0.79
0.6	499	0.87
0.8	256	1.02
1.0	104	1.21

Mining operations

The study considered the development of the West and Central Zones as an open pit (Fig. 7). Initially, production from the open pit would supplement production from the present operation but will later replace underground production as the reserve becomes exhausted. A summary of the operation is listed in Table 5.

Table 5. *Summary of mining operations.*

	Expansion option		
	1 15 000 tpd	2 8500 tpd	3 8500–15 000 tpd
Tonnes (millions)			
Pre-stripping	13.1	8.9	8.6
Waste	183.6	141.9	185.3
Low grade stockpile (above 0.7%, below cut off)	31.8	24.5	34.1
Ore	144.9	81.7	144.9
Total	360.2	249.4	364.3
Average head grade % Cu	1.05%	1.09%	1.05%
Waste to ore ratio	1.49	2.04	1.51

Table 6. *Capital costs for years 1–6 of the expansion options.*

Year	8500 tpd option				15 000 tpd option				8500/15 000 tpd option			
	U/G	Pit	Plant	Total	U/G	Pit	Plant	Total	U/G	Pit	Plant	Total
1	3.5	27.7	4.0	35.2	3.5	28.8	1.9	34.2	3.5	31.0	3.4	37.9
2	3.3	10.7	9.2	23.2	3.3	16.8	36.7	56.8	3.3	13.0	9.6	25.9
3	2.6	1.1	0.0	3.7	2.6	8.3	36.9	47.8	2.6	4.9	61.4	68.9
4	0.9	0.0	5.8	6.7	0.9	0.0	0.0	0.9	0.9	0.0	0.0	0.9
5	0.0	1.4	0.0	1.4	0.0	0.0	0.0	0.0	0.0	0.0	0.0	0.0
6	0.0	0.0	0.0	0.0	0.0	0.0	0.0	0.0	0.0	0.0	0.0	0.0
Net present value*	8.5	35.7	15.2	59.4	8.5	46.3	59.8	114.6	8.5	42.6	57.2	108.2

*With 10% rate of discount

Table 7. *Summary of operating costs. Operating costs for years 1–6 of the expansion options.*

Year	8500 tpd option				15 000 tpd option				8500/15 000 option			
	U/G	Pit	Plant	Adm*	U/G	Pit	Plant	ADM*	U/G	Pit	Plant	Adm*
1	5.00	0.00	3.74	2.40	5.00	0.00	3.74	2.40	5.00	0.00	3.74	2.40
2	5.00	5.85	3.57	1.78	5.00	0.00	3.74	2.40	5.00	5.85	3.57	1.78
3	5.00	5.85	3.41	1.41	5.00	4.27	3.16	0.80	5.00	7.86	3.41	1.41
4	5.00	6.59	1.41	1.42	5.00	3.62	3.16	0.80	5.00	3.64	3.16	0.80
5	5.00	3.82	3.41	1.41	5.00	2.94	3.16	0.80	5.00	2.91	3.16	0.80
6	0.00	2.74	3.41	1.41	5.00	2.45	3.16	0.80	0.00	2.37	3.16	0.80
Net present value†	8.5	35.7	15.2	59.4	8.5	46.3	59.8	114.6	8.5	42.6	57.2	108.2

* Includes general management, operation management, administrative and finance management and engineering and serive costs.
* With a 10% rate of discount

Plant

It is proposed to use the same layout and process as is presently used in the operation. The 15 000 tpd case requires some improvements to the crushing installation, a new grinding circuit, a new pulp line and major changes to the concentrator. The 8500 tpd case is only a marginal increase of present facilities and requires a new ball mill and minor changes in the concentrator. The 15 000 tpd case needs additional capacity for tailings disposal in year 13 while the 8500 tpd case needs no such new facility. General infrastructure is adequate for a 15000 tpd expansion.

Capital and operating costs

Tables 6 and 7 give summaries of the capital and operating costs for years 1 to 6 of the proposed expansion.

Conclusions

The feasibility study indicates that an expansion of the concentrator facilities at Los Pelambres together with development of an open pit on the West and Central Zones of the deposit is technically feasible, can be achieved without interruption to the present operation and is commercially attractive. Fluor concluded in 1993 that the stepped expansion gives the higher Net Present Value and Internal Rate of Return while minimizing the financial risk. There are a number of concepts such as the use of contractors in the mining operation which have not been considered and which may improve the economics of project.

References

ANACONDA MINERALS COMPANY 1983. *Los Pelambres Feasibility Study.* Unpublished internal report.

FAUNES, A., FORKES, J. & O'LEARY, J. Ore Reserve estimation at Los Pelambres, A Chilean porphyry copper. *In:* ANNELS, A.E. (ed.) *Case Histories and Methods in Minerals Resource Evaluation.* Geological Society, London, Special Publications, **63**, 277–288.

RTZ CONSULTORES 1988. *Review of Los Pelambres Project.* Unpublished internal report.

Index

Abbeyleix 233–4
Abernant Middlings 226
accuracy 46, 118, 121, 130, 195
acquisition costs 126
ADDMOD program 161
aggregates 5, 191, 202, 221
  lightweight 225–6
  silica content 226
aluminium, as impurity 59
andesites 254
Anglian Boulder Clay 196
Anglian Glaciation 196, 221
anisotropy 49–50, 72, 74, 77, 79, 91, 164–5, 203
apatite 54, 56–7, 59–60
aquifer heads 217
aquifer recharge 214
aquifer recovery 216
area errors 112
Argentina 253–4
assay values, Lisheen 237
Association of Professional Engineers of Ontario 11, 13–14, 132
assurance definition 132
attribute accuracy 46
Australian IMM/AMIC classification 6, 11–12, 132, 135
avalanches 254, 256, 258–9

back-estimation 167
backfill sloughing 102
backfilling 103
bailer 192
ball mills 256, 263
Bama mine 177
Barham Sands and Gravels 196
basalt, andesitic 207
Bayes' theorem 90
Belfry quarry 222, 224–5
bell pits 221, 226, 232
beneficiation 54–6, 59–61, 157, 258
Bergslagen ore province 172–3
bimodal distribution 115, 120, 162
Birimian Group 159
blast damage 97, 100
blast hole drilling 103
blast hole sampling 68, 72, 75, 77
blast vibration 101, 103
blasting overbreak 100
block confidence limits 133
block errors 197
block grade models 89, 94, 161
block grades 93
block models 73, 75, 77, 82, 84, 87, 92, 112–16, 121, 144, 254
block revenue models 147
block sizes 87–9, 91, 93, 116
borehole assays 72

borehole databases 48, 191–2
borehole layout diagrams 30, 80
borehole sampling 98, 192
borehole simulation 72
bornite 254
boulder clay 196
boundary element methods 100
boundary errors 112
boundary location 111
boundary recognition 111
boundary variation 111
braided rivers 191
Brandelos mine 177
breccias 234, 237, 254
British Coal Opencast 37–8, 45, 48, 51, 226, 231–2
British Geological Survey 1, 8, 241
British Standards, aggregates 191, 202
bulk properties, kaolin 244
buried canyons 213
buried channels 195
burnt coal 50
Bushveld Complex 16

calc-silicate rocks 174
calcination 61–2
Calgon 243, 245
Canadian IMM 6
carbonates, as impurities 61
carbonatites 53–4
cash flows 132–4, 138–9, 146
cavity surveying 106
central tendency, measures of 72, 75
ceramics 243, 246
Chacay 253, 258
chalcocite 254
chalcopyrite 254
channel bedloads 204
chemical barrier 162
Chile
  government and laws 250–1
  investment in 123–8, 251
Chilulwe 241–3, 245–7
chippings 224
chlorine, as impurity 62
Choapa 253
chromite 16
Chuquicamata 250
churn drilling 161–2
classification 3
  alternatives 13, 16
  international 14
  quantitative 16
clay minerals 242
Clee Group 220
Clee Hill Quarry 219–32
Clee Hills 219, 221
coordinate systems, unfolded 186–9

INDEX

coal, Welsh 226
Coal Measures 220–2, 224–6, 228, 231
coal quality 231
coal reserves 226–7
coal seam thickness 38, 40–1, 50, 231
coding, nested 237–8
La Coipa 127
Collahuasi 123, 127
communications, underground 95, 103, 106
completeness 46
computer aided design 141
computer based pit design 147
computer errors 111
computer integrated mining system 104, 107
computer models 109, 121
computer networks 236–7
concrete production 191
conditional simulation 91
confinement effects 100
contamination 222
continuity, geological 13–15, 176
continuous assessment 130
control points 181–8
control sampling 93
copper, grades 252, 258
Copper Cliff North mine 106
copper projects, Chile 123–8, 249–63
Copperbelt 242
core drilling 222, 225, 227, 231
corporate borrowing 133
cost benefit analysis 212–13
Council of Mining and Metallurgical Institutions 6
crater blocks 103
cratonic blocks 157
cross-validation 201–4
crushing 255–6
currency 47
cut and fill mining 102, 173, 177
cutoff grades 89, 96–7, 161, 166, 168

dam, tailings 258–9
data, uses 3
data analysis 2, 37–43
data capture errors 111
data collection 2–3, 6, 105
  Zinkgruvan 177
data errors 111
data flow model 47
data modelling 29–35
  see also data analysis
data organization 159
data point coordinates 179–80, 199
data preparation 48
data quality 45–6
data set, statistical assessment 69, 202
data storage 29
database management 233–9
DATAMINE program 164, 186–7
debt swaps 250
deconvolution 164
definitions, nomenclature 7
deformed orebodies, reconstruction of 171, 180
Delaunay triangulation 50
depletion 70

depletion schedules 147
deregularization 164–5
Derrykearn 234
Desert Valley 207–9, 214, 216
development data, Chile 125–6
deviation, drilling 103, 106
dewatering 207–17
  costs of 212–13, 217
  environmental effects of 214, 217
  simulated 212
dewatering stresses 211
Dhustone 221–2, 224, 226, 231
diamond drilling 161–2, 172, 233, 252, 254
digital terrain model 30–1, 34
dilution 96, 98, 255
  backfill 102
  control of 103, 105–6
  definition 96
  measures 95
  planned 96–8, 100, 107
  total 97
  unplanned 97, 100–2, 107
diopside 175
dip contours 32, 34
dip sections 33–4
discovery costs 126
discovery to startup times 127
disseminated deposits 141, 144, 174
Ditton Series 220
dolerite 219–22, 224–5, 231
dolomite 61, 233–4, 236–8
dolostone 237
downhole assays 161, 177
Downton Series 220
drawdown 211–12, 214
dredging 55
drift and bench mining 173
drill drifts 100
drillhole sections 117, 121, 197
drilling
  for dewatering 208
  monitoring 106
drilling accuracy 103
drilling and blasting 100, 105
drilling grids 87–90, 144, 159–60, 177, 191–2, 197, 235–6
dry holes 193
duplicate sampling 192

economics, Chilean mining 123–8, 251
edge blocks 93
en echelon faults 236
end-user involvement 3, 8
enrichment 70
entry costs 124
environmental impact 259
erosional scours 196–8
error estimation 75, 83, 85, 109, 130
error quantification 112, 121
error types 111, 194–5
La Escondida 123, 127, 251
estimates, precision of 25
estimation 3
  of variance 110, 134

Europe, Eastern 1, 5
exploration costs 127
exploration quality 98
explosives 100, 103, 106
exposure time 103
extension variance 197
external environment factors 3, 5
extraction reserves 97
extractive equipment 54

fault blocks 185
fault interpretation 29–35
faulted orebodies 171, 180
faulting
   Clee Hills 224, 228
   Lisheen 236
   Zinkgruvan 173
feasibility studies 6, 8, 12, 24–5
feldspar, in kaolin 244–6
ferroan dolomite 234
financial conditions, Chile 251
financial forecasting 134
financial returns 133
financial risk 132, 263
fines, loss of 192–3
finite element methods 100, 208, 210
fireclay 232
fixed heads 211
flint 197
flotation 59, 61–2
flowsheets 257
fluorine, as impurity 62
fluorspar veins 143–4
fluvial systems 197
folding
   complex 184
   isoclinal 176
   minor 180
   synclinal 224
footwall prediction 118, 141
fragmentation sensing 106
fuller's earth, UK 5

galena 174–5
Galmoy 233
Gaussian distribution 70, 115
geographic information systems 45–6, 106, 111, 237
geological distances 182, 185–6
geological interpretation 229–30
geological reserve 96
Geological Survey of Canada 98
geometric modelling 109, 120
GeoMODEL 39–40, 48
geophysical logging 98, 227, 231
geostatistical noise 204
geostatistical techniques 15, 50, 110, 147
geostatistics 98, 109, 157–69, 171, 191, 198, 200
German (DDR) classification 135
German Mining Engineers and Metallurgists Society 136
Ghana 157
GIS *see* geographic information systems
Glenover 60
global estimates 77

global positioning systems 46
gold mine, Nevada 207–17
gold mineralization 91, 250
gold projects, Chile 123–8
Graça orebody 117–18
grade contours 73
grade control 177
grade estimation 67–87, 109, 129
   procedures 69
   statistics 80–1
grade reduction 107
grade shells 70, 72, 74, 86
grade variability 71
grade zones 73–4
graphs, directed 89–90, 145
Graunt, John 2
gravel gradings 192
gravel thickness 197–201
gravels, Beestonian 191
Great Dyke 16
gridded seam model 112–16
grinding 256
ground-penetrating radar 98
groundwater 208, 210, 212, 216–17

hanging-wall prediction 119, 141
haulage 106
head deposits 221
heuristic algorithms 145
El Hueso 127
hydrocyclones 242–3, 247
hydrogeology 208–9, 216–17
hydrothermal alteration 254

Illapel 258
IMM classification 6, 11, 13, 26–7, 132, 137
implementation 3
impurities, kaolin 244
in-situ resources 131
Incline quarry 222, 224, 226
indicator maps 75
El Indio 127
Industrial Minerals Map, UK 8
information effects 89
infrared studies 58
Institution of Mining and Metallurgy *see* IMM
   interceptor model 208–11
INTMOV 37–43
intrusives 159
investment criteria 24, 130
investment risks 14
iron, as impurity 59
iron ores 60
irrigation 217, 258
isopach map 31
ITH drilling 100

Jacupiranga 61

kaolin 241–7
kaolinite 241–3, 245–7
Kesgrave Gravels 196–7
Khibiny intrusion 59
Killoran fault 234, 236

INDEX

Kiruna 60
Knalla mine 173, 175–6
kriging 74–5, 77, 90–1, 134, 164–8, 172, 197, 199, 201, 203, 254

Lake Lahontan 207, 216
laterite 159, 242–3
leaching 159
lead–zinc mining, Poland 21
legislation, mining 250
Lerchs–Grossmann algorithm 90–1, 94, 145, 147–55
limestones, Lisheen 233–4, 236
lineage 47
Lisheen 233–9
load-haul-dump 106
loading equipment 105
loading performance 106
Lobo 127
locational errors 111, 121
Loch Borralan 60
log-normal distribution 203–4
log-normal shortcut 91
logical consistency 46
Lombador deposit 112
London Basin 197
London Clay 195
longitudinal retreat 103
low grade deposits 130
Lowestoft lodgement till 191
Ludlow 221

macros 238
magnetic surveys 222
Main Ore 174–5
Malmberget mine 98
management systems 105
manganese mines 157–69
Mathew's method 101
meta-tuffs 159
metadata 46
metallogenesis 237
Metallogenic Map of the UK 8
metatuffite 174–5
mine design 34, 48, 50, 95, 97, 100
mine models 104–5
mine planning 71, 77
mineable reserves 96
Mineral Occurrence Database 8
Mineral Planning Guideline 5
mineral potential 13, 21, 137
mineral reserves, definition 1, 137
mineral resources, definition 1
mineralization envelopes 161
mineralization zones
    Chile 255–6, 262
    Lisheen 234
mineralogy 243
mining costs 144
mining depths 142
mining dilution 13
mining efficiency 95
mining feasibility 134
mining lines 96–8, 100
mining reserves 97

mining resources 131
mining sequence planning 145–6
mining trials 104
model structures 110
model validation 167
modelling, three-dimensional 34
MODFLOW 208, 217
molybdenite 254
molybdenum 250, 252
moving cone method 144–5
moving-windows statistics 37–9, 72, 77
mucking 103
muckpiles 105
mudstone 225–7
multi-probe logging 98
multiple-unit deposits 144
MWINDOW 37

Nauru 62
nepheline 59
networks 236–7
Neves–Corvo mine 109, 112, 117
nomenclature 6, 9
    definitions 7
normal distribution 198, 200, 203–4
Nsuta mine 157–69
nugget effect 165, 180, 200–1, 203
numerical modelling 100
Nygruvan 172, 174–6, 189

Old Red Sandstone, Lower 220
old workings 51
open hole drilling 222, 224, 227, 231
open pit design 87–94, 141, 145, 209, 261
open pit mining 54, 207–17, 250, 255
ophiolite complex 177
optimal pits 87, 90–3, 141, 145
ore blocks 75, 185
ore homogeneity 163
ore intersections 98
ore lobes 70, 72
ore loss 96, 98, 100
ore parcels, values of 141
ore passes 255
ore reserves, 12, 96, see also reserves
ore zone shortening 176
ore zones, Zinkgruvan 174
ore–skarn contact 176, 180, 185
orebodies
    boundaries 109, 141
    delineation 98–9, 107
    orientation 183
    outlines 117
    sections 159
    shapes 109, 119, 161
    thickness 115–16, 119–20, 161, 180, 185
organic matter, as impurities 62
Oron 61
output forecasts 50–1
overburden ratios 54
overburden removal 145, 208, 216, 222, 226, 232
overburden thickness 193, 222–4, 226, 231
overprinting 70

overpull 97, 103, 106

El Pachon 254
palaeocurrents 195
palinspastic maps 171
paper-coating 243–4, 246
Parallel Ore 174–6
passive exploration 128
pegmatites 241–2, 247
Los Pelambres 249–63
percussion drilling 222
permeability 214, 216
petrography 56, 197
Phalaborwa 60
phosphate rock 53–65
   analysis 57
   characterization 55–6
   extraction methods 54–5
   impurities 56, 58
   infrared analysis 58
phosphatic limestones 53–4
phosphogypsum 61
Phosphoria Formation 57
phosphorites 53
   Saudi Arabia 6–8
phyllite 159
pillar recovery 97–8
pit dimensions 209–10
pit expansion 217, 261–3
pit floor elevations 210, 212–14, 216
pit optimization 143
pit shapes 142
planned dilution 96–8, 100, 107
planning applications 231–2
planning consents 5
planning tools 130
Pocuro Fault Zone 254
Polish classification 23
porphyry copper 21, 254
positional accuracy 46
power supplies 209, 214–16, 258
pre-resource mineralization 13
precision 46, 118, 121, 192, 194
   estimates 25
prediction, bias in 120
primary stoping 97–8
probable ore reserves 12
problem structuring 2
production control 95
production data, Chile 125
production management 103–7
production requirements 130
production schedules 147
productivity 103
profits, from quarries 144
project costs, Chile 124
project financing 129–39
proved ore reserves 12–13
psilomelane 159
pumping rates 209, 211–14, 216–17
pumps, sump 208
pushbacks 146, 214
pyrite 236, 254
pyrolusite 159, 163

pyroxenite 60
pyrrhotite 175

quantitative studies 1–2
quarry design 141–55
quarry wall slopes 144
quartz veins 159
Quaternary deposits 191, 221

radiometric ages, Chile 254
Rathdowney Trend 233
re-evaluation 104
recoverability 88–90
recovery 98, 100, 103, 130
recovery measures 95
redrilling 191, 193
reference planes 178
Refugio 127
regression analysis 204
removal cones 144–5
research drilling 191
reserve base 130
reserve matrix 260, 262
reserves
   assessment 45–52, 129–39
   classification 5–6, 11, 23, 130
   Comecon 17
   demonstrated 134
   detected 27
   documented 18, 24, 26
   economic 20, 25
   estimation 3, 172, 177
   indicated 12, 26, 118, 134, 136
   inferred 12, 27, 134, 136
   inventories 17–18
   measured 12, 132, 134
   mineable 96, 255
   mining 97
   ore 12, 96
   possible 130, 135
   potential 18, 26
   probable 6, 13, 26–7, 130, 135–6
   prognostic 19, 23, 26
   proven 6, 13, 26, 130, 135–6
   theoretical 19, 21
   uneconomic 20
   unworkable 20
   workable 20
resistivity surveys 222
resolution 47
resource assessment 3, 9
resource estimation 34
resources
   classification 4–6, 11, 130, 136
   identified 12
   indicated 6, 12–13, 27
   inferred 12
   measured 12–13, 27
   mining 131
   potential 131
   subeconomic 12
   undiscovered 12
restoration 232
revenue blocks 93, 100

revenue calculations 143
rewatering 216
reworked gravels 197
rhyolite 173, 207
risk analysis 147
risk categories 132
roadstone 224
rock fragmentation 97, 105
rock quality 102

safety levels, dewatering 212
Salamanca 253
sample composites 160, 162, 166, 192, 194
sample contamination 193
sampling 16, 172, 177
sampling density 22
sampling errors 111, 192–3, 195
sampling grids 90, 177
sand and gravel deposits 191–205
saturated thickness 212–14, 216
Saudi Arabia, phosphorite 6, 8
scale, economies of 100
schedule of estimated quantities 45, 50
screening 242–5
scrubbing 242
seam correlation 227–8
seam modelling 34, 48, 50
secondary phosphates 60
Securities and Exchange Commission 6, 11, 132
sedimentation, rhythmic 225
seismic surveys 222, 224–5
seismic tomography 98
seismicity, Chile 254, 258
selective mining units 94
semi-variograms 70, 72–3, 160, 164–8, 180, 197–8, 201, 203–4
sensitivity analysis 147
sensors 104, 106–7
sequential gaussian simulation 90–1
Serenje 241
shear-zone deposits 16
shell and auger drilling 192
Sherwood Sandstone Group 197
sieve sampling 193–4, 202–3
silica, as impurity 58–9
silica content, aggregates 226
sills 220–1
silver mining 67, 250
simulation, geostatistical 90
sintering 226
site investigation databases 48
site surveys 39
size fractionation 242
skarn 175, 180, 185
Sleeper mine 207–17
slot raises 100
sludge drilling 177
slurrying 55
smoothing effects 161
Society for Mining, Metallurgy and Exploration 6, 11
software packages 29–35, 37–43, 48, 141, 161, 164, 186–7, 195, 197, 208, 217, 233, 236, 238
solifluction 221
Soviet Union, former 1, 5, 17

sparse data 87
spatial data 37
specific capacity 213–14
sphalerite 174–5
stability graphs 101
stationarity 42, 204
stationary random functions 37
statistical assessment 69
statistical techniques 2, 37, 80
stock market, Chile 252
stockpiles 256, 259
stockworks 177, 234, 254
stope boundaries 96
stope cavities 105
stope design 95, 100
stope grades 96
stope mapping 177
stoping methods 100, 105
stoping practice 102–3, 107
stratification 174
stratiform deposits 87, 141, 171
stratigraphic units 142
streamlines 211
Stretton Series 221
strike lengths 183
stripping ratios 147, 192, 222, 226, 231, 256
stromatolites 59
structural unrolling 171–2, 175–7, 180–5, 189
subjectivity 15
sublevel caving mining 256
sulphide deposits 112, 174, 234
supergene enrichment 159, 162
superquarries 5
support effect 89
support installation 105
SURPAC2 29–35, 141, 195, 197
Sveco-fennian orogeny 173

tailings 258–9, 263
Tarwaian Group 159
TECHBASE 233, 236, 238
El Teniente 250
Thiessen polygons 48, 50
Thompson mine 98, 103
three-dimensional models 74, 141, 157, 165, 186, 195, 217
Thurles 233
till 191, 197, 221
Titterstone quarry 222
tomography, seismic 98
Tonduff 234
topographic surfaces 111
total dilution 97
trace elements, as impurities 62
trial pits 193
Trinity Silver Mine 67
truck haulage 106
tube drilling 100
turbulent flow 213
turning bands method 90

underground drilling 117, 120
underground exploration 104
uniform grade deposits 142

United Nations 252
United Nations classification 131, 136
United States Geological Survey 208
univariate statistics 69–70, 197
unplanned dilution 97, 100–2, 107
unrolling, structural 171–2, 175–7, 180–5, 189
unrolling grids 185
Uranium Resource Appraisal Group 136
USBM/USGS classification 6, 8, 11–12, 26, 130–2, 134

variable grade deposits 142
variation errors 111
variograms 91
Ventanas 258
viability, economic 132
Viscaria mine 98
viscosity, kaolin 243
volume estimation 111–12
volumetric modelling 110

wall geometries 98
wallrock sloughing 97, 100–1
washouts 48, 50
waste blocks 75

Waulsortian facies 233–4, 236, 238
weathering, alteration by 57, 221–2, 224
wellfield design 209–10, 212–15, 217
wells
 exploration 213
 interception 208–9, 212–14
 peripheral 208, 212–13
Wendy Norte 127
Wenlock 220
Westphalian 221
wetland project 214, 217
Whalesback mine 100
wireframe models 157, 161
workability, criteria 24–5
workforce 106

X-ray analysis 57–8, 243, 245

Zaldivar 127
Zambia 241–7
zinc–lead–silver ores 173, 233
Zinkgruvan mine 98, 171–89
 geological map 173